INVASIVE PLANTS

0 11557 00284 3

INVASIVE PLANTS

A Guide to Identification, Impacts, and Control of Common North American Species

Sylvan Ramsey Kaufman
and
Wallace Kaufman

STACKPOLE
BOOKS

*This book is dedicated to Drs. Peter Smouse and Jean Marie Hartman
of Rutgers University, who started Sylvan's pursuit of invasive plant species,
and to Dr. Dan Livingston of Duke University, who taught her father
not just zoology but the discipline of science.*

Copyright 2012 by Dr. Sylvan Ramsey Kaufman and Wallace Kaufman

Published by
STACKPOLE BOOKS
5067 Ritter Road
Mechanicsburg, PA 17055
www.stackpolebooks.com

Printed in China

10 9 8 7 6 5 4 3 2 1

Second edition

Cover design by Wendy A. Reynolds
*Cover photos by Barry Rice/sarracenia.com (top), J. S. Peterson @USDA-NRCS PLANTS
Database (middle), and Sylvan Ramsey Kaufman (bottom)*

Library of Congress Cataloging-in-Publication Data

Kaufman, Sylvan Ramsey.
    Invasive plants : a guide to identification, impacts, and control of common North
American species / Sylvan Ramsey Kaufman and Wallace Kaufman. — 2nd ed.
        p.   cm.
    Includes bibliographical references and index.
    ISBN 978-0-8117-0284-3
    1. Invasive plants—United States.   2. Invasive plants—Canada.   3. Invasive
plants—United States—Identification.   4. Invasive plants—Canada—Identification.
I. Kaufman, Wallace.   II. Title.
SB612.A2K38   2013
581.6'2—dc23

                                                                                2012034473

CONTENTS

ACKNOWLEDGMENTS

This book would not be possible without Adkins Arboretum and its director Ellie Altman, who generously gave Sylvan time and resources. Arboretum staff and volunteers Carol Jelich, Maureen MacFarland, Mary Anne Hartman, Michelle Lawrence, Gayle Jayne, Laura Sanford, and Erica Weick helped review pages of the guide and were enthusiastic supporters. Jil Swearingen and Kerrie Kyde of the Maryland Invasive Species Council offered useful advice and resources toward the development of the book.

Many thanks to the staff of The Nature Conservancy's Global Invasive Species Initiative for their informative workshops and stewardship abstracts on invasive species and for contributing photographs. We are deeply grateful for the understanding and collegial help of the many other photographers who also donated photographs. Katelin Mielke helped locate and organize photographs of many plant species.

Thanks to our agents at New England Publishing Associates for their usual high-class attention to a humble project and for placing us with Stackpole. We have not worked with more cooperative and efficient editors than Ken Krawchuk, Mark Allison, and Kathryn Fulton at Stackpole Books, and Stackpole's designers were undaunted by the many problems of organization and layout to make this book practical and comfortable to look at.

Finally, every writer must acknowledge the support that cannot be neatly described but extends far beyond any one book or subject—the patience and encouragement of our friends and family, including Nelli Uteyeva, Sarah Ramsey, Robert Kelly, and Jake Barnes.

PART I

About Invasive Plants—Definitions, Issues, History, and Management Options

The Invasive Species Challenge

Nothing unites a country like an invasion, and the war against invasive species has created rare common ground for forest owners, homeowners, farmers, ranchers, liberal environmentalists, and free market environmentalists.

In 2002, President Bush's Agriculture Undersecretary Mark Rey called invasive species the most underappreciated problem affecting national forests, noting that "populations of nonnative invasive species in the U.S. are expanding annually by 7 to 14 percent." That same year, U.S. Bureau of Land Management Director Kathleen Clarke warned, "Invasive plant species are estimated to cause more than $20 billion annually in economic damage and affect millions of acres of private and public lands."

In a well-publicized and often-quoted speech before the Izaak Walton League in July 2003, U.S. Forest Service Chief Dale Bosworth asked the nation to appreciate the latest great environmental threat to its forests. "The second great threat [after fire]," Bosworth said, "is the spread of unwanted invasive species . . . Nationwide, invasive plants now cover an area larger than the entire Northeast, from Pennsylvania to Maine. Each year, they gobble up an area larger than the state of Delaware . . . All invasives combined cost Americans about $138 billion per year in total economic damages and associated control costs." Bosworth also cited studies that estimate invasives "have contributed to the decline of almost half of all imperiled species."

In the literature of environmental groups, the alert has a doomsday ring: "An invasion is under way that is undermining our economy and endangering our most precious natural treasures" begins the invasive species page of Nature-Serve, an international database supported by natural heritage programs. Environmental groups first began to coalesce around this issue in the late 1990s. Representatives of groups that fund environmental programs, meeting as the Consultative Group on Biological Diversity in 1999, commissioned a review of the invasive species threat that linked the issue to both economics and biodiversity. Among the groups that have made invasives a top priority are Defenders of Wildlife, the Union of Concerned Scientists, The Nature Conservancy, World Resources Institute, Conservation International, the Wilderness Society, the Environmental Defense Fund, Natural Resources Defense Council,

Sierra Club, and Audubon Society. The U.S. federal government's role is guided by an interagency National Invasive Species Council (NISC) created in February 1999 when President Bill Clinton issued Executive Order 13112 (available online at www.invasivespecies.gov). Canada issued an "Invasive Alien Species Strategy for Canada" in 2004 as part of its commitment to the United Nations Convention on Biological Diversity. The leaders of 150 nations signed the convention, which recognizes invasive species as a major threat to biodiversity globally (http://www.cbd.int/invasive/problem.shtml).

The NISC now estimates that "just 16 invasive plants alone infest over 126 million acres of range and pasture lands. They are spreading at a rate of 1.3% to 25% annually. Nationwide aquatic weeds are estimated to cost the economy from $1 to 10 billion annually. The State of Florida spends $30 million anually to control invasive aquatic weeds alone." The NISC underscores the importance of invasives in natural areas with this assertion: "Nationwide, 42 percent of the species listed under the Endangered Species Act are at risk primarily because of invasive species."

This kind of notice has put the invasive species issue on center stage and has brought with it a cast that demands new laws, regulations, and funding. Secretaries of Agriculture under both Presidents Bush and Obama won more and more funding for the fight against invasive species. Speaking for President Bush on October 22, 2003, and announcing $1.5 million in grants to universities in eight states, Agriculture Secretary Ann Veneman said, "Protection of the nation's agriculture and natural resources from invasive pests is a top priority for the Bush Administration." In 2002 an interagency task force of federal government scientists issued a report whose alarm was as intense as that of any environmental group. Its executive summary began, "America is under siege by invasive species of plants and animals, and by diseases. The current environmental, economic, and health-related costs of invasive species could exceed $138 billion per year—more than all other natural disasters combined" (USGS 2002).

This is the kind of language used to underline the huge size of assets at risk and thus to justify large public programs and expenditures. Even before the Bush administration signed on, total federal spending to fight invasive species had risen to some $600 million in fiscal year 2000 (Tate 2002). Spending by the Animal and Plant Health Inspection Service (APHIS) rose from some $556 million in fiscal year 2000 to $987 million in the 2005 U.S. Department of Agriculture budget. The USDA's 2011 budget for invasive species topped $1.3 billion (ISAC 2011).

A Nature Conservancy task force in 2001 recommended that The Nature Conservancy "[e]levate the political profile of the invasive alien species issue to establish new funding and policy support for invasive species management in the U.S. and internationally" (The Nature Conservancy 2001). They proposed to spend $10 million in research. The Union of Concerned Scientists credits itself with rallying more than 300 experts on invasives to its "Sound Science Initiative" (UCS 2001).

The Nature Conservancy, which owns hundreds of thousands of acres of land, began working on invasive species in the late 1980s, but farms, forests, highway departments, parks, and homeowners have been fighting invasive species for over a century. Farm stores and the garden section of every department store have offered a variety of remedies to attack invasive species, from herbicides to traps and bullets. Individual species like the snakehead fish, sudden oak death and giant hogweed have sometimes made national news. In 2003 the government agencies joined environmentalists, greatly expanding its presence in an increasingly popular war against invasives in general. Why did this old, ever simmering guerilla war suddenly become a major battlefront for the environmental movement?

First, the problem is real, it is big, and it is both an economic issue and an environmental issue. Since the advent of European settlement in North America over 50,000 species of plants alone have been introduced. While many enhance our landscaping and others provide 98 percent of our crops, some 5,000 have gone wild to compete with some 17,000 native plants. (Morse et al. 1995; Morin 1995). Estimates of how fast and how extensively they are replacing natives vary, but no one who has seen the blooms of garlic mustard in eastern forests, the broad yellow fields of star thistle in the West, the hair-thick stands of melaleuca in the Everglades, or the impenetrable mats of water chestnut on northern lakes and rivers can doubt that change is everywhere.

Often a plant that is a normal part of its native environment becomes a domineering force in its new home. Australian melaleuca grows much more densely in the Everglades than in its native habitat, and has spread at a rate of 29,700 acres a year (Campbell 1994). It has real costs to both wildlife and to the free flowing water regime that filters and provides much of Florida's water. In Utah's Great Basin, European cheatgrass has accelerated fire frequency from every 60 to 110 years to every 3 to 5 years. This volatile invader has come to dominate some 5 million acres in Idaho and Utah (Whisenant 1990).

Many environmental groups and journalists have eagerly dramatized the invasives problem, aided by scary names like bushkiller, skunk vine, fire ant, African killer bee, mile-a-minute vine, fishhook water flea, dog strangling vine, and sudden oak death. In April 2012, the Florida Exotic Pest Plant Council and the state chapter of the Wildlife Society called their joint conference, "Invasion of the Habitat Snatchers." The National Forest Foundation starts the scare early when their classes for children say of invasive plants, "They're really 'mean' plants; they're playground bullies that put native plants in a headlock and give them a noogie."

While the problems are often large and even scary (e.g., West Nile virus), the negatives are not the whole story. A complete balance sheet would also note that many introduced plants, including some that invade natural areas, have had economic and social benefits. In fact, many species, like kudzu (used for erosion control in the Southeast), were introduced for their benefits and have provided those benefits even as escapees. In this sense, many species are unwelcome only in a superabundance or in the wrong place or because they serve no important economic need or because the media ignores their services.

The zebra mussel, notorious for clogging power plant intakes, also provides water filtering and clarifying that benefits some plant and fish populations. Louisiana has been trying to create a market for the large muskrat-like nutria, a good source of meat and fur. Tamarisk, or salt cedar, was introduced in the early 1800s for its ability to grow rapidly (up to 12 ft. a year), provide dense windbreaks, and colonize heavily saline soils where little else will grow. It turns out salt cedar invasions have been a boon to populations of the endangered willow flycatcher, which prefers salt cedars for nesting (Zavaleta et al. 2001). Japanese barberry, besides being an attractive ornamental, provides an abundance of fruits for wildlife. Judgments about some invasives, like salmon in the Great Lakes, are a matter of environmental preference, while the European honeybee, a continuing boon to farms and gardens across America, appears to have no organized opposition. As we note in our habitat descriptions for individual plants, a great many colonize abandoned and barren lands and this is often a real service.

Nevertheless, the damages and the costs in billions of dollars required to control unwanted and overabundant invasives is one more proof for environmental pessimists that humankind has ruined nature and should not disturb nature's landscape plan. Some have projected present rates of spread into the future ad infinitum without allowing for saturation, the development of natural controls, or other vectors that might slow or stop an invasive. The invasive species issue also has a convenient link to one of the great bugaboos of social activists of all sorts—globalization. Increased global trade has indeed accelerated the movement of biological agents between countries and radically accelerated its ancient role in the spread of invasive species. Some environmentalists have already nominated free trade as the primary villain. Forestry activist and respected plant ecologist Dr. Jerry Franklin has declared, "It's time to stop moving green plants and raw wood between continents" (McClure 2003).

While farmers are well aware of the costs of invasive species, they are also frightened by the potential for eco pessimists to capture the issue. "Unless farmers and ranchers become active in their approach to this issue now, due to heavy environmental influence, federal controls could far surpass the type of abuses of power already experienced with the Endangered Species Act," says Michele Dias, California Farm Bureau Federation attorney. She noted that the 108th Congress was considering more than 50 bills addressing invasive species (Dias 2003).

To paint the invasive species issue as a choice between the native environment and alien species, between preservation and human meddling, obscures the real issue. Ecosystems change over time with shifts in climate, nutrient inputs, rainfall patterns, and the relative abundance of species. Human choices about energy sources, farming and ranching practices, natural resource extraction, urban planning, wildlife management, and conservation increasingly guide what our ecosystems will look like in future. How should we manage the new competition between the native plants that existed here before the European settlement or since the last ice age and those native and alien species that seem well-adapted to current environmental conditions? Two subsidiary questions are: 1. What can we mange successfully? and 2. Do the real benefits justify the full costs? The invasive species issue is real, and environmentalists can take a major part of the credit for bringing it onto the public stage. The heart of the matter, however, is what it means to restore an ecosystem. We will make intelligent decisions only when the debate shifts from the notion that "native" is always better to the all-important question of how we should manage change in that dynamic system of tradeoffs that is our natural economy.

Ashe, D. (chief of the National Wildlife Refuge System). April 19, 2001. Appearing before the Fisheries Conservation, Wildlife, and Oceans Subcommittee regarding Invasive Species Control within The National Wildlife Refuge System.

Bosworth, D. (chief of U.S. Forest Service). July 17, 2003. "We need a new national debate." Pierre, SD: Izaak Walton League, eighty-first annual convention.

Campbell, F. T. 1994. "Killer pigs, vines, and fungi: Alien species threaten native ecosystems." *Endangered Species Technical Bulletin* 19(5):3–5.

Center for Invasive Species and Ecosystem Health. http://www.bugwood.org

Cusak, C., M. Harte, and S. Chan. 2009. "The economics of invasive species." SeaGrant Oregon. http://www.oregon.gov/OISC/docs/pdf/economics_invasive.pdf.

Dias, M. (attorney, California Farm Bureau Federation). August 6, 2003. "Farmers must be involved in invasive species debate."

Dudley, D. 2000. "Wicked weed of the west—yellow star-thistle: Scourge of the golden state." *California Wild* (fall).

Hebert, G. October 21, 2003. "Aquaculture project aims to alleviate hunger, promote sustainability in poverty-stricken countries." *Marine Biological Laboratory.*

ISAC. 2011. Invasive Species Advisory Council December meeting minutes. http://www.invasive species.gov/global/ISAC/ISAC_Minutes/2011/Minutes_ISAC_Dec_6_8_1_FINAL.pdf

Jetter, K. M., J. Hamilton, and J. H. Klotz. 2002. "Red imported fire ants threaten agriculture, wildlife, and homes." *California Agriculture* (January–February). http://ucanr.org/repository/cao/landingpage.cfm?article=ca.v056n01p26&fulltext=yes.

Ludke, L., F. D'Erchia, J. Coffelt, and L. Hanson. 2002. *Invasive Plant Species: Inventory, Mapping, and Monitoring—A National Strategy.* U.S. Geological Survey Information and Technology Report 2002-0006. http://www.fort.usgs.gov/Products/Publications/21272/21272.pdf

McClure, R. 2003. "Debate over forests is a difference in priorities." *Seattle Post Intelligencer* (October 14).

Morin N. 1995. "Vascular plants of the United States." In *Our Living Resources: A report to the nation on the distribution, abundance, and health of U.S. plants, animals, and ecosystems,* edited by E. T. LaRoe, G. S. Farris, C. E. Puckett, P. D. Doran, and M. J. Mac, 200–205. Washington, DC: U.S. Department of the Interior, National Biological Service.

Morse, L. E., J. T. Kartesz, and L. S. Kutner. 1995. "Native vascular plants." In *Our Living Resources: A report to the nation on the distribution, abundance, and health of U.S. plants, animals, and ecosystems,* edited by E. T. LaRoe, G. S. Farris, C. E. Puckett, P. D. Doran, M. J. Mac, 205–209. Washington, DC: U.S. Department of the Interior, National Biological Service.

National Forest Protection Alliance. http://www.forestadvocate.org/news/RESTOR.FS.WEB.pdf.

National Forests Foundation. http://www.nationalforests.org/blog/post/40/weeds-kids-and-bugs-at-gold-creek-pond#post.

National Invasive Species Council. Issues Overview Factsheet. http://www.invasivespecies.gov/Factsheets/Issue_Overview.pdf.

Nature Conservancy, The. 1996. *America's least wanted: Alien species invasions of U.S. ecosystems.* Arlington, VA.

———. 2001. *Abating the threat to biodiversity from invasive alien species: A business plan for engaging the core strengths of the Nature Conservancy.* Arlington, VA.

NatureServe. 2003. http://www.natureserve.org/conservation/invasivespecies.jsp.

Office of Technology Assessment. 1993. *Harmful non-indigenous species in the United States.* Washington, DC: U.S. Congress.

Olson, L. J. 2006. "The economics of terrestrial invasive species: a review of the literature." *Agricultural and Resource Economics Review* 35(1):178–194. http://ageconsearch.umn.edu/bitstream/10181/1/35010178.pdf.

Pimentel, D. L. Lach, R. Zuniga, and D. Morrison. 2005. "Update on the environmental and economic costs associated with alien-invasive species in the United States." *Ecological Economics* 52(3):273–288.

Silliman, H. (assistant city editor). 1998. "Prickly star thistle invasion spreads." *Montana Democrat* (September 9).

Tate, J., Jr. (science advisor, U.S. Department of the Interior). October 2, 2002. Appearing before the House Agriculture Subcommittee on Department Operations, Oversight, Nutrition, and Forestry.

Union of Concerned Scientists. *Annual Report,* 2001. Cambridge, MA. http://www.ucsusa.org/assets/documents/ucs/annrepfinal-1.pdf.

U.S. Bureau of the Census. 1998. *Statistical abstract of the United States, 1996.* 200th ed. Washington, DC: U.S. Government Printing Office.

U.S. Department of Agriculture. "National Agricultural Library." http://www.invasivespeciesinfo.gov/economic/main.shtml.

U.S. Department of Interior. 1998. *U.S. Geological Survey, 1998: Status and trends of the nation's biological resources.* Washington, DC: U.S. Government Printing Office. http://www.nwrc.usgs.gov/sandt.

———. April 29, 2003. "The growing problem of invasive species." Joint oversight hearing. U.S. House of Representatives. http://www.gpo.gov/fdsys/pkg/CHRG-108hhrg86708/pdf/CHRG-108hhrg86708.pdf.

Whisenant, Steven G. 1990. "Changing fire frequencies on Idaho's Snake River plains: ecological and management implications." In: *Proceedings of a symposium on cheatgrass invasion, shrub die-off, and other aspects of shrub biology and management,* edited by E. D. McArthur, E. M. Romney, S. Smith, and P. T. Tueller, 4–10. Ogden, UT: U.S. Department of Agriculture Forest Service, Intermountain Research Station.

Wilcove, D. S. D. Rothstein, J. Bubow, A. Phillips, and E. Losos. 1998. "Quantifying threats to imperiled species in the United States." *BioScience* 48(8):607–615.

———. 2000. "Leading threats to biodiversity: What's imperiling U.S. species." In *Precious heritage: The status of biodiversity in the United States,* edited by A. S. Bruce, L. S. Kutner, and J. S. Adams, 242. Oxford, UK: Oxford University Press.

Zavaleta, E.S., R.J. Hobbs and H.A. Mooney. 2001. "Viewing invasive species removal in a whole-ecosystem context." *Trends in Ecology and Evolution,* 16:454–459.

Using This Book

If you encounter a plant that appears to be spreading rapidly in a preexisting plant community or a weed in your garden that suddenly appears to be taking over, chances are you can learn more about it in this book. In the following pages we describe more than 200 plant species that occur from the southernmost states of the United States to northernmost Canada and Alaska. These plants grow in all kinds of habitats, from rivers to marshes to forests to prairies.

This book classifies invasive plants first by whether plants occur on land (terrestrial) or in water (aquatic) and second by type of plant (tree, shrub, vine, herbaceous, grass or sedge, or fern). Within each plant type, the species are further divided by characteristics that are relatively easy to distinguish with a little practice, such as whether the leaves alternate along the stems or occur opposite to each other and whether the leaves are entire or made up of many leaflets. If you are unsure of any of the botanical terms, check the glossary toward the end of the book for an explanation. For readers more familiar with botany and the scientific names of plants, the plants are arranged alphabetically by Latin name within each grouping.

USING THE KEY: WHAT IT DOES AND DOES NOT DO

Many field guides include keys to help identify plants without having to flip through all the pages of photographs. Our key (starting on page 36) helps you narrow your search, but because we cover so many plants (all of them nonnative) throughout North America, the key will not take your identification to an individual species but will direct you to a group of similar species that is small enough that you can quickly survey it and see if your target is there. Within each species account we list other plants that may look similar, and we list special characteristics to help you distinguish close relatives or native plants that may be similar in appearance.

You may want to confirm your identification using a regional field guide for a specific group of plants. Such guides will have more comprehensive keys, allowing for comparison of both native and nonnative species that occur in a specific region.

MANAGING INVASIVE SPECIES: A START

Strategies and tactics for managing or eradicating invasive species have complex ecological, social, geographical, economic, and even political dimensions. We list general options in the chapter on management, particularly focusing on manual and chemical means of controlling invasive plant species. In the field guide pages, we list recommended approaches for each species, including biological, manual, and chemical methods. Because chemical controls can have unwanted consequences for non-target species, readers will need more detailed information on rates and means of applications and the range of plants and animals affected. Readers who want to undertake active management should follow our references and explore the dimensions of each option.

FOLLOW UP WITH REFERENCES

This book is your portal to enter the world of invasive plants and invasive plant science. Our references—general, chapter related, and species particular—take you deeper into the subject. With many references we give website addresses. These addresses may or may not last as long as this edition of the book. If the full address does not work, we advise the reader to shorten it to the root part of the address, go there, then search for the species of plant to which the reference refers. The references at the end of the book also list many regional guides to invasive plant species, some of which are available online.

ONE

The Aliens Landed Long Ago and Keep Arriving

"Alien species" means, with respect to a particular ecosystem, any species, including its seeds, eggs, spores, or other biological material capable of propagating that species, that is not native to that ecosystem.

"Invasive species" means an alien species whose introduction does or is likely to cause economic or environmental harm or harm to human health.
—Both quotes from the President's Executive Order 13112, 1999

The government's distinctions in the quotes above can be summarized as follows: All invasives are aliens but not all aliens are invasives. Aliens are also called nonnative, exotic, or nonindigenous. All are life forms moved purposely or accidentally to a new location where they did not evolve. Today a debate rages about how to deal with invasive species or whether to deal with most of them at all. Is "harm" often nothing more than change? The debate is new but the phenomenon is old.

Some 180 million years ago, the breaking apart of Pangaea, the single earth landmass from whose division modern continents began to form, decisively separated evolving species and allowed or forced them to take very different paths. Mountains, deserts, and climate zones further divided the geographical stages for evolution. Over millions of years the species that evolved in each territory established relationships that were sometimes cooperative, sometimes adversarial. Within certain limits, and subject to many accidents of climate and geology, most species established a durable place within a landscape or ecosystem. These systems were not static but dynamic—populations waxed and waned, exploded and collapsed. New species occasionally appeared or disappeared, but except for radical changes caused by global warming or cooling or sudden catastrophes, the plants and animals typical of the ecosystem remained in place.

As that ingenious animal *Homo sapiens* began to multiply its abilities and powers with technology, it also began to reshape the landscapes wherever it settled. Fire, hunting, flooding, roadways, and cultivation upset ecosystems. Ever more mobile human populations also took with them, deliberately and by accident, plants and animals that would compete with the native plants and animals in the places they were transported to. The human economy changed the natural economy, giving some species an advantage and others a disadvantage in their competition for resources. Humans also changed the mix of competitors vying for sunlight, nutrients, and real estate in the natural economy.

The natural economy, unlike the human economy, evolves very slowly; most species lack the intellect and imagination that make quick decisions and adaptations possible (exceptions being microbes whose size and reproductive rate allow for rapid changes). Nature's way is more conservative and lower risk, but it also means that in nature, unlike in many human habitats, room for one more does not always exist. A newcomer must muscle out, in whole or part, those natives who use the resources it needs. Those resources often include water, sunlight, and important soil nutrients.

The long history of invasive species in America is like many other histories of environmental issues—one in which the main character (for us, invasive plants) has played a variety of roles. For instance, the automobile, recently the cause of so much air pollution, a century ago was the machine that helped replace horses as a source of power and transportation that had consumed 2 to 10 acres per animal for fuel and left the streets of American towns rank with manure and buzzing with flies. Coal once replaced the clearing of forests for fuel wood and the killing of whales for lamp oil, but then became the source of major air pollutants. So too, many of today's invasive plants came to America as welcome guests, often the guests of a government program or of brilliant people like plant breeder Luther Burbank or naturalist John Bartram.

European colonists who found themselves in what Massachusetts governor John Winthrop called a "howling wilderness" and Shakespeare might have called a "brave new world," found themselves surrounded by strange plants and animals. They immediately began importing their familiar home country species for agriculture, medicine, and decoration.

Spanish missionaries in coastal California first planted the 3-foot-tall black mustard (*Brassica nigra*) that now grows wild and paints many meadows and old fields yellow in spring. The use of mustard in food and medicine dates back over 2,000 years, and some European monasteries had developed a good business growing mustard. It was a welcome cure for headache and flu as well as a spice. Spanish missionaries also introduced the bamboo-like giant reed (*Arundo donax*) for use in light construction. The reed quickly escaped to colonize thousands of acres of riverbanks once dominated by cottonwoods and other native plants. The result has been the replacement of natural firebreaks with the very flammable giant reed.

Spanish missionaries, of course, had lots of company in the colonial era. By 1727 English ivy appears in American history, and before long it had established itself in moist landscapes from Massachusetts to the Pacific Northwest. English ivy had been established in European herbal medicine for many centuries. The best universities of early America adorned their brick buildings with the vine, which will grow up to 90 feet long. By the end of the twentieth century, however, the ivy had taken over so much forest habitat that citizens of the Pacific Northwest formed the No Ivy League, organizing work parties to comb woods and neighborhoods, pulling out ivy by its roots.

Plants like kudzu were sometimes promoted to solve environmental crises. Agricultural experts encouraged the planting of this Asian broad-leafed vine that can grow 6 inches in a day as a way of controlling soil erosion caused by intensive farming in the South. Kudzu has several beneficial functions besides stabilizing clay soils. The Chinese use an extract of kudzu isoflavones to control the desire for alcohol and the physiological damage from alcohol. The Japanese batter and fry the young leaves, and a starch made from the roots is rich in isoflavones thought by some to be effective against prostate trouble and some cancers. The Japanese also use the fibrous vine in making cloth.

Many prominent Americans in colonial times and the first years of the republic actively planted alien species. George Washington in 1786 noted in his journals that he had planted several European plants given to him as a present by the French botanist André Michaux. The notes reveal he had already planted other aliens—pistachio nuts, Spanish chestnuts, Chinaberry tree, and buckthorn from Europe. The famed Quaker botanist John Bartram proudly cultivated European and Middle Eastern plants in his garden. He is also responsible for introducing some 200 species of American plants to Europe in his long exchange with English botanist Peter Collinson.

So far as we know, no one has introduced an invasive plant maliciously, with the intent of damaging the American landscape or economy. Plants that were not once welcome immigrants almost always came in accidentally. The song that celebrates "rolling along with the tumbling tumbleweed" seems quintessentially American, but the round, dry, bushy plant that winds dislodge and blow across the high plains and deserts of the West is actually Russian thistle (*Salsola tragus*). It is native to the Ural Mountains that divide European Russia from Siberia. Its seeds came to South Dakota mixed with flax seed imported by Ukrainian immigrant farmers. The frequent droughts of the plains often devastated wheat, corn, and flax, but tumbleweed needs very little soil or water.

While the introduction of alien plants into North America had its initial explosion in colonial times, a second great wave began with the advent of modern shipping and the globalization of trade. Again, aliens arrived by both intent and accident. Thousands of plants and animals have arrived in the ballast of ships, on their hulls, as contaminants in cargoes of seeds, or in packing materials made from plant materials. Landscapers, nurserymen, and homeowners imported many plants that continue to decorate our yards and public spaces even while they invade our wild places. The most familiar names include Japanese barberry, burning bush, privet, water hyacinth, purple loosestrife, and that staple of shady tree-lined suburban streets, the Norway maple. Just as Americans develop cravings for foreign foods, they have had love affairs with exotic plants. The nursery industry raises millions of barberry and burning bush plants that are planted across the United States in front of fast-food restaurants, schools, and suburban homes, and yet these same species invade the surrounding forests in the northeastern United States.

Government environmental policy also played a role in importing invasives. Our third president, Thomas Jefferson, wrote, "The greatest service which can be rendered any country is to add a useful plant to its culture" (Randall 1994). Our sixth president, John Quincy Adams, established as national policy that "[t]he United States should facilitate the entry of plants of whatever nature whether useful as a food for man or the domestic animals, or for purposes connected with . . . any of the useful arts" (Hyland 1977).

Introducing new plants to America was once so popular that the activity was encouraged by the United States Office of Plant Introduction whose officials once boasted of introducing 200,000 species and varieties of nonnative plants. In the 1930s the U.S. Soil Conservation Service promoted multiflora rose for erosion control, highway dividers, and as a living fence, but it spread so rapidly that several states now list it as a noxious weed.

By 1900 the dangers of importing plants deliberately or accidentally had become so obvious that Congress passed the first federal laws to control alien species. The Lacey Act applied only to plants and animals that threatened agriculture, but it was soon followed by the Plant Pest Act, the Plant Quarantine Act, and the Noxious Weed Act. The primary promoters of these acts were farmers and their goal was to protect their crops and animals from wild species.

In the 1970s, with the advent of modern environmentalism and its concomitant enthusiasm for things wild, natural, and native, exotic species began to lose much of their prestige and attraction. In areas where water shortages occurred, homeowners and landscapers often responded by looking for native species adapted to the local water regime. A new environmental awareness, coupled with rising costs of labor, fertilizers, and pesticides, has created a demand for landscapers and extension services offering advice on plantings that conserve water and require little fertilizer or pesticide.

Some government-supported programs that encouraged species like kudzu and multiflora rose were indicted by costly unintended consequences. Kudzu was killing trees, and multiflora rose was growing in impassible thickets that ruined pastureland. Today the scope of government action has been expanded to protecting America's wild lands from both wild and domesticated aliens.

By Executive Order 13112 in 1999, President Bill Clinton created the interagency National Invasive Species Council, which includes most cabinet secretaries. On January 18, 2001, the Council issued their *Invasive Species Management Plan*. The plan summarized in startling numbers the government's view of invasive plants: "Invasive plants are estimated to infest 100 million acres in the United States. Every year, they spread across three million additional acres, an area twice the size of Delaware. Every day, up to 4,600 acres of additional Federal public natural areas in the western continental United States are negatively impacted by invasive plant species" (National Invasive Species Council 2001).

Canada issued its National Invasive Species Strategy in 2004, attributing $7.5 billion in damages to the impacts of invasive plants on the agriculture and forestry industries, social costs to rural Canadian and aboriginal populations, and environmental costs to Canada's biodiversity. The country's first nationwide ecosystem evaluation of biodiversity in 2010 listed invasive species as a major threat to Canada's ecosystems, with 24% of Canada's flora nonnative and 486 nonnative plants identified as weedy or invasive.

President Barack Obama issued Executive Order 13514, "Federal Leadership in Environmental, Energy and Economic Performance," in 2009, leading to guidelines for sustainable landscaping at federal government's 429,000 buildings on 41 million acres. These guidelines emphasize removing invasive ornamental plants and selecting noninvasive plants to replace them. Most states and provinces now have invasive species councils that exchange information about new occurrences of invasives, regulatory issues, and plant, agriculture, and pet industry concerns. A few states, including Florida, Massachusetts, Connecticut, and Maryland, now regulate the production and sale of some invasive ornamental plants. Some towns and homeowner associations have removed invasive plants from their approved planting lists.

Scientists and economists are perfectly capable of waging hot and complex debates among themselves, but whenever government devotes a large part of its energy and taxpayers' resources to an issue, the debate becomes public and often very divisive. Money is part of the reason; the other part is distrust of political intervention into a scientific and legal dispute. The Clinton administration in the late 1990s committed the government to a major effort to control and/or eradicate many invasive species. The Bush administration did not seriously question that initiative. As we noted in the introduction, the chief of the U.S. Forest Service, Dale Bosworth, called invasive species the second greatest threat to national forests, and Agriculture Secretary Ann Venneman said in 2003, "Protection of the nation's agriculture and natural resources from invasive pests is a top priority for the Bush Administration" (Venneman 2003).

The case that America is overreacting to invasive species has been summed up by Professor Mark Sagoff, Pew Scholar in Conservation and the Environment at the University of Maryland. Sagoff argues that "the concept of 'harm to the environment' may not be definable in scientific terms," and that "introduced species typically add to the species richness of ecosystems; studies suggest, moreover, that increased species richness correlates with desirable ecosystem properties, such as stability and productivity." He also argues that the level of alarm about extinctions is overblown, saying "there is no evidence that nonnative species, especially plants, are significant causes of extinction, except for predators in certain lakes and other small island-like environments." Sagoff argues that excluding nonnative species from definitions of biodiversity or ecosystem integrity feeds the idea that these species are harmful when in fact they might be innocuous or even beneficial (Sagoff 2005). The World Conservation Union's (IUCN) Invasive Species Specialist Group, however, contends that "habitat alteration and invasive species impacts have been the major cause of species extinctions over the past few hundred years, increasing the rate of extinction by about 1,000 percent."

A guidebook like this one cannot and should not take sides in this debate because we cannot present the issues in the detail they deserve, but we would be social delinquents if we omitted any mention of the debate, since it is one of the most important reasons for writing this book. Our notes on each species describe its role in the environments it invades as well as its most significant impacts and benefits. Our bibliography and list of websites include references for any reader who wants to explore the arguments about the costs and benefits of invasive species and how we should think about them.

Canadian Councils of Resource Ministers. 2010. "Canadian Biodiversity: Ecosystem Status and Trends 2010." http://www.biodivcanada.ca/ecosystems.

Clavero M, and E. Garcia-Berthou. 2005. "Invasive species are a leading cause of animal extinctions." *Trends in Ecology and Evolution* 20: 110.

Environment Canada. 2004. "An Invasive Alien Species Strategy for Canada." http://www.ec.gc .ca/Publications/default.asp?lang= En&xml=26E24C67-2299-4E7A-8014-9FB6B80695C5.

Hyland, H. L. 1977. "History of U.S. plant introduction." *Environmental Review* 4:26–33 (National Agricultural Library Q125.E5).

Invasive Species Specialist Group (of the IUCN). 2012. "About Invasive Species." http://www .issg.org/about_is.htm.

National Invasive Species Council. 2001. *National management plan: Meeting the invasive species challenge.* http://www.invasivespecies.gov/main_nav/mn_NISC_Management Plan.html.

Randall, W. S. 1994. *Thomas Jefferson: A life.* New York: Harper Collins.

Sagoff, M. 2005. "Do non-native species threaten the natural environment?" *Journal of Agricultural and Environmental Ethics* 18:215–236. doi 10.1007/s10806-005-1500-y.

TWO

Invasives Changing Wild America

Our reference point for understanding what alien plants have done to America's landscapes has to be the pre-Columbian environment of North America. Indigenous people who began spreading across the continent in large numbers some 12,000 years ago, of course, changed the natural landscape forever by hunting some species to extinction and by hunting others with habitat-changing fire. As far as we know, however, their migration route across the ice age land bridge joining Alaska and Russia and their slow spread south prevented them from bringing in plants that would establish themselves in the New World. The natives of the Bering Straits region were never agriculturalists. Evidence for early migration across the Atlantic and possibly across the Pacific may offer greater potential for pre-Columbian plant introductions, but scientists have found no evidence for wide-scale additions to the flora. Pre-European Americans did introduce some plants to places where they had never grown before—corn being but one example. So far as we know, these economic relocations never became naturalized. While the history of aboriginal changes to the landscape is fascinating, our topic is the change brought by alien and invasive plants. So far, the known story begins only with European settlers.

By far the greatest changes in the American land made before or after Columbus were physical changes made by the spread of human settlement and its infrastructure of housing, roads, canals, and cultivated lands, and the resources exploited to build them. The story of invasives is part of this change. Scientists debate the exact size of changes wrought by the invasive species, but not the expanding presence and influence of invasives on our landscape and ecosystems. The fact that invasives are increasingly present is one of the principal reasons for this guidebook.

Common barberry was the first plant to win notoriety as an invader because it harbored wheat stem rust. In 1726 the General Court of Connecticut noted: "The abounding of barberry bushes is thought to be very hurtful, it being by plentiful experience found that, where they are in large quantities, they do occasion, or at least increase, the blast on all sorts of English grain." The court decided that, from then on, town meetings in the state could pass measures enforcing barberry eradication and fining violators. By 1779 several

New England colonies had laws that allowed anyone to eradicate barberry wherever it occurred (Fulling, 1943).

Another invasive species that arrived in the early 1900s changed American public opinion forever because it devastated the much-loved dominant tree of mature eastern forests. A fungus from Japan that probably piggybacked on resistant Asian trees appeared in the the New York Botanical Garden and killed its American chestnut trees. It quickly escaped into the wild. Chestnut blight spread at the rate of 25 miles a year, relentlessly killing the chestnuts, trees which could grow up to 200 feet tall and 14 feet in diameter (yielding a great deal of prized lumber) and which made up 25 to 50 percent of the eastern tree population (Ashe 1911; Buttrick 1925). The chestnuts' death became a national disaster. Oaks were the most frequent replacements for chestnuts, and many eastern forests have large oaks that date back to the demise of the chestnut. The awakening of the American public with the fall of the chestnuts illustrates that until invasive species problems express themselves in a dramatic way on a large and appealing species, rousing the interest of the public or government is difficult.

No introduced plant has come near rivaling chestnut blight in changing the North American landscape or attracting public attention. The most intensive efforts to control invasive species still focus on alien pathogens and insects. Various boring insects now threaten large areas of evergreens and hardwoods. A fungus-like alga that kills oaks has been found on both coasts where "sudden oak death" has become a well-known alarm. Aquatic invaders like the zebra mussel and Asian carp also receive significant national attention. Although invasive plants do not get the same attention as these more virulent invaders, they are making noticeable changes in both the economy and natural history of North America.

The U.S. government's Bureau of Land Management (BLM), which manages more than 250 million acres, estimated in 2000 that 35 million acres of its lands in the West were infested with invasive species. They expect that estimate to double in a new survey completed in 2010 (BLM 2010). The U.S. Department of Agriculture's National Invasive Species Strategy reports that "the Natural Resource Information System documents approximately 6 million National Forest System acres infested with invasive plants." The National Park Service estimates that 2.6 million acres of park land are dominated by invasive plants.

Government itself, of course, has brought in many invasive species, with good intentions. In the Southeast government planners deliberately spread kudzu vine and subsidized its planting in a scheme to control erosion and feed cattle. The government in the early 1900s paid farmers to plant kudzu, then in the Great Depression its Civilian Conservation Corps workers planted kudzu roots over thousands of acres. Today kudzu thrives on over a quarter million acres from southern Ontario and Illinois in the north to Florida and Texas in the south and has been found recently in Oregon and Arizona.

Compared with kudzu, cheatgrass (*Bromus tectorum*) is invisible to the average nature watcher, although anyone who has walked through western meadows and prairies in the summer may have spent a long time picking the sharp seed spikes out of socks. Imported in 1898 from Eurasia as a promising dry land cattle forage, cheatgrass quickly spread across all of the lower 48 states and proved it feeds range fires better than it feeds cattle. Its seeds germinate in the fall and by spring its early-developing root system uses most of the winter and spring moisture before native grasses get started, thus setting the stage for a longer fire season and fewer native species.

Few ecosystems in the United States play a more important role for their urban and agricultural neighbors as Florida's Everglades, which is also one of the richest wildlife preserves in America. Miami and other Florida cities depend on the Everglades for water. Yet in the first half of the 1900s both the government and private landowners thought the Everglades had too much water. One of the methods employed to drain it was the planting of Australian melaleuca trees, sometimes known as the "tea tree" or "paperbark tree."

Private landowners seeded melaleuca from the air while the U.S. Army Corps of Engineers planted it in 1940 and 1941 to stabilize levees protecting fields and towns from floods. Unlike many stands in its native regions, melaleuca in Florida did not develop open groves of thick-trunked trees. It grows in hair-like thickets often with more than 1,500 trees per acre (30 trees in the space of an average living room). Patches up to 5 acres large eliminate most native plants and habitat. On millions of Everglades acres they now absorb water four times faster than the native sawgrass they replace, increasing fire danger and blocking normal water flow so vital to the Everglades ecosystem.

Another almost unnoticed invasion has radically changed wetlands in more northern states. A tall grass called phragmites (*Phragmites australis*) has been native to freshwater and brackish wetlands for thousands of years and more. During the past century phragmites stands seemed to grow larger and much denser, outcompeting other wetland plants, destroying aquatic habitat, and filling in many shallow ponds. Recently scientists studying the genetic makeup of phragmites recognized that the denser stands are not a native genetic strain of phragmites but a European strain that is indistinguishable to the untrained eye (Saltonstall 2003). The extent of the invasive strain has not been well measured, but the value of the habitat it is invading is clear. Once-open beaches on the shores of the Chesapeake Bay are now solid stands of phragmites that terrapins cannot dig through to lay their eggs.

Since change is always present in natural systems and large changes are not uncommon, the question raised by the changes invasive plants make in communities and ecosystems is "so what?" Apart from the troubles these changes cause people and their economy, do invasives actually harm a natural system any more than noninvasives, or do they merely change communities and ecosystems? We've already noted that Mark Sagoff of the University of Maryland argues that "harm" is not a term that can easily be defined by science (Sagoff 2005). He points out that often invasives are defined as harmful simply because someone has decided that any change caused by a nonnative species damages the native ecosystem. If harm is defined as change, then the question is why change caused by an alien is harmful and change caused by a native is natural?

Other scientists point out that ecologists, conservationists, and land managers often believe invasives are a leading cause of species extinctions, because "[n]ative species' declines often occur simultaneously and in the same place as invasion by nonnative species," but "[e]xisting data on causes of extinctions and threats are, in many cases, anecdotal, speculative, or based upon limited field observation" (Gurevitch and Padilla 2004). Because so many forces are acting at one time in any ecosystem, no one can say for certain that invasive plants have caused extinctions of native species.

Nevertheless, since two species cannot occupy the same ground at the same time, simple logic tells us invasives usually displace natives—with the exception of an invasive that might occupy ground that no native can tolerate. In the cases of melaleuca, kudzu, English ivy, and European phragmites, the invasives often grow so densely or cover the ground so completely that they either crowd out natives or cut off vital sunlight. Others drop leaves or have roots that release chemicals that prevent competitors from growing (some natives like black walnut also do this, of course). Invasives that establish themselves in the territory of a threatened or endangered species become an important part of the forces pushing that species closer to the edge of extinction.

In other cases invasive species are close enough genetically to a native species that the invasive can dilute the gene pool through hybridization. The climbing vine American bittersweet (*Celastrus scandens*) has long been noted for its graceful and bright orange-red berries favored by birds. The more aggressive Asian or oriental bittersweet (*C. orbiculatus*) has been imported by the nursery business and it establishes itself on forest edges. From there it invades as trees mature and die or as storms make forest openings. Once in the forest, the vine twines around trees and can overtop a tree, stealing the sunlight from the canopy. Oriental bittersweet may be hybridizing with American bittersweet, and because the oriental is more common in the landscape and produces more fruits, the genes of American bittersweet are likely to be lost over time.

Garlic mustard (*Alliaria petiolata*) not only crowds out native plants but may threaten the endangered West Virginia white butterfly. Early settlers cultivated this biennial herb for its vitamin C–filled leaves that were available early in spring, but it is seldom collected anymore and has spread into eastern forests. Over 2,000 seedlings can grow in a square yard. This leaves little space for other plants to establish, and the dense shade from the leaves robs light from native spring wildflowers that have important environmental functions in addition to delighting the human eye and curiosity. The caterpillars of the West Virginia white butterfly usually feed on one of several native mustard species, but cannot survive on a diet of garlic mustard. Unfortunately adult butterflies cannot distinguish garlic mustard from native mustards, and because garlic mustard is increasingly common, they often lay their eggs on the garlic mustard instead of on the less common native mustard.

In the northeast, Autumn olive, Japanese barberry, shrubby honeysuckles, privets, and burning bush tend to form a dense understory layer that precludes other species from establishing; over time they outshade existing species. Autumn olive and barberry also alter the chemistry of the soil, potentially making restoration more difficult. Autumn olive fixes nitrogen and thus increases the fertility of the soil. Barberry leaves alter the pH of the soil. All of these species were introduced because they are hardy and produce fruits for wildlife and are attractive in landscapes. By displacing native shrubs, however, they change the quality of the food and the types of shelter available to animals. Often the fruits of the invasive shrubs differ in quality (lipid content, carbohydrates, etc.) from those of displaced or diminished native shrubs like spicebush and viburnums, thus favoring a different set of bird species. Exotic shrubs also differ in architecture, sometimes making nests more accessible to predators.

The economic costs of invasives are clearer and easier to quantify than their environmental impact, though often the two forces overlap or are the same. In the Great Plains, leafy spurge (*Euphorbia esula*), a Eurasian invasive, has spread across some 5 million acres. It often dominates livestock pasture and native grasses, but cattle and other grazers do not eat it. The U.S. Department of Agriculture estimates leafy spurge in just the Dakotas, Montana, and Wyoming costs ranchers more than $144 million a year in lost forage and control expenses. Knapweed and thistle are other invasives that have displaced native rangeland plants and increased fire danger.

Distinguishing between the "good" and "bad" nonnative species is not simple. Some species produce both positive and negative consequences, depending on the location and who is judging. Purple loosestrife, for example, is an attractive nursery plant but a major wetland weed. Tamarisk, or salt cedar, came into the Southwest in the nineteenth century and has established itself as the dominant species along creek and riverbeds of the arid West. It has often replaced native willow and cottonwood and outcompetes natives for water. In these tamarisk stands, however, about 25 percent of the endangered southwestern willow flycatchers find shelter for breeding. This combination of an endangered species and an invasive challenges state and federal land managers who want both to restore native vegetation and to protect the endangered bird.

Some scientists argue that nonnative species are creating novel plant communities that still perform basic ecological functions such as slowing stormwater, intercepting pollutants from rainfall, storing carbon, and taking up excess nutrients. Particularly in urban environments where soils and hydrology have been significantly altered by development, these communities that are a mix of native and nonnative species might be important in maintaining some natural processes (DelTredici 2010). Eighteen scientists from several countries argued in the prestigious journal *Nature* that "it is time for scientists, land managers, and policy-makers to ditch this preoccupation with the native–alien dichotomy and embrace more dynamic and pragmatic approaches to the conservation and management of species—approaches better suited to our fast-changing planet" (Davis et al. 2011).

What to do about invasive plant species will vary not only by species but by the perspective of the observer. It takes a lot of time to research the ecological impacts of each species in different environments and with many nonnative species we simply do not know yet what environmental impact they might have. We are only beginning to get adequate information on the long-range dynamics of many invasives—how rapidly they will spread, if their populations will peak and decline, if natural predators will spread to control them, their net impact on biodiversity and animal populations, and so on.

Because the invasive species issue is relatively new and the parameters are constantly changing, scientists and land managers need all the information they can get. So does the public. Whether readers of this book are environmental activists or just interested observers, we hope this will be their portal into a new area of natural history. In many cases, readers who discover new populations of an invasive species could play an important role in helping scientists and their communities understand and manage the species and the landscape. Be it for better or worse, "seek and ye shall find."

Ashe, W. W. 1911. *Chestnut in Tennessee.* Bulletin of the Tennessee State Geological Survey. Nashville.

Bureau of Land Management. 2010. "Scale of weed invasions on public lands." http://www.blm.gov/wo/st/en/prog/more/weeds/scale_of_problem.html.

Buttrick, P. L. 1925. "Chestnut in North Carolina." Economic paper 56. North Carolina Geological and Economic Survey.

Davis et al. 2011. "Don't judge species on their origins." *Nature* 474(7350):153–154.

DelTredici. 2010. *Wild Urban Plants of the Northeast: A Field Guide.* Ithaca, NY: Comstock Publshing Associates.

Gurevitch, J., and D. K. Padilla. 2004. "Are invasive species a major cause of extinctions?" *Trends in Ecology and Evolution* 19(9):470–474 (September).

Fulling, E.H. 1943. "Plant life and the law of man IV. barberry, currant and gooseberry, and cedar control." *The Botanical Review* 9(8):484–540.

National Park Service. 2012. *Exotic Plant Management Teams.* http://www.nature.nps.gov/biology/ invasivespecies/documents/EPMT_PrgmInfoSheet_20080214.pdf.

Sagoff, M. 2005. "Do non-native species threaten the natural environment?" *Journal of Agricultural and Environmental Ethics* 18:215–236. doi:10.1007/s10806-005-1500-y.

Saltonstall, K. 2003. "A rapid method for identifying the origin of North American phragmites populations using RFLP analysis." *Wetlands* 23(4):1043–1047.

USDA Forest Service. 2010. Research and Development National Invasive Species Strategy. http://www.fs.fed.us/research/invasive-species/docs/RD_National_InvasiveSpecies Strategy.pdf.

THREE

Managing the Good, the Bad, and the Ugly

Here are five basics of prevention and control of invasive plants:
1. Prevent introduction of invasive species. Discourage the use of invasive species in landscaping and erosion control and maintain the quality of natural areas by limiting excessive disturbance and maintaining a cover of native species.
2. Learn to recognize invasive plants and monitor natural areas regularly to catch invasions early.
3. Prevent invasive plants from reproducing vegetatively or by seed.
4. Thoroughly clean all tools and equipment used in managing invasive plants to avoid spreading plants to new areas.
5. Share information with other landowners and land managers and work cooperatively to prevent and control invasive species.

The easiest course of action is to prevent invasions in the first place. If you are landscaping, learn what plants are invasive in your region and avoid planting them. Clean your boats and fishing gear to prevent accidental aquatic invasions and clean hiking boots, tires, paws, and hooves to prevent terrestrial invasions. Maintaining a healthy landscape will reduce the ability of invasive plants to establish because a diverse, vigorous community of native plants competes better with invasive plants. Monitor your property for the appearance of new populations of an invasive species. It is always easier to remove something as soon as it appears than to wait until it is established and reproducing. Many invasive plants get their start on disturbed ground and in areas where the native plant cover has been damaged. Be particularly careful to monitor these areas and to restore native plant communities if needed. Monitoring is especially important after earth moving, hurricanes and tornadoes, floods, fires, and grazing. In some remote areas with rough ground, people are using trained dogs to sniff out certain invasive plants or employing satellite imagery to detect new infestations.

If you are faced with an area full of invasive plants, don't despair! Map out which invaders occur and where, then categorize the level of infestation (light, moderate, heavy). Next determine which areas have top priority for removal of invaders. These might be areas with rare plants or animals, heavily infested areas near high-quality uninvaded habitat, areas with new, still-small invasions, or places like streambanks and trail edges from which invaders are likely to spread quickly.

Next determine when and how you want to control each species. Most control techniques can be categorized as mechanical, biological, or chemical methods. Mechanical controls vary from hand pulling plants to mowing to cutting them down with axes or chainsaws. Generally, mechanical controls are cheap but labor-intensive. Biological controls rely on an animal eating the plant or a disease infecting it. Chemical controls use herbicides that either kill plants on contact or are absorbed into the plants. There are many ways to apply herbicides that will target specific plants.

MECHANICAL CONTROLS

Wear gloves when handling plants as some people are allergic to toxins contained in certain plants, and thorny plants can injure anyone. Hand pulling is often effective for controlling young plants or plants without extensive root systems. For larger plants, tools like a weed wrench, root talon, or mattock can be used to pull up or dig up plants. The disadvantage of pulling or digging plants is that it often causes soil disturbance, which can create new sites for the invader's seeds to germinate.

Cutting down plants is often a central objective for invasive plant control. In some situations, brush cutters or brush mowers or weed eaters can be used to mow down small shrubs, tree seedlings, grasses, and taller herbaceous plants. Otherwise, chainsaws, axes, loppers, and other cutting tools can be used to cut down one tree or shrub at a time. Sometimes cutting invasive plants will allow surrounding desirable vegetation to gain an advantage and outcompete the invasives. Other times, you will need to plant a temporary cover crop or use mulch to hold soil in place and then replant desired plants.

For aquatic plants, many small infestations can be removed by hand or with a pond rake. For larger infestations there are floating mechanical harvesters that rototill or rake and collect plants underwater. They may be called rotovators or hydrorakes. Another way to collect plants is by using a hose attached to an engine that creates suction like a dredge. A diver holds the hose and vacuums plants out of the water. The apparatus is attached to a barge that collects the plant material for disposal.

Both aquatic and terrestrial herbaceous plants can be smothered by covering the soil with material such as plastic, thick layers of newspaper and mulch, burlap, or geotextile fabric. The material chosen depends on the terrain and the type of plant. Plastic traps heat and when used in summer will kill plants with the heat as well as by depriving them of water. Other fabrics allow water to pass through but reduce the supply of light. For aquatic plants, the materials used are called benthic barriers and are usually plastic tarps or geotextile fabric made of plastic, nylon, fiberglass, or natural materials. They are held down with pvc pipe or rocks. Smothering is usually cost-effective only for small sites.

Controlled fires (prescribed burns) can be used in some ecosystems to kill emerging seedlings or reduce a seed bank. But some invasive plants are well adapted to fires and may even benefit from controlled burns. It is important to understand the life history and adaptations of the plant you are trying to control and the ecology of the plant community it is growing in. Controlled burns usually require permits and trained personnel.

Flame weeders are propane-powered torches that can be used to kill seedlings and thin-barked trees and shrubs. For woody plants, the flame is used to girdle branches or small trunks near the root crown. Flame weeding can be a quick and inexpensive technique, but of course you have to take care not to set surrounding vegetation on fire! A couple of commercial machines produce steam to kill the above-ground plant parts. One study found that steam suppresses weeds but does not provide long-term control (Barker and Prostak 2008).

Many plants can send out roots from cut plant parts. Others have seeds that mature even after the plant is cut down. You will need to consider your options for disposing of plant material removed mechanically. If plants are removed prior to flowering and fruiting they can be left to air-dry on sheets of cardboard or plastic. The dried material can then be composted as long as the plants don't have parts like rhizomes, runners, or bulblets that might have survived. Woody plants can be made into brush piles for wildlife habitat or mulched. Plant material can be taken off-site for disposal in landfills or incinerators.

BIOLOGICAL CONTROL

Grazing by animals is essentially a means of cutting plants down and can be quite effective in certain situations. Different grazing animals have different food preferences, and some, like goats, will eat almost any woody plant whether native or not. Goats can quickly clear vine- and brush-choked forest edges and fields. They feed by stripping leaves off of stems, eating small stems, and sometimes eating bark off of trees. Sheep and cattle eat herbaceous plants and graze very close to the ground. Different breeds also feed differently, and even within a breed some animals are likely to prefer certain plants or to browse differently from other individuals. It is important to time grazing to when the target plant is most palatable. Grazers, however, may eat the plant and excrete the seeds. The science and art of choosing the right herd of grazers at the right time for the right plants is known as targeted grazing. Some companies will rent out herds of goats or sheep and monitor their grazing for you.

Another common form of biological control is insects that have been carefully selected or bred and deliberately introduced into an area to control a particular invasive plant species. Many biological control agents have been introduced, and many more are under investigation by the U.S. Department of Agriculture's Agricultural Research Service and by Agriculture and Agri-Food Canada. These groups work to find insects or diseases that will attack particular invasive species without posing harm to native species or economically important species like agricultural or horticultural crops. Often they go back to the invader's native range to find what insects attack it at home, then they do extensive laboratory and field trials to determine how effective that particular species will be on controlling a particular invader.

When using biological control agents, you have to be careful to not introduce an invasive species of your own. A vegetarian fish called the grass carp is sometimes used to control aquatic plants, but fertile fish that were released have proliferated, happily eating native aquatic vegetation. Distinguishing sterile and fertile fish requires a blood test. Some states do allow sterile triploid grass carp to be used under certain conditions, but a permit must be obtained first. The fish have strict preferences for certain plants. They seldom completely eliminate the invasive plant.

The advantages of biological control agents are that they are relatively inexpensive, they should be able to spread to new populations of the invader on their own and remain active, and they are less harmful to the environment than chemical controls if they remain specific to the invasive species. Generally, biological control agents do not eliminate a species; they simply keep it in check and prevent it from spreading. If you want to obtain a biocontrol agent for a project, talk to your state or provincial agricultural or natural resource agency.

CHEMICAL CONTROL

Chemical control means using herbicides. Herbicides vary considerably in price, effects, safety, and method of application. Different states and provinces have different rules about herbicide use. Although many people prefer not to use herbicides at all, in some cases only herbicides can be effective against species that are extremely difficult to control or in situations where mechanical control would cause greater environmental damage. Every control technique has its drawbacks, and many times herbicides can be applied with less disturbance to soils and surrounding vegetation than mechanical controls. Some plant species, particularly annual plants repeatedly exposed to a single type of herbicide, have evolved resistance to particular herbicides. Using a combination of control strategies, both chemical and nonchemical, will help reduce the risk of herbicide resistance.

Herbicides are grouped generally by how they kill or suppress the growth of a plant. Some herbicides act on certain types of plants such as grasses only or woody plants only. Other herbicides known as nonselective or broad-spectrum herbicides kill almost any plant.

Nonselective herbicides are grouped into two categories—those that kill on contact (contact herbicides) and those that are absorbed into the plant (systemic herbicides). Contact herbicides kill the foliage they touch, but not necessarily the plant's root system. Therefore they are most effective on seedlings. Nearly all organic herbicides currently on the market are contact herbicides. Vinegar (acetic acid) can be used as a contact herbicide, but it is a strong acid that will damage surrounding plants and could kill organisms living in the soil. Pelargonic acid is used in the herbicide Scythe and provides short-term weed suppression. Herbicides containing natural oils such as citrus oil, clove oil, cinnamon oil, or pine oil also act as contact herbicides. They are most effective on young herbs and grasses; their effectiveness is often increased when they are mixed with an organic additive called an adjuvant. Because these organic herbicides need to be used in higher volumes and in repeated applications, they tend to cost more than synthetic herbicides. A few organic herbicides are soaps that contain fatty acids. One brand (A.D.I.O.S.) contains sodium chloride as the active ingredient. EcoSense Weed B Gon uses chelated iron as an active ingredient and is marketed for killing broadleaf weeds in lawns because some grasses are less susceptible to the iron.

Most selective herbicides are also systemic herbicides. Selective and nonselective systemic herbicides may inhibit the production of certain amino acids, disrupt cell membranes, inhibit the synthesis of lipids, or have other effects on a plant's metabolism. Some systemic herbicides are applied to the soil before the seeds germinate and are called "preemergent" herbicides. They prevent seed germination or inhibit the formation of plant tissue. Corn gluten is a nontoxic preemergent, but it also acts as a fertilizer which can encourage perennial weed growth. "Postemergent" herbicides are applied to the growing plants. Most modern systemic herbicides have little effect on animal species because they target chemical pathways specific to plants, but sometimes they are mixed with other chemicals that can harm animals (including people).

The differences among brand name herbicides often relate to active chemicals in different quantities and dissolved in different solvents (e.g., water, oil, or alcohol). Many will also contain what are called "adjuvants" or "surfactants." An adjuvant may make the herbicide more effective under particular conditions, reduce foaming or drift, increase the activity of the herbicide, make it more waterproof, or enable it to penetrate into the plant more easily. Herbicide labels will specify which adjuvants are recommended for use with that particular herbicide. Adjuvants are not regulated by the federal government, so research the possible side effects of the adjuvant. Compare the quantity of the active ingredients in different herbicides to determine value.

Surfactants, also called surface-active ingredients, are a specific type of adjuvant that help the herbicide stick to the leaf or penetrate the leaf's outer layer by reducing the surface tension of water. Dishwashing liquid is a surfactant, but it can precipitate solids when mixed with some herbicides or hard water and the solids can clog spray equipment. There are three basic types of surfactants. Emulsifying agents mix oil and water. These are used to make oil-soluble herbicides work as a water-based spray, to control drift, to make the herbicide more resistant to rain, or to enhance the activity of the herbicide. Wetting agents, or spreaders, decrease the surface tension of the herbicide, allowing it to spread over the surface of the plant. Surfactants that contain silicone also act as wetting agents. Stickers help the herbicide adhere to plant surfaces, prolonging contact with the target plant. Although the active ingredient in the herbicide may be considered nontoxic to animals, the surfactant or adjuvant may not be so benign, so caution should be used in handling any herbicide or adjuvant. Each herbicide will have a label saying which plants it is effective on and in what quantities it should be used.

The table below lists the different systemic herbicides mentioned in the management section of each species page with their mode of action and some currently available brand names. Many of these herbicides are available only to licensed professional herbicide applicators.

SYSTEMIC HERBICIDES

Active Ingredient	Sample Brands	Mode of Action	Notes
Nonselective systemic			
Glyphosate (dry sites)	Roundup, Glyphos, Glypro, Glyphomax, Kleen-up, ClearOut, Razor Pro	Amino acid inhibitor	Short length of activity in soils
Glyphosate (wet sites)	Rodeo, AquaMaster, Accord, Aqua Neat	Amino acid inhibitor	Short length of activity in soils
Imazapyr	Chopper, Stalker, Arsenal	Amino acid inhibitor	Persists in soil
Oryzalin (preemergent)	WeedImpede, Surflan AS	Inhibits cell division	Prevents seed germination, toxic to fish

SYSTEMIC HERBICIDES *continued*

Active Ingredient	Sample Brands	Mode of Action	Notes
Selective systemic			
2,4-D, 2,4-DP	Patron 170, Navigate, Class, Weed-Pro, Justice, Weed B Gon, Weedone, Weedar	Auxin mimic	Targets broadleaf plants
Aminopyralid	Milestone	Auxin mimic	Targets broadleaf weeds, less persistent than clopyralid, picloram
Acrolein	Magnicide	Degrades cellular structure	For submerged aquatics
Clethodim	Select, Centurion, Compass	Lipid inhibitor	Targets grasses
Clopyralid	Reclaim, Curtail, Transline, Stinger	Auxin mimic	Targets plants in aster, buckwheat, and pea families; can persist in soil
Dicamba	Vanquish, Banvel, Veteran CST	Auxin mimic	Targets broadleaf plants; persists in soil and can leach into groundwater
Fosamine	Krenite	Mitosis inhibitor	Affects woody plants
Fluazifop-p-butyl	Fusilade DX, Fusion, Ornamec	Lipid biosynthesis inhibitor	Affects grasses
Fluridone	Sonar	Inhibits pigments	Affects grasses and broadleaf weeds

SYSTEMIC HERBICIDES *continued*

Active Ingredient	Sample Brands	Mode of Action	Notes
Selective systemic			
Hexazinone	Velpar, Pronone	Inhibits photosynthesis	Affects grasses, broadleaf plants and some woody species. Can contaminate groundwater
Imazapic, Imazameth (pre- and postemergent)	Plateau, Cadre	Amino acid inhibitor	Controls some broadleaf plants and annual grasses; persists in soil
Isoxaben (preemergent)	Gallery	Inhibits cellulose formation	Persists in the environment, primary effective-ness on annual broadleaf plants
MCPA	Weedar, MCPA, Weed B Gon Max	Inhibits cell division and respiration	Targets broadleaf plants
Metsulfuron methyl	Ally, Escort	Inhibits cell division	Does not effect most grasses; can contaminate groundwater, persists in soil
Picloram	Tordon, Pathway, Grazon	Auxin mimic	Kills broadleaf plants; persists in the environ-ment
Prodiamine	Barricade, Endurance	Inhibits cell division	Applied to soil to kill grasses and broadleaf plants

SYSTEMIC HERBICIDES *continued*

Active Ingredient	Sample Brands	Mode of Action	Notes
Selective systemic			
Sethoxydim	Poast, Torpedo, Bonide Grass-Beater	Inhibits lipid synthesis	Kills grasses, low toxicity and persistence
Sulfosulfuron and halosulfuron-methyl	Sedgehammer, NuFarm Halosulfuron Pro, Certainty	Amino acid inhibitor	Affects sedges and some broadleaf plants
Triclopyr	Garlon, Remedy, Brush-B-Gon, Brush Killer, Turflon	Auxin mimic	Targets broadleaf plants

TECHNIQUES FOR HERBICIDE APPLICATION
Spraying
Many herbicides are designed to be sprayed onto foliage. These are usually mixed with water and sometimes a surfactant; a dye can be added to distinguish plants that have and have not been sprayed. The disadvantage of spraying is that surrounding plants are likely to be killed, but it works well where there is a solid stand of something you want to kill. Spraying is also safest when plants are low growing since holding a spray nozzle over your head is both tiring and dangerous to the person spraying.

Most plants should be sprayed between midsummer and late fall, before the leaves turn, so that the plant's process of storing nutrients for winter will also aid herbicide absorption. In some cases, however, you will want to spray in winter or early spring when other plants are dormant. Check the herbicide's label, but generally speaking you want to spray when the temperature is above 55° F and below 80° F (13 to 27° C). Spray until the leaves are wet but not dripping. Do not spray on windy days, as the herbicide may drift onto other plants. Often, early mornings provide cooler temperatures and less wind, but don't spray when plants are wet with dew.

Various types of spray equipment are available. Hand pump sprayers work well for small sites. Backpack sprayers hold more liquid, and some have motors to regulate the volume of spray. Larger spray equipment with tanks can be mounted on tractors, trucks, or off-road vehicles.

Hack-and-Squirt

The hack-and-squirt method involves cutting into the bark of a tree or large shrub with a hatchet, drill, or other tool and then squirting herbicide into the cut. This allows the herbicide to get directly into the plant's circulatory system with little risk of contaminating surrounding desirable vegetation. Care should be taken not to girdle a tree when making the cuts as that can cause some species to send up root suckers or resprouts. One cut for every 4 inches in diameter of trunk should do. Small hand sprayers can be used as squirt bottles. The bottles used in labs for squirting liquid also work well. Generally, the best time of year for hack and squirt is late in the growing season when plants are beginning to translocate nutrients down to their roots. Triclopyr and glyphosate are most commonly used for the hack-and-squirt method. Triclopyr can be used when temperatures are colder but can volatilize (turn into a gas) when temperatures are over 80° F.

Cut-Stump

Cut trees or shrubs close to the ground, then paint or squirt herbicide onto the cut tissue. Handheld spray bottles, sponge paintbrushes, and squirt bottles work well for this. Similar to the hack-and-squirt method, this method minimizes risk to surrounding vegetation. This works best on species unlikely to send up root suckers and can be done throughout most of the year if temperatures are favorable. Glyphosate and triclopyr are the herbicides most frequently used for this kind of treatment.

Basal Bark

For basal bark application, the herbicide is generally mixed with oil so that it will have time to penetrate the bark of the tree or shrub. Herbicide is sprayed or painted onto the bark at the base of the tree in a 6- to 15-inch-wide (15 to 38 cm) band. Basal bark treatments are usually done in cooler weather since the herbicides volatilize at higher temperatures. The advantage of this method is that a large number of plants can be treated fairly quickly. After trees die they can be cut down or left to fall on their own.

Safety Precautions

Anyone using pesticides is required by law to read and follow the directions on the pesticide container. The label lists information on proper use, rate and timing of application, storage, cleanup, and emergency procedures. Use of some herbicides requires a pesticide applicator's license, usually available through your state's Cooperative Extension office or Agriculture Department. Wear chemical-resistant gloves and safety glasses when handling and applying herbicides and wear protective clothing. Clothing that may have herbicide residue on it should be washed separately from other clothing and should be discarded if it came in contact with concentrated herbicide.

Product labels and Material Safety Data Sheets (MSDS) are also available online to provide you with emergency measures in case of accidental exposure. Some herbicides are more toxic than others, particularly those that are easily absorbed through the skin or inhaled. These include some possible carcinogens.

Store herbicides away from food and animal feed, preferably in a locked cabinet. Keep them in their original containers with the labels on them.

When mixing herbicides and loading them into sprayers, tanks, bottles, and so on, wear protective clothing. Place the equipment you will be using for application into a chemical-resistant tub to contain any spills and do not mix within 100 feet of surface water, a well, or a storm drain. Try to mix only as much herbicide as needed to avoid having to dispose of unused herbicide mixtures. When replenishing a pressurized sprayer, first release pressure gradually through the spray wand with tank turned upside down so only air exits the nozzle. When opening the tank beware that often some residual pressure might blow pesticide spray out of the mouth of the tank.

INTEGRATED PEST MANAGEMENT

Integrated pest management requires monitoring pests so that you can catch them when populations are small and when they are at their most vulnerable. Early detection of plant pests is extremely important.

Many times you will find that a combination of the techniques described is most effective for controlling invasive plants. Cutting plants to remove biomass forces a plant to put more resources into regrowing. Spraying the regrowth catches the plant at a time when it is vulnerable. Combining targeted grazing with spot-spraying of plants the grazers missed can increase chances of success. For plants with a large seed bank, using grazers or a brush mower to cut down larger plants and then using a flame weeder to burn off seedlings will kill two generations of plants.

RESTORATION

Once invasives are removed from an area, it will be necessary to restore the area. This may require replanting, controlling erosion until new plantings are established, reducing soil fertility or changing soil ph (acidity), mowing or burning on a certain schedule, and, of course, continuing to monitor for new invasives. Think about your long-term goals for the site before you start removing invasive plants, so that you can plan your restoration strategy accordingly. Use nearby sites in good condition as reference sites that you can use to compare progress on the site undergoing restoration. For more information on the basics of ecological restoration visit the website of the Society for Ecological Restoration, http://www.ser.org.

MARKET CONTROLS

We would be remiss if we did not mention a number of proposals in recent years to use market forces to eliminate or reduce invasive plant and animal populations. It stands to reason that if market forces sometimes endanger or eliminate desirable species, they can do the same to undesirable species. If fancy East Coast restaurants have successfully turned snakehead fish into gourmet meals, certainly invasive plants can also have a market. Many invasive plants have economic values that could lead to their exploitation. Harvesting Himalayan blackberries may be a delicious pastime even if it won't control the invasion—but harvesting invasive pear trees will have an impact if the stumps are prevented from regrowing. The wood of these trees makes excellent turning material, firewood, and fuel for smoking and barbecue. Some aquatic weeds are harvested for livestock feed. Other invasives have edible leaves, fruits, and seeds. Taras Grescoe the author of *Bottomfeeder: How to Eat Ethically in a World of Vanishing Seafood,* says, "it is high time we developed a taste for invasive species" (2008). Creating a market for an invasive species, of course, carries with it the risk that demand will result in its further spread under cultivation.

Barker, A.V. and R. G. Prostak. 2008. *Herbicides alternatives research.* Amherst, MA: University of Massachusetts Transportation Center. http://www.mhd.state.ma.us/downloads/manuals/rpt_herbicides_alternative.pdf.

Czaraparta, E. J. 2005. *Invasive Plants of the Upper Midwest.* Madison: University of Wisconsin Press.

Environmental Protection Agency, Office of Pesticide Programs. http://www.epa.gov/pesticides.

Grescoe, T. 2008. *Bottomfeeder: How to Eat Ethically in a World of Vanishing Seafood.* New York: Bloomsbury U.S.A.

Langeland, K., J.Ferrell, B. Sellers, G. E. MacDonald and R. K. Stocker. 2011. *Integrated management of non-native plants in natural areas of Florida.* University of Florida IFAS SP-242. http://edis.ifas.ufl.edu/wg209.

Launchbaugh, K. (editor). 2006. *Targeted grazing: A natural approach to vegetation management and landscape enhancement.* American Sheep Industry Association. http://www.cnr.uidaho.edu/rx-grazing/handbook.htm.

National Pesticide Information Network. Call: (800) 858-7378. http://npic.orst.edu.

Ross, M. A., and D. J. Childs. 1996. "Herbicide Mode of Action Summary WS-23-W." Purdue University Cooperative Extension Service. http://www.extension.purdue.edu/extmedia/ws/ws-23-w.html.

Society for Ecological Restoration International Science & Policy Working Group, 2004. *The SER International Primer on Ecological Restoration.* Tucson: Society for Ecological Restoration International. http://www.ser.org/content/ecological_restoration_primer.asp.

Tu, M., C. Hurd, and J. M. Randall. 2001. *Weed Control Methods Handbook: Tools and Techniques for Use in Natural Areas.* The Nature Conservancy, Wildland Invasive Species Team. http://www.invasive.org/gist/handbook.html.

FOUR

A Quick Key to Identifying Major Species of Invasive Plant Groups

Far North
Pacific West
Central West
Great Lakes
Mid-Atlantic
Northeast
Southeast
Southwest

There are far too many species covered in this book for us to give a full description of each in this key. However, we have classified them by major characteristics—habitat, type of plant, leaf arrangement, size, and so on; once you've narrowed down your identification to a particular group of similar species, you can turn to the species accounts for more specific information. If a species can occur both on land and in water, or if it has another combination of characteristics, it is cross-listed in both groups in the key.

This key is color-coded by region, corresponding to the eight regions on the map. Each plant in the chart is coded with the colors of all the regions where it currently grows. You can use this as a quick reference to find out what plants are invasive in your region.

SPECIES KEY

REGION		LATIN NAME	COMMON NAME	PAGE
Terrestrial—Trees and shrubs—Evergreen—Needles or inconspicuous leaves—Trees				44
	*	Casuarina spp.	Australian pine	44
		Pinus sylvestris	Scots pine	46
Terrestrial—Trees and shrubs—Evergreen—Needles or inconspicuous leaves—Shrubs				49
		Tamarix spp.	salt cedar	49
		Ulex europaeus	common gorse	51
Terrestrial—Trees and shrubs—Evergreen—Alternate leaves—Trees				55
	*	Bischofia javanica	bishopwood	55
		Cinnamomum camphorum	camphor tree	57
	*	Cupaniopsis anacardiodes	carrotwood	58
	*	Ficus microcarpa	laurel fig	60
	*	Manilkara zapota	sapodilla	62
	*	Melaleuca quinquenervia	melaleuca	64
		Schinus molle	pepper tree	66
Terrestrial—Trees and shrubs—Evergreen—Alternate leaves—Shrubs				69
	*	Ardisia elliptica	shoebutton and coral ardisia	69
	*	Colubrina asiatica	latherleaf	71
		Mahonia aquifolium	Oregon grape holly	74
		Mahonia bealei	leatherleaf mahonia	74
	*	Scaevola sericea	scaevola	76
		Schinus terebinthifolius	Brazilian pepper tree	78
	*	Solanum viarum	tropical soda apple	81
	*	Thespesia populnea	seaside mahoe	83
Terrestrial—Trees and shrubs—Evergreen—Opposite or whorled leaves—Trees				86
	*	Casuarina spp.	Australian pine	44
	*	Schefflera actinophylla	schefflera	86
Terrestrial—Trees and shrubs—Evergreen—Opposite or whorled leaves—Shrubs				88
		Cytisus scoparius	Scotch broom	88
		Euonymus fortunei	winter creeper euonymus	91
		Lantana camara	lantana	93
		Ligustrum spp.	privets	95
		Nandina domestica	nandina	98
	*	Psidium spp.	common and strawberry guava	100
		Tamarix spp.	salt cedar	49
		Ulex europaeus	common gorse	51
Terrestrial—Trees and shrubs—Deciduous—Alternate leaves—Trees				104
		Ailanthus altissima	tree of heaven	104
		Albizia julibrissin	mimosa	107
		Alnus glutinosa	black alder	110
		Brousonnetia papyrifera	paper mulberry	116
		Leucaena leucocephala	leucaena	112
		Melia azedarach	chinaberry tree	114
		Morus alba	white mulberry	116
		Populus alba	white poplar	119
		Prunus avium	bird cherry	121
		Pyrus calleryana	Callery pear	123
		Quercus acutissima	sawtooth oak	126
		Robinia pseudoacacia	black locust	128
		Salix spp.	willows	131
		Triadica sebifera	Chinese tallow tree	134
		Ulmus pumila	Siberian elm	137

* = Florida only

SPECIES KEY *continued*

REGION	LATIN NAME	COMMON NAME	PAGE
Terrestrial—Trees and shrubs—Deciduous—Alternate leaves—Shrubs			140
	Alhagi maurorum	camel thorn	140
	Berberis spp.	Japanese and common barberry	142
	Elaeagnus angustifolia	Russian olive	145
	Elaeagnus umbellata	autumn olive	145
	Hippophae rhamnoides	sea buckthorn	148
	Rosa multiflora	multiflora rose	150
	Rosa rugosa	rugosa rose	153
	Rubus spp.	blackberries	154
	Rubus phoenicolasius	wineberry	158
	Sesbania punicea	rattlebox	160
	Spiraea japonica	Japanese spiraea	162
Terrestrial—Trees and shrubs—Deciduous—Opposite or whorled leaves—Trees			164
	Acer platanoides	Norway maple	164
	Paulownia tomentosa	royal paulownia	166
	Phellodendron amurense	Amur cork tree	169
Terrestrial—Trees and shrubs—Deciduous—Opposite or whorled leaves—Shrubs			172
	Berberis spp.	Japanese and common barberry	142
	Buddleja davidii	butterfly bush	172
*	Eugenia uniflora	Surinam cherry	175
	Euonymus alatus	burning bush	177
	Lonicera spp.	shrub honeysuckle	179
	Lonicera fragrantissima	fragrant honeysuckle	182
	Rhamnus spp.	common, glossy, dahurian buckthorn	184
	Rhodotypos scandens	jetbead	187
	Viburnum dilatatum	linden viburnum	187
	Vitex rotundifolia	beach vitex	192
Terrestrial—Vines—Evergreen—Alternate leaves			194
	Akebia quinata	five leaf akebia	213
	Delairea odorata	cape ivy	194
	Hedera helix	English ivy	196
Terrestrial—Vines—Evergreen—Opposite leaves			200
	Euonymus fortunei	winter creeper euonymus	91
*	Jasminum spp.	jasmines	200
	Lonicera japonica	Japanese honeysuckle	201
	Macfadyena unguis-cati	cat's claw vine	204
	Paederia foetida	skunk vine	206
	Vinca minor	common periwinkle	208
Terrestrial—Vines—Deciduous—Alternate leaves			211
	Abrus precatorius	rosary pea	211
	Akebia quinata	five leaf akebia	213
	Ampelopsis brevipedunculata	porcelainberry	215
	Cayratia japonica	bushkiller	217
	Celastrus orbiculatus	Oriental bittersweet	219
	Convolvulus arvensis	field bindweed	221
	Dioscorea bulbifera	air potato	224
	Dioscorea polystachya	cinnamon vine	226
	Persicaria perfoliata	mile-a-minute vine	228
	Pueraria montana var. lobata	kudzu	231
	Wisteria spp.	Chinese and Japanese wisteria	234

Far North   **Pacific West**   **Central West**   **Great Lakes**
Mid-Atlantic **Northeast**   **Southeast**   **Southwest**

SPECIES KEY *continued*

REGION	LATIN NAME	COMMON NAME	PAGE
Terrestrial—Vines—Deciduous—Opposite leaves			237
	Clematis terniflora	sweet autumn virgin's bower	237
	Clematis vitalba	old man's beard	239
	Cynanchum spp.	swallow-worts	241
	Glechoma hederacea	ground ivy	244
	Humulus japonicus	Japanese hop	246
	Vitex rotundifolia	beach vitex	192
Terrestrial—Herbaceous plants—Rosette or basal leaves only			249
	Colocasia esculenta	wild taro	260
	Ficaria verna	lesser celandine	249
	Hemerocallis fulva	orange daylily	252
	Hieracium aurantiacum	orange hawkweed	254
	Hieracium pillosella and *H. caespitosum*	mouse ear and meadow hawkweed	254
	Iris pseudacorus	yellow flag iris	257
Terrestrial—Herbaceous plants—Alternate leaves—Leaves smooth edged			260
	Colocasia esculenta	wild taro	260
	Cynoglossum officinale	houndstongue	263
	Euphorbia esula	leafy spurge	265
	Fallopia japonica	Japanese knotweed	268
	Isatis tinctoria	dyer's woad	271
	Kochia scoparia	kochia	273
	Linaria dalmatica	Dalmatian toadflax	275
	Linaria vulgaris	yellow toadflax	275
	Murdannia keisak	marsh dewflower	277
	Polygonum spp.	smartweeds	278
	Salsola tragus	tumbleweed	281
	Tradescantia spp.	small leaf spiderwort and boat lily	283
	Tragopogon dubius	yellow salsify	285
	Verbascum thapsus	woolly mullein	287
Terrestrial—Herbaceous plants—Alternate leaves—Leaves toothed or lobed			290
	Acroptilon repens	Russian knapweed	290
	Aegopodium podagraria	goutweed	292
	Alliaria petiolata	garlic mustard	294
	Arctium minus	common burdock	297
	Artemisia absinthium	absinthe	299
	Artemisia vulgaris	mugwort	301
	Brassica tournefortii	Asian mustard	303
	Cardaria draba	white top	305
	Carduus nutans	musk thistle	307
	Centaurea calcitrapa	red star thistle	309
	Centaurea solstitialis	yellow star thistle	312
	Centaurea stoebe	spotted knapweed	315
	Chelidonium majus	celandine	318
	Cirsium arvense	Canada thistle	320
	Crupina vulgaris	common crupina	322
	Heracleum mantegazzianum	giant hogweed	324
	Hesperis matronalis	dame's rocket	326
	Hyoscyamus niger	black henbane	328
	Impatiens glandulifera	Himalayan balsam	330
	Jacobaea vulgaris	tansy ragwort	332
	Lactuca serriola	prickly lettuce	335
	Lepidium latifolium	perennial pepperweed	337
	Leucanthemum vulgare	ox-eye daisy	339
	Onopordum acanthium	scotch thistle	341

* = Florida only

SPECIES KEY continued

■ **Far North**   ■ **Pacific West**   ■ **Central West**   ■ **Great Lakes**
■ **Mid-Atlantic**   ■ **Northeast**   ■ **Southeast**   ■ **Southwest**

SPECIES KEY *continued*

REGION	LATIN NAME	COMMON NAME	PAGE
Terrestrial—Grasses and sedges—Leaves not angular—Plants are not in distinct clumps			
or are growing along creeping stems—Under 4 ft. (1.2 m) tall			416
	Aegilops triuncialis	barbed goat grass	416
	Agrostis stolonifera	creeping bentgrass	419
	Ammophila arenaria	European beach grass	382
	Arthraxon hispidus	small carpgrass	421
	Brachypodium sylvaticum	false brome	422
	Bromus inermis ssp. *inermis*	smooth brome	424
	Bromus tectorum	cheatgrass	426
	Cenchrus ciliaris	buffel grass	398
	Cynodon dactylon	Bermuda grass	429
	Microstegium vimineum	Japanese stilt grass	431
	Osplimenus hirtellus ssp. *undulatifolius*	wavyleaf basketgrass	434
	Panicum repens	torpedo grass	435
	Schismus arabicus	Mediterranean grass	437
	Schedonorus phoenix	tall ryegrass	439
	Taeniatherum caput-medusae	medusahead	442
Terrestrial—Ferns			444
	Lygodium japonicum	Japanese climbing fern	444
*	*Lygodium microphyllum*	Old World climbing fern	444
	Nephrolepsis spp.	sword ferns	447
Aquatic—Herbaceous—Rosette or basal leaves only			449
	Colocasia esculenta	wild taro	260
	Eichornia crassipes	water hyacinth	449
	Hydrocharis morsus-ranae	European frog-bit	452
	Iris pseudacorus	yellow flag iris	257
	Pistia stratiotes	water lettuce	453
Aquatic—Herbaceous—Alternate leaves			456
*	*Ipomoea aquatica*	water spinach	456
	Ludwigia hexapetala	water primrose	457
	Potamogeton crispus	curly pondweed	460
	Murdannia keisak	marsh dewflower	277
	Trapa natans	water chestnut	462
Aquatic—Herbaceous—Opposite or whorled leaves			465
	Alternanthera philoxeroides	alligator weed	465
	Egeria densa	Brazilian elodea	467
	Hydrilla verticillata	hydrilla	469
	Lythrum salicaria	purple loosestrife	364
	Myriophyllum spicatum	Eurasian water milfoil	472
	Najas minor	brittle naiad	475
Aquatic—Algae, ferns, and grasses			477
	Arundo donax	giant reed	385
	Butomus umbellatus	flowering rush	477
	Carex acutiformis	European lake sedge	377
	Caulerpa taxifolia	killer algae	479
	Cyperus entrerianus	deep rooted sedge	380
	Didymosphenia geminata	rock snot	481
	Glyceria maxima	reed sweetgrass	401
	Phalaris arundinacea	reed canarygrass	403
	Phragmites australis	phragmites	406
	Salvinia spp.	salvinia	483
	Spartina alterniflora	smooth cordgrass	485
	Urochloa mutica	Pará grass	414

* = Florida only

PART II

A Field Guide to Individual Species

Terrestrial Plants—Trees and Shrubs— Evergreen—Needles or Inconspicuous Leaves—Trees

Australian Pine   *Casuarina* spp.

NAME AND FAMILY
Australian pine, horsetail pine, beach sheoak, casuarina, scaly bark oak, common ironwood, swamp oak, whispering pine (*Casuarina equisetifolia* L.), Casuarina family (Casuarinaceae). Other invasives in this family include Brazilian oak (*C. glauca*) and Australian river oak (*C. cunninghamiana*). Hybrids among the species also occur, making them difficult to distinguish.

IDENTIFYING CHARACTERISTICS
It is not a pine and the green foliage is not needles but rather very slender green twigs with small nodes where six to eight tiny scalelike leaves grow. Trees grow rapidly and can reach 80 to 100 ft. (24 to 30.5 m) tall with a feathery appearance. At the twig tips, male flowers form as thin cylinders, while the red female flower heads are attached to small branches and mature into 0.5 in. (1.3 cm) diameter green, then brown fruits that look like rounded cones and contain 70 to 90 light, flat seeds easily released and carried by the wind. The bloom peaks

Australian pine dominating a sand dune.

in April with a lesser bloom in September and some fruits present at any time. The root system is shallow but dense. Brazilian oak and Australian river oak are not as salt tolerant as Australian pine and are dioecious, with separate male and female trees. Species can be distinguished based on the number of leaves at each node.

HABITAT AND RANGE

Intolerance for cold temperatures limits the range of Australian pine to southern and central Florida, but its smaller relative Australian river oak tolerates short periods of cold temperatures to 15° F (–9.4°C) and is found in north Florida as well. Australian pine prefers sandy soils near the coast and tolerates salt well. In Florida it grows mostly south of Orlando on both coasts and in the Everglades.

Australian pine's branches give the tree a feathery appearance.

WHAT IT DOES IN THE ECOSYSTEM

Australian pine seeds quickly germinate, especially in disturbed soils. A tree can grow 10 ft. (3 m) a year, and after five years it is producing thousands of seeds. It may change soil nitrogen levels and be more competitive by cooperating with soil bacteria (actinomycetes) to manufacture its own nitrogen. Its growth on sand dunes inhibits the lower-growing native plants and eliminates their stabilizing effects; when violent storms blow down the shallow-rooted trees, the dunes, devoid of native stabilizers, erode quickly. In hurricanes, fallen Australian pines have also been a major obstacle, blocking escape routes. Hurricane Donna in 1960 spread the tree into the Everglades. Australian pine in beach areas reduces nesting area for American crocodiles and loggerhead and green sea turtles. The copious pollen produced by Australian pine triggers allergic reactions in many people.

HOW IT CAME TO NORTH AMERICA

Casuarinas are native to Australia. The U.S. Department of Agriculture brought in Australian pine seeds from France in 1898, and other Australian species made their appearance by the 1920s, probably from the Caribbean, where they were introduced about 1870. Promoters of the tree thought it would provide paper pulp and lumber and a source for tannins, but it is not particularly good for any of these uses. It was used extensively as an ornamental tree for landscaping, but the Florida Department of Agriculture has now banned its sale, propagation, and transportation.

MANAGEMENT
For small stands, cut the trees or pull out saplings. Cutting may need to be repeated until the root system is exhausted. The most effective method of attacking large trees and extensive stands is a combination of cutting and a systemic herbicide. Trees can be girdled and resprouts sprayed. The hack-and-squirt method, cut-stump treatment, injection method, and basal bark treatment are also effective. Burning works if new seedlings are yanked early after they appear.

FOR MORE INFORMATION
Klukas, R. W. 1969. *The Australian pine problem in Everglades National Park: Part 1. The problem and some solutions.* Internal report, South Florida Research Center, Everglades National Park.

Langeland, K. A., J. A. Ferrell, B. Sellers, G. E. MacDonald, and R. K. Stocker. 2011. "Integrated Management of Nonnative Plants in Natural Areas of Florida." SP 242. Gainesville, FL: University of Florida IFAS. http://edis.ifas.ufl.edu/wg209.

Morton, J. F. 1980. "The australian pine or beefwood (Casuarina equisetifolia L.): An invasive 'weed' in Florida." *Proceedings of the Florida State Horticultural Society.*

Wheeler, G.S., G. S. Taylor, J. F. Gaskin, and M. F. Purcell. 2011. "Ecology and Management of Sheoak (*Casuarina* spp.) and invader of coastal Florida, U.S.A." *Journal of Coastal Research* 27(3):485–492.

Woodall, S. L., and T. F. Geary. 1985. "Identity of Florida casuarinas." Research Note SE-332. U.S. Department of Agriculture Southeast Forest Experiment Station.

Scots Pine   *Pinus sylvestris*

NAME AND FAMILY
Scots pine, Scotch pine (*Pinus sylvestris* L.); pine family (Pinaceae).

IDENTIFYING CHARACTERISTICS
Many people have bought Scots pine as a Christmas tree because of its ample branching and dense needle growth. This is the only pine that bears blue-green to slate-green needles (1.5 to 4 in. [3.75 to 10 cm] long) that grow in pairs from their fascicles. Needles usually twist a full 360 degrees. New buds are orange brown and may have a resinous cover. The 2.5 to 7 in. (6.2 to 18 cm) long cones start green and darken to grayish, maturing in November to December. Trees can grow to 100 ft. (30 m) and live several hundred years. Trees whose lead buds are not damaged grow straight trunks that are topped by a mass of foliage. Many varieties of Scots pine exist, often distinguishable from each other only by genetic analysis or resin chemistry.

Scots pine cones mature, open, and drop seeds in winter.

HABITAT AND RANGE

This hardy native of northern Europe can grow from sea level to 8,000 ft. (2,400 m). It must have open sunlight to grow but grows on many soil types, from sand to peat. It has naturalized in Canada's eastern provinces, in New England, across the United States to Wisconsin, and in cooler areas of New York and Pennsylvania.

WHAT IT DOES IN THE ECOSYSTEM

Most authorities do not consider Scots pine an aggressive competitor except in open areas normally dominated by herbaceous plants and shrubs that grow on poor soils. In such situations Scots pine seedlings form thick mats, growing in stands that shade out the lower-growing native plants. Scots pine has colonized heathlands, bogs, and dunes. Because of its ability to grow in poor soils and reproduce at an early age, it has been used to reclaim mining sites and for windbreaks. Moose browse on it, but it is a low-

Scots pine trees grow tall and straight in good soil.

Blue-green needles occur in pairs and twist.

priority browse for deer. Grosbeaks eat the buds and porcupines eat the bark. It can be used for pulpwood but has little value as lumber.

HOW IT CAME TO NORTH AMERICA

Scots pine is the national tree of Scotland and once dominated thousands of acres of the now almost vanished Great Wood of Caledon. It is the world's most widely distributed pine, ranging from Scotland, across northern Europe to the east coast of northern Asia. Colonial settlers probably brought the first specimens to North America for windbreaks, erosion control, herbal medicine, or for planting in poor soils. Its needles were once used in bedding and called "pine wool." It may constitute as much as 30 percent of all Christmas trees bought in North America.

MANAGEMENT

Scots pine is not an aggressive invader of woodlands because it does not tolerate shade. Hand pulling seedlings and cutting larger trees at ground level are effective controls.

FOR MORE INFORMATION

Catling, P.M. and S. Carbyn. 2005. "Invasive Scots Pine, *Pinus sylvestris,* replacing Corema, *Corema conradii,* heathland in the Annapolis valley, Nova Scotia, Canada." *Canadian Field-Naturalist* 119(2):237–244.

Skilling, D. D. 1990. "*Pinus sylvestris* L.: Scotch pine." In *Silvics of North America.* Volume 1. Conifers. Agriculture Handbook 654. R. M. Burns and B. H. Honkala, technical coordinators, 489–96. Washington, DC: U.S. Department of Agriculture, Forest Service. http://na.fs.fed .us/pubs/silvics_manual/Volume_1/pinus/sylvestris.htm.

Terrestrial Plants—Trees and Shrubs— Evergreen—Needles or Inconspicuous Leaves—Shrubs

Salt Cedar   *Tamarix* spp.

NAME AND FAMILY
Salt cedar, tamarisk (*Tamarix africana, T. aphylla, T. chinensis, T. gallica, T. parvi-flora,* and *T. ramosissima*); tamarisk family (Tamaricaceae). Taxonomists disagree on which species of salt cedar are present in the United States, and species probably hybridize. Management of the different species is the same, however.

IDENTIFYING CHARACTERISTICS
The salt cedars distinguish themselves by their dense growth, spreading, multi-trunked form, and tiny, overlapping scalelike leaves about 1/16 in. (0.15 cm) long. These characteristics create the typical feathery appearance. They grow 10 to 15 ft. (3 to 4.6 m) high and form hairlike thickets. *Tamarix aphylla* can grow to 50 ft. (15 m) and is evergreen while others drop their leaves. Young bark is smooth and reddish brown and becomes dark brown to purplish in maturity. From early spring to early fall, pink to white four- or five-petaled flowers blossom at the tips of branches in dense clusters of 2 in. (5 cm) long spikes. Seeds shelter within a small capsule that usually has a tuft of hair to catch the wind.

HABITAT AND RANGE
Salt cedar has colonized moist soils from the Southwest north to Montana and across the United States to Connecticut. It has also become common in parts of southern British Columbia, Canada. While it will grow

On salt cedar's feathery stems, leaves appear like small scales.

Left: *Salt cedar has a feather appearance and flowers from spring through fall.*
Right: *Flowers cluster tightly at the tips of the branches.*

in many moist soils, it has become densest and most troublesome along stream- and riverbanks, especially in the Southwest, where its high salt tolerance helps it colonize land where evaporation has left salt deposits.

WHAT IT DOES IN THE ECOSYSTEM

Salt cedars have several adaptations that allow them to form dense colonies in difficult soils. They have long taproots for seeking scarce moisture; they can spring back from roots after fires; they tolerate submersion for longer than 2 months; and they grow up to 12 ft. in one season. Because salt cedars have one of the highest rates of evapotranspiration in semiarid and arid climates, they rob other plants of water. They also accumulate salt in their leaves, and fallen leaves increase soil surface salinity, making it difficult for native plants to establish. Since their introduction in the early 1800s, they have become the dominant plant in many riparian areas of the Southwest. Salt cedars now occupy over 1 million acres, replacing cottonwood, willows, and other bottom- land vegetation. The dense stands change water flow, widening floodplains by catching sediment and lowering water tables. While they offer little food for wildlife, they do provide important shelter for doves and plentiful nectar for honeybees. The endangered southwestern willow flycatcher (*Empidonax trailii extimus*) also uses the salt cedar for shelter.

HOW IT CAME TO NORTH AMERICA

Its showy flower clusters and feathery appearance made salt cedar a favorite among nineteenth-century landscapers who introduced it to North America in the 1820s. Several other varieties followed for ornamental use and as windbreaks and streambank erosion stabilizers. By the end of the nineteenth century, it had begun to spread rapidly in natural areas.

MANAGEMENT

Young first-year salt cedars can be pulled by hand. Once the plants are established, with deep taproots, machines and chemicals are the only effective tools, although some research is being done on biological controls. A leaf-feeding beetle, *Diorhabda elongata*, has been released as a biocontrol agent in the U.S. Burning only sets back growth in this fire-adapted species. Herbicides are effective in dense stands and can be sprayed on foliage or painted on cut stumps, but only chemicals approved for use in aquatic environments should be used. Goat browsing has also shown some results in controlling salt cedars.

FOR MORE INFORMATION

Nissen, S., A. Sher, and A. Norton. 2010. *Tamarisk Best Management Practices in Colorado Watersheds.* Ft. Collins, CO: Colorado State University.

Dudley, T. L., and C. J. DeLoach. 2004. "Saltcedar (*Tamarix* spp.), endangered species, and biological weed control—can they mix?" *Weed Technology* 18:1542–1551.

Lindgren, C., C. Pearce and K. Allison. 2010. "The biology of invasive alien plants in Canada. 11. *Tamarix ramosissima* Ledeb., *T. chinensis* Lour. and hybrids." *Canadian Journal of Plant Science* 90(1):111–124.

Common Gorse   *Ulex europaeus*

NAME AND FAMILY

Common gorse, European gorse, furze, prickly broom (*Ulex europaeus* L.); pea family (Fabaceae). Brooms (*Cytisus* and *Genista* spp.) may look similar to gorse but lack the spines.

IDENTIFYING CHARACTERISTICS

An evergreen, densely branched shrub, gorse has spine- or scalelike leaves that are only 0.25 to 0.75 in. (0.6 to 1.9 cm) long. Unlike most legumes, the typical leaf made up of three leaflets appears only in seedlings and on plants growing in rich soils. Spines up to 1 in. (2.5 cm) long cover the branches of

The dense branches of gorse are heavily armed.

mature shrubs, which grow to 3 to 6 ft. (1 to 2 m) in height. The twigs are covered in grey to reddish brown hairs. Shiny, yellow, pealike flowers about 1 in. (2.5 cm) long bloom singly or in groups from the leaf axils and ends of the branches in winter to early spring. Seeds mature inside very hairy pods, 0.5 to 1 in. (1.3 to 2.5 cm) long. The brown seeds have a white, fleshy appendage called an aril. Gorse tends to have a dense root mat at the surface and some roots that hang down from the stems; it also has a taproot.

HABITAT AND RANGE

Gorse's distribution is restricted by its intolerance of extremely dry conditions and of extremes in temperature. It tends to colonize disturbed areas such as sand dunes, gravel bars, overgrazed pastures, or logged areas. Gorse occurs along the west coast of North

Yellow, pealike flowers bloom in winter.

America from British Columbia south

to California and in several eastern states from Massachusetts to Virginia; however, it is considered invasive principally on the West Coast.

WHAT IT DOES IN THE ECOSYSTEM
Gorse forms dense stands made impenetrable by its thorns. A large amount of litter, made up mostly of old stems, falls to the ground and tends to change soil chemistry. As a legume, gorse fixes nitrogen, potentially increasing soil nitrogen levels. Because of the oils contained in the plant, gorse also poses an increased risk of fire. Seeds can remain dormant in the soil for more than 30 years. Although the primary method of seed dispersal is ejection from the seedpod, seeds may be further transported by ants, in soil, and by water. Quail may also eat and disperse the seeds.

HOW IT CAME TO NORTH AMERICA
Gorse was introduced as an ornamental plant, probably in the mid-to-late 1800s, from central or western Europe. Chemical lectins from gorse have been tested as biomarkers for some cancers.

MANAGEMENT
Gorse is difficult to manage due to the long life of the seed bank and the ability of stumps to resprout after cutting or fire. Planting fast-growing trees toler-

Gorse's hairy seedpods.

Dense stands of gorse are formidable competitors.

ant of acidified soils among the gorse may eventually lead to shading out the shrubs and any new seedlings. Seedlings can be hand pulled or pulled out using tools. Angora goats have also been used to control gorse. Shrubs can be cut before seeds are set. A weevil introduced from France in 1953 feeds on gorse and has been partially effective where the climate is not too cold for it. Applying lime to pasture and meadow areas has decreased germination rates. Herbicides can be effective but often require more than one application. Picloram will kill young plants, while glyphosate gives at least partial control on established stands. Spraying is best done on young leaves or in early fall, when the waxy coatings of the leaves have worn off some.

FOR MORE INFORMATION

Hoshovsky, M. 1989. *Element stewardship abstract for* Ulex europaeus. The Nature Conservancy. http://www.invasive.org/weedcd/pdfs/tncweeds/ulexeur.pdf.

Rees, M., and R. L. Hill. 2001. "Large-scale disturbances, biological control, and the dynamics of gorse populations." *Journal of Applied Ecology* 38:364–377.

Terrestrial Plants—Trees and Shrubs— Evergreen—Alternate Leaves—Trees

Bishopwood   *Bischofia javanica*

NAME AND FAMILY
Bishopwood, Javawood, toog (*Bischofia javanica* Blume); spurge family (Euphorbiaceae).

IDENTIFYING CHARACTERISTICS
This evergreen tree has shiny, bronze-hued leaves made up of three leaflets. Leaves are arranged opposite along the branches. Each leaflet is 6 to 8 in. (15 to 20 cm) long with toothed edges. Leaves and twigs contain a milky sap. Branches are smooth and dense, giving the canopy a rounded shape. Tiny greenish-yellow flowers without petals bloom in clusters in the leaf axils, with male and female flowers on separate trees. On female trees, pea-sized fruits ranging in color from reddish brown to blue black hang in grapelike clusters. Trees grow 30 to 70 ft. (9 to 21 m) tall. Researchers in Japan have noted that the trees change sex, with larger trees becoming female.

HABITAT AND RANGE
Common in disturbed wetlands and old fields, but also invading hardwood islands (hammocks) and cypress forests. Prefers the moist soils of south Florida.

A single bishopwood leaf is made up of three leaflets.

Mature trees have reddish bark and a rounded crown.

WHAT IT DOES IN THE ECOSYSTEM

Birds disperse the abundant fruits, and seedlings can germinate in full sun to shade, sprouting in abundance in hardwood stands. Bishopwood changes the composition of forest communities and threatens the already-rare remaining hardwood hammock and cypress communities of south Florida.

HOW IT CAME TO NORTH AMERICA

In 1912 nursery grower E. N. Reasoner introduced this native of Asia and the Pacific islands to the west coast of Florida. It is frequently planted as an ornamental street tree in south Florida. In Asia the wood is used for furniture, and oil is extracted from the seeds.

MANAGEMENT

Remove female trees first to discourage seed dispersal. A mixture of triclopyr and oil sprayed around the lower trunk of the tree will kill it. Trees will sprout from roots if cut, but painting triclopyr on cut stumps may discourage resprouting. Seedlings can be hand pulled.

FOR MORE INFORMATION

Horvitz, C. C., and A. Koop. 2001. "Removal of nonnative vines and post-hurricane recruitment in tropical hardwood forests of Florida." *Biotropica* 33:268–281.

Langeland, K. A., H. M. Cherry, C. M. McCormick, and K. A. Craddock Burks et al. 2008. *Identification and Biology of Nonnative Plants in Florida's Natural Areas,* 2nd ed. SP 257. Gainesville, FL: University of Florida IFAS.

Langeland, K. A., J. A. Ferrell, B. Sellers, G. E. MacDonald, and R. K. Stocker. 2011. "Integrated Management of Nonnative Plants in Natural Areas of Florida." SP 242. Gainesville, FL: University of Florida IFAS. http://edis.ifas.ufl.edu/wg209.

Yamashita, N., and T. Abe. 2002. "Size distribution, growth, and inter-year variation in sex expression of *Bischofia javanica*, an invasive tree." *Annals of Botany* 90:599–605.

Camphor Tree   *Cinnamomum camphora*

NAME AND FAMILY
Camphor tree, camphor laurel (*Cinnamomum camphora* [L.] J. Presl); laurel family (Lauraceae).

IDENTIFYING CHARACTERISTICS
The odor of crushed leaves and stems gives camphor tree its name. The evergreen tree grows 50 to 100 ft. (15 to 30 m) tall and the spread of limbs is often twice as wide as the tree's height. It has green to reddish twigs. Leaves are simple. They are oval, sometimes with wavy edges, 1.5 to 4 in. (3.7 to 10.1 cm) long and 0.8 to 2 in. (2 to 5 cm) wide and reddish when young, then glossy green on the upper side and waxy green below. On new growth, 3 in. (7.6 cm) long spikes of white- to cream-colored small flowers bloom in spring. Small, round, black fruits form in summer and persist into winter.

HABITAT AND RANGE
Most common in northern and central Florida. Also grows from North Carolina to Texas. Invades drier areas with sandy soils such as upland pine forests, scrubland, roadsides, and field edges.

WHAT IT DOES IN THE ECOSYSTEM
Forms dense stands and can prevent regeneration of natives or shade out native plants. In Florida scrublands it is displacing the endangered Florida jujube (*Ziziphus celata*). Seeds are bird dispersed and are a winter food source for birds.

HOW IT CAME TO NORTH AMERICA
Introduced in 1875 to Florida as an ornamental shade tree. Until camphor began to be produced artificially in 1920, the tree was used for the pro-

Camphor trees prefer dry soils and warm climates.

Glossy green leaves are reddish when young.

duction of medicinal camphor and mothballs. Also planted for windbreaks. Native to eastern Asia.

MANAGEMENT
Seedlings and saplings can be hand pulled. Larger trees can be cut and treated with triclopyr, or triclopyr mixed with oil can be sprayed on the base of the trunks as a basal bark treatment.

FOR MORE INFORMATION

Langeland, K. A., H. M. Cherry, C. M. McCormick, and K. A. Craddock Burks et al. 2008. *Identification and Biology of Nonnative Plants in Florida's Natural Areas,* 2nd ed. SP 257. Gainesville, FL: University of Florida IFAS.

Murray, A., and V. Ramey. 2003. "Camphor tree." University of Florida IFAS, Center for Aquatic and Invasive Plants. http://aquat1.ifas.ufl.edu/camphor.html.

Carrotwood   *Cupaniopsis anacardioides*

NAME AND FAMILY
Carrotwood (*Cupaniopsis anacardioides* [A. Rich.] Radlk.); soapberry family (Sapindaceae).

IDENTIFYING CHARACTERISTICS
Called carrotwood because of their orange inner bark, trees grow to 35 ft. (10.7 m) and have a broad, irregular crown with dark gray outer bark. The leaves of this evergreen tree are made up of 4 to 10 oblong, shiny leaflets, 4 to 8 in. (10 to 20 cm) long and attached to the stem by a swollen stalk. The tips of the leaflets are rounded and sometimes indented. In winter, branched clusters of small, greenish-white, five-petaled flowers bloom on stalks that can be more than a foot long and that emerge from the leaf axils. Yellow-orange seed capsules open into three sections, revealing three shiny black seeds covered in orange arils (fleshy tissue).

HABITAT AND RANGE
Grows in southern Florida following the distribution of mangrove forests, but also grows in dunes, marshes, riverbanks, pinewoods, cypress swamps, and hardwood islands. Grows in sun or shade and is salt tolerant.

Left: *Carrotwood trees are common ornamentals in southern Florida.* **Right:** *Carrotwood leaves are divided into several leaflets.*

WHAT IT DOES IN THE ECOSYSTEM

Thick stands of carrotwood displace native species, outcompeting them for light, nutrients, and space. Birds and probably small mammals disperse the seeds to new locations. Carrotwood is of particular concern in mangrove forests because mangroves serve as nurseries to many fish and crustaceans and as habitat for threatened bird species.

HOW IT CAME TO NORTH AMERICA

Carrotwood was introduced to Florida from Australia as early as 1955 as an ornamental plant. Large scale propagation of carrotwood began in Sarasota in 1968 and by the early 1990s carrotwood was found in natural areas. The apricot-colored wood is prized by some wood turners and sculptors.

MANAGEMENT

Seedlings and young plants can be hand pulled. Larger plants can be killed through basal bark or cut-stump application of triclopyr or glyphosate.

FOR MORE INFORMATION

Langeland, K. A. 2003. "Natural area weeds: carrotwood (*Cupaniopsis anacardioides*)." University of Florida IFAS. http://edis.ifas.ufl.edu/ag111.

Lockhart, C. S., D. Austin, L. Downey, and B. Jones. 1999. "The invasion of carrotwood (*Cupaniopsis anacardioides*) in Florida natural areas (U.S.A.)." *Natural Areas Journal* 19(3):254–262.

Laurel Fig   *Ficus microcarpa*

NAME AND FAMILY
Laurel fig, Chinese banyan, curtain fig, Indian laurel (*Ficus microcarpa* L. f. or *F. thonningii* Blume); mulberry family (Moraceae). Two other introduced figs have naturalized in areas of Florida, lofty fig (*F. altissima* Blume) and banyan fig (*F. benghalensis* L.). The native strangler fig (*Ficus aurea* Nutt.) is also similar in appearance to laurel fig.

IDENTIFYING CHARACTERISTICS
Figs often begin their lives growing on other trees as epiphytes. Seedlings germinate in the crotches of trees, on overpasses, and in walls. Aerial roots hang down from the branches and often root in the ground, thickening into pillars around the original host. They can also germinate on the ground and grow as an independent tree. The leathery evergreen leaves are arranged alternately and are variable in size and shape. Generally they are oblong, 2 to 3 in. (5 to 7.6 cm) long, 1 to 2 in. (2.5 to 5 cm) wide, smooth, and with smooth edges. Tiny 0.25 to 0.5 in. (0.6 to 1.3 cm) red to yellow fruits form that hang down from the leaf axils. The fig fruit (a synconium) contains the flowers inside it and after pollination by a tiny wasp many seeds form with a fleshy coating on the outside. Grayish white, smooth bark covers the branches, which reach out broadly. Branches and leaves have a milky sap. Trees can reach 90 ft. (27 m) in

Fig roots wrap around a dead tree trunk.

The spreading branches of a mature laurel fig tree.

height. Lofty fig and banyan fig tend to have much larger, more rounded, leaves, 10 in. (25 cm) long and 6 in. (15 cm) wide. The leaves of the native strangler fig have more distinct veins than those of the introduced figs.

HABITAT AND RANGE
Laurel fig grows in southern Florida in cypress swamps, pine rocklands, hardwood hammocks, and urban and suburban areas.

WHAT IT DOES IN THE ECOSYSTEM
Laurel fig grows over other trees, eventually killing them through its constricting roots, shade, and competition for nutrients. Birds disperse seeds when they eat the fruits, and seeds are also dispersed by ants when fruits fall to the ground. Laurel fig can cause structural damage if it grows too close to cement or rock structures.

HOW IT CAME TO NORTH AMERICA
Introduced to Florida from southern Asia in 1912 as an ornamental plant and street tree, laurel fig wasn't invasive until its wasp pollinator, *Parapristina verticillata*, was introduced accidentally to Florida sometime in the 1970s. Thomas Edison is supposed to have planted the first banyan fig at his home in Ft. Myers, Florida, a gift to him from industrialist Harvey Firestone.

MANAGEMENT

Small plants can be hand pulled. Figs are particularly susceptible to the herbicide triclopyr. Larger plants can be cut and the stumps treated with triclopyr, but care must be taken not to get herbicide on the host plant.

FOR MORE INFORMATION

Hammer, R. L. 1996. "*Ficus altissima, F. benghalensis, f. microcaspa.*" In *Invasive Plants: Weeds of the Global Garden,* edited by J. M. Randall, and J. Marinelli, 33–34. Brooklyn, NY: Brooklyn Botanic Garden.

Langeland, K. A., H. M. Cherry, C. M. McCormick, and K. A. Craddock Burks et al. 2008. *Identification and Biology of Nonnative Plants in Florida's Natural Areas,* 2nd ed. SP 257. Gainesville, FL: University of Florida IFAS.

Nadel, H., J. H. Frank, and R. J. Knight. 1992. "Escapees and accomplices: The naturalization of exotic *Ficus* and their associated faunas in Florida." *Florida Entomologist* 75(1):29–38.

Sapodilla   *Manilkara zapota*

NAME AND FAMILY

Sapodilla, chicle-gum tree (Manilkara zapota [L.] van Royen); soap berry family (Sapotaceae).

IDENTIFYING CHARACTERISTICS

This evergreen tree can grow to be 60 to 100 ft. (18 to 30 m) tall. The stiff leaves cluster at the ends of the shoots, emerging a light green to pinkish color and maturing to dark green. They grow 2 to 5 in. (5 to 13 cm) long with pointed ends. Bell-shaped flowers only 0.38 in. (0.9 cm) in diameter bloom in the leaf axils. Round to oval fruits, 2 to 4 in. (5 to 10 cm) in diameter, can mature year-round but are most abundant from May to September. They are covered with a hairy brown peel that hides the very sweet, light-brown to reddish brown pulpy flesh. Each fruit contains 0 to 12 flattened shiny black seeds each 0.75 in. (1.9 cm) in diameter. On larger trees, the bark has an attractive red-brown, flaky appearance.

HABITAT AND RANGE

Sapodilla grows in central and southern Florida, often establishing in more elevated parts of hardwood hammocks. It prefers calcareous, well-drained soils. Can withstand temperatures to 26°F (–3°C).

WHAT IT DOES IN THE ECOSYSTEM

These trees cast dense shade, making it difficult for other plants to survive. Seedlings can also grow very densely, potentially inhibiting establishment of other plant species.

Left: *Fruits begin to form from the five-petaled flowers.* Right: *Mature brown fruits surrounded by simple, evergreen leaves.*

HOW IT CAME TO NORTH AMERICA

Sapodilla fruits are eaten throughout much of Central America, where it originated, and it may have been introduced to Florida as early as the 1500s. It is also a source of chicle, the original base of chewing gum (and the once popular brand, Chiclets), made from the latex-like sap.

MANAGEMENT

Seedlings can be pulled up by hand. Triclopyr mixed with oil can be applied as a basal bark treatment.

FOR MORE INFORMATION

Balderi, C. F., and J. H. Crane. 2000. "The Sapodilla (*Manilkara zapota* van Royen) in Florida." Fact Sheet HS-1. University of Florida IFAS. http://edis.ifas.ufl.edu/MG057.

Melaleuca   *Melaleuca quinquenervia*

NAME AND FAMILY
Melaleuca, paperbark tree, punk tree, cajeput tree (*Melaleuca quinquenervia* [Cav.] Blake); myrtle family (Myrtaceae). Melaleuca is a close relative of eucalyptus. *M. quinquenervia* is not the tea tree (*M. alterniflora*) from which oils and other herbals are made. The related weeping bottlebrush tree (*Callistemon viminalis* (Sol. x Gaertn.) Cheel) is reported as invasive in parts of California and occasionally naturalizes in Florida.

IDENTIFYING CHARACTERISTICS
This fast-growing evergreen tree is covered with white and cinnamon-colored, peeling, spongy bark. It can grow to be 80 ft. (24 m) tall with a substantial trunk. Narrow elliptical leaves, 1 to 2 in. (2.5 to 5 cm) long and gray green in color, grow alternately along the stems. Leaves smell like eucalyptus when crushed. Throughout much of the year, white, bottlebrush-like clusters of flowers bloom

at the ends of the branches. Rounded seed capsules are held tightly against the stems, and stems continue to elongate after producing seeds. Seed capsules generally remain closed until they are dried by fire or a branch is injured by breaking, herbicide, or frost. Each capsule contains hundreds of seeds. Seeds are dispersed by wind and water. Red bottlebrush, *Callistemon viminalis*, bears some resemblance to melaleuca in bark, leaves, and seed pods and is used extensively in landscaping, but its flowers are red and it is not considered invasive except in parts of California

HABITAT AND RANGE
Grows throughout southern Florida in wet areas, including the Everglades, cypress swamps, and along canals and lake edges.

WHAT IT DOES IN THE ECOSYSTEM

Melaleuca features peeling bark and evergreen leaves.

Trees can form extremely dense stands that block light to species in

Bright-green new leaves mature to gray green.

Seed capsules cluster at branch tips.

the understory and prevent establishment of other vegetation. It often converts marshes into tree-dominated swamps, changing habitat for wildlife. Melaleuca took over hundreds of thousands of acres in the Everglades before a massive control program began in the 1990s. Trees are fire adapted and can cause very hot crown fires. Honey bees use melaleuca flowers as a source of nectar. The pollen causes allergies in some people.

HOW IT CAME TO NORTH AMERICA
Melaleuca was introduced from eastern Australia as an ornamental tree and spread in the 1930s and 1940s in an attempt to afforest marshy "wastelands." It was introduced at least twice to Florida, to the east coast of Florida in 1907 by Dr. John Gifford and to the west coast of Florida in 1912. The Koreshan Unity sect of

utopians adopted it as a fast-growing substitute for the native trees they had cut. The Koreshans planted it in nurseries and sold saplings to the public. The bark has been used for mulch and the wood for pulpwood.

MANAGEMENT
Seedlings can be hand pulled. Larger trees are often cut or girdled and treated with the herbicide imazapyr. Three biocontrol insects, the melaleuca snout beetle or melaleuca weevil (*Oxyops vitiosa*), melaleuca psyllid (*Boreioglycaspis melaleucae*), and a gall midge *(Lophodiplosis trifida)* from Australia have been successfully released in the Everglades to combat the trees. A rust fungus, *Puccinia psidii*, probably introduced through the nursery trade, also attacks melaleuca trees.

FOR MORE INFORMATION
Kaufman, S. R., and P. E. Smouse. 2001. "Comparing indigenous and introduced populations of *Melaleuca quinquenervia* (Cav.) Blake: Response of seedlings to water and pH levels." *Oecologia* 127:487–494.

Mazzotti, F. J., T. D. Center, F. A. Dray, and D. Thayer. "Ecological consequences of invasion by *Melaleuca quinquenervia* in South Florida Wetlands: Paradise damaged, not lost." University of Florida IFAS. http://edis.ifas .ufl.edu/UW123.

Center, T. D., M. D. Purcell, P. D. Pratt, M. B. Rayamajhi, P. W. Tipping, S. A. Wright, and F. A. Dray, Jr. 2011. "Biological Control of *Melaleuca quinquenervia*, an Everglades invader." *Biological Control*. doi 10.1007/s10526-011-9390-6

Turner, C. E., T. D. Center, D. W. Burrows, and G. R. Buckingham. 1997. "Ecology and management of *Melaleuca quinquenervia*: An invader of wetlands in Florida, U.S.A." *Wetlands Ecology and Management* 5:165–178.

Pepper Tree   *Schinus molle*

NAME AND FAMILY
Pepper tree, California pepper tree, Peruvian pepper tree, false pepper (*Schinus molle* L.); sumac family (Anacardiaceae).

IDENTIFYING CHARACTERISTICS
The hanging clusters of purplish to pinkish-red peppercorn-sized fruits that mature in fall earn the pepper tree its name. These evergreen trees grow to 40 ft. (12 m) tall with a short trunk and slender, weeping branches. The bark of mature trees is deeply fissured, flaky, and aromatic. Compound leaves up to a foot long that alternate along the stems are divided into 14 to 40 lance-shaped leaflets, 0.75 to 1.5 in. (1.9 to 3.7 cm) long, that zigzag along the leaf stem (rachis). Leaflet edges are usually toothed. Male and female flowers occur on separate plants, so only female plants bear fruit. Yellowish-white flowers bloom in hanging clusters from the leaf axils and ends of the branches in spring to early summer.

HABITAT AND RANGE

Found in natural areas in California. Tends to colonize disturbed areas such as desert washes, dry shrub communities, grasslands, and old fields. Trees are very drought tolerant.

WHAT IT DOES IN THE ECOSYSTEM

Plants spread by root sprouts and seeds to form stands of pepper trees. Currently a localized problem. Trees are susceptible to scale and root rot. In Australia, plants shade out other vegetation, which can lead to erosion on slopes. Fruits are eaten by birds.

Right: *Pink-red fruits give pepper tree its name.* **Below:** *The weeping branches of a large pepper tree.*

Tiny flowers bloom in spring to early summer.

Plants can cause a poison ivy–like dermatitis in some people.

HOW IT CAME TO NORTH AMERICA
The mission of San Luis Rey de Francia in Oceanside, California, claims that the first pepper trees in California were planted there by Father Antonio Peyri, who had received the seeds from a sailor returning from South America in the early 1830s. Pepper tree was widely planted as an ornamental tree. Pepper tree is native to Peru and surrounding South American countries. Dried fruits are sometimes sold as pink pepper but if eaten in large quantities can be toxic.

MANAGEMENT
Remove female trees first to limit seed production. Smaller saplings and seedlings can be pulled or dug up. Larger trees can be cut down and the stumps treated with herbicide glyphosate or triclopyr.

FOR MORE INFORMATION
Howard, L. F., and R. A. Minnich. 1989. "The introduction and naturalization of *Schinus molle* (pepper tree) in Riverside, California (U.S.A.)." *Landscape and Urban Planning* 18(2):77–96.
Nilsen, E. T., and W. H. Muller. 1980. "A comparison of the relative naturalization ability to two *Schinus* species in southern California. I. Seed germination." *Bulletin of the Torrey Botanical Club* 107:51–56.
"San Luis Rey de Francia." http://www.athanasius.com/camission/luis.htm.
Weber, E. 2003. *Invasive Plant Species of the World.* Cambridge: CABI Publishing.

Terrestrial Plants—Trees and Shrubs— Evergreen—Alternate Leaves—Shrubs

Shoebutton Ardisia   *Ardisia elliptica*

NAME AND FAMILY
Shoebutton ardisia, seashore ardisia (*Ardisia elliptica* Thunb.); Myrsine family (Myrsinaceae). It can be confused with the native marlberry (*A. escalloniodes*). Farther north in Florida and Louisiana, coral ardisia (*A. crenata*) is invasive.

IDENTIFYING CHARACTERISTICS
Shoebutton ardisia grows as a tall shrub or small tree, up to 20 ft. (6.1 m), with leathery evergreen leaves alternating along the stems. The leaves are 3 to 6 in. (7.6 to 15.2 cm) long, about 1 in. (2.5 cm) wide, and elliptical, with smooth edges. New leaves are often reddish in color. Star-shaped, five-petaled, mauve-colored flowers hang in clusters from the branches, attached where the leaves meet the stems. Pink fruit clusters ripen to shiny black to dark purple with 1 seed in each fruit. Plants flower and set fruit year-round, but peak flowering occurs in summer. Native marlberry can best be distinguished from shoebutton

Five-petaled flowers cluster in the leaf axils.

Left: *Ardisa's leathery green leaves.* **Right:** *Fruits change color from pink to dark purple.*

ardisia because fruits and flowers occur only at the ends of the branches in marlberry. Coral ardisia is a smaller shrub and has coral to red fruits.

HABITAT AND RANGE
Shoebutton ardisia is a common invader in southern Florida in shady, moist areas. It grows on tree islands in the Everglades and in other wet, lowland areas and in old fields. It spread into formerly agricultural areas within Everglades National Park after being mistaken for the native species and being planted around the new visitors center around 1947. Coral ardisia is becoming a problem plant farther north in Florida where it has escaped cultivation.

WHAT IT DOES IN THE ECOSYSTEM
Shoebutton ardisia forms dense colonies that displace native plants. Birds eat the fruits and disperse them to new tree islands throughout the Everglades. Raccoons and opossums also eat the fruit.

HOW IT CAME TO NORTH AMERICA
This native of India and Southeast Asia was introduced as an ornamental plant in the late 1800s. Coral ardisia, native from Japan to India, was introduced as a more cold-tolerant ornamental than shoebutton ardisia.

MANAGEMENT
Seedlings can be pulled up by hand. Stands can also be sprayed with glyphosate, or to minimize damage to surrounding plants, herbicide can be applied to cut stumps (glyphosate or 2,4-D) or to the bark (triclopyr mixed with oil).

FOR MORE INFORMATION
Francis, J. K. Ardisia elliptica *Thunb.* U.S. Forest Service. http://www.fs.fed.us/global/iitf/pdf/shrubs/Ardisia%20elliptica.pdf.

Koop, A. L. 2004. "Diferential seed mortality among habitats limits the distribution of the invasive non-native shrub *Ardisia elliptica.*" *Plant Ecology* 172: 237–249.

Langeland, K. A., H. M. Cherry, C. M. McCormick, and K. A. Craddock Burks et al. 2008. *Identification and Biology of Nonnative Plants in Florida's Natural Areas,* 2nd ed. SP 257. Gainesville, FL: University of Florida IFAS.

Latherleaf   *Colubrina asiatica*

NAME AND FAMILY
Latherleaf, Asiatic colubrina, Asian snakewood, hoop withe (*Colubrina asiatica* [L.] Brongn.); buckthorn family (Rhamnaceae). Can be confused with native colubrinas.

DESCRIPTION
Latherleaf grows as a low shrub but has climbing, vine-like branches that can reach 30 ft. (9 m) long. Thin, shiny green egg-shaped leaves come to a long point and occur alternately along the stem. Leaves are 1.5 to 5.5 in. (3.7 to 13.7 cm) long with toothed edges. Small greenish flowers bloom in clusters from where the leaves meet the stems in midsummer, but it can flower year-round.

Small gray seeds are held in 0.5 in. (1.3 cm) greenish capsules that turn dark brown with age, maturing by early fall. Plants can begin to produce seeds after only one year. Seeds float and are salt tolerant, so they disperse along ocean currents. Native colubrinas grow as trees and have hairy shoots as opposed to the sprawling branches of latherleaf and its smooth shoots.

HABITAT AND RANGE
Latherleaf occurs along the east and west coast of Florida and in the Florida

Latherleaf fruits will turn brown as they mature.

Shiny green leaves come to a long point.

Keys on upland sites like coastal hardwood forests, dunes, roadsides, and elevated ridges in mangrove forests. Prefers areas with full to part sun.

WHAT IT DOES IN THE ECOSYSTEM

Mats of stems several feet thick prevent light from reaching the forest floor and impede germination of nearly all plants. Vines can climb in a dense wall up native foliage, growing as much as 30 ft. (9 m) a year. Because hardwood coastal forests are increasingly uncommon and many species in them are considered endangered, latherleaf may threaten the species in those forests. Seeds can survive three to five years in the soil and plants also spread vegetatively from trailing stems and by resprouting from cut stems. Vines can climb upwards, then fall back, produce roots where they contact the ground, and send up new vines.

Leaves alternate along long branches.

Sprawling growth of latherleaf.

HOW IT CAME TO NORTH AMERICA

East Asian immigrants probably brought latherleaf first to Jamaica around 1850 because of its many traditional uses as a soap substitute (thus "latherleaf"), fish poison, food, and medicine. It is native to tropical Asia. It was first collected in the wild from the Florida Keys in 1937.

MANAGEMENT

Seedlings and young plants can be hand pulled. Larger plants can be killed through basal bark or cut-stump application of triclopyr or by spraying glyphosate on leaves.

FOR MORE INFORMATION

Jones, D. T. 1997. "Ecological consequences of latherleaf (*Colubrina asiatica*) in southern Florida." *Wildland Weeds* (winter): 11–12.

Langeland, K. A., H. M. Cherry, C. M. McCormick, and K. A. Craddock Burks et al. 2008. *Identification and Biology of Nonnative Plants in Florida's Natural Areas,* 2nd ed. SP 257. Gainesville, FL: University of Florida IFAS.

McCormick, C. M. 2007. Colubrina asiatica *(Latherleaf) management plan*. South Florida Water District. http://www.fleppc.org/Manage_Plans/CA%20Mngt%20Plan.pdf.

Mahonia   *Mahonia* spp.

NAME AND FAMILY
Oregon grape holly, Oregon grape, tall Oregon grape (*Mahonia aquifolium* [Pursh.] Nutt.) and leatherleaf mahonia, Asian mahonia (M. *bealei* [Fortune] Carriére); barberry family (Berberidaceae). Some cultivars of Oregon grape may be hybrids with other native mahonias.

IDENTIFYING CHARACTERISTICS
Although sometimes called "tall" Oregon grape, M. *aquifolium* seldom grows to more than 6 ft. (2 m); however, plants can occasionally reach 16 ft. (5 m). The dark-green, shiny, leathery, spiked leaflets suggest it is related to American holly, but it is not. The five to nine individual leaflets are part of a larger leaf 6 to 12 in. (15 to 30 cm) long. The leaves alternate along the stems and are evergreen, although they can turn yellow or reddish on occasion, depending on temperature, age, and soil conditions. Fragrant yellow flowers 0.25 in. (0.5 cm) long bloom in clusters at the ends of branches in spring. The flowers are followed by greenish berries ripening to a deep blue black with a powdery surface suggesting grapes. The shoots or trunks are usually thin and branch irregularly, and the inside bark has a distinct yellow color (used for dying) caused by the alkaloid berberine.

Leatherleaf mahonia has grey green leaves with 9 to 13 leaflets held on strong, upright stems. It begins flowering in winter and its fruits are similar in appearance to those of Oregon grape, with a bluish color when ripe. Leatherleaf mahonia generally grows 5 to 10 ft. (1. 5 to 3 m) tall.

Oregon grape leaves have five to nine leaflets.

HABITAT AND RANGE

Oregon grape is native to the areas of British Columbia, Washington, Oregon, and northern California that have wet winters and dry summers; it is associated with Douglas fir as an understory plant but is also present in other areas. Does well in rocky to gravelly poor soils, although it is most often found naturalizing in sites with some moisture. Outside its native range, Oregon grape is now found in Georgia, Michigan, Montana, New Jersey, New York, Ohio, Ontario, and Quebec. Leatherleaf mahonia grows from Maryland south to Florida in open woodlands.

WHAT IT DOES IN THE ECOSYSTEM

So far, mahonias have not had serious consequences for native plants. In Europe, Oregon grape can produce dense stands that outcompete native plants, so many land managers are

Flowers of leatherleaf mahonia bloom in late winter.

being proactive in removing mahonias when they appear in eastern parks. Oregon grape seems to be more invasive in its introduced range because the seeds came from cultivated selections that were chosen for their hardiness and vigorous growth. Mahonia flowers attract insect pollinators and the fruits are dispersed by birds. Leatherleaf mahonia can hybridize with North American mahonias. Mahonias are used as dye plants and some herbalists say the alkaloid berberine is anti-inflammatory and antibacterial.

HOW IT CAME TO NORTH AMERICA

Native to the Pacific Northwest, Oregon grape has been introduced to other areas as a drought-resistant, slow-growing ornamental. Leatherleaf mahonia is native to China, Japan, and Taiwan, and was introduced as an ornamental plant in 1845.

MANAGEMENT

Oregon grape in limited quantities is easy to control by cutting and pulling. Gardeners who have it as an ornamental can control its spread by clipping off the fruits before they fully ripen. Cut stems can also be treated with herbicides containing glyphosate or triclopyr.

FOR MORE INFORMATION

Allen, C. R., A. S. Garmestani, J. A. LaBram, A. E. Peck and L. B. Prevost. 2006. "When landscaping goes bad: the incipient invasion of *Mahonia bealei* in the southeastern United States." *Biological Invasions* 8(2):169–176. http://digitalcommons.unl.edu/cgi/viewcontent.cgi?article=1017&context=ncfwrustaff.

Ross, C. and D. Faust, and H. Auge. 2009. "Invasions in different habitats: local adaptation or general-purpose genotypes?" *Biological Invasions* 11(2):441–452.

Scaevola   *Scaevola sericea*

NAME AND FAMILY

Scaevola, Hawaiian half-flower, beach naupaka, sea lettuce (*Scaevola sericea* Vahl); Goodenia family (Goodeniaceae). There are two varieties of scaveola; *S. sericea* var. *sericea* is the more widespread, but *S. sericea* var. *taccada* is also present in Florida. Can be confused with native inkberry or gullfeed (*S. plumieri*), which is a threatened species in Florida.

IDENTIFYING CHARACTERISTICS

Evergreen glossy, succulent, spoon-shaped leaves cover this rounded shrub or small tree that grows to 16 ft. (4.8 m). Leaves taper gradually toward the base, growing 5 to 9 in. (12.7 to 23 cm) long and 1 to 4 in. (2.5 to 10 cm) wide, and the leaf edges usually have indentations. *S. sericea* var. *sericea* has silky hairs on the leaves and stems whereas var. *taccada* is hairless. Clusters of two to four white to pinkish flowers appear year-round from the leaf axils. Flowers growing from leaf axils have five white to pale lilac petals, fused at first, then separate and arranged in a semicircle, less than 1 in. (2.5 cm) across. Fleshy rounded fruits

Scaevola grows along the water's edge in coastal area of Florida.

Spoon-shaped leaves are glossy and succulent.

begin green and mature white, 0.5 to 0.75 in. (0.6 to 1.9 cm) wide. Inkberry has smaller leaves (to 4 in. [10 cm]), smooth leaf edges, and black fruits.

HABITAT AND RANGE
Grows along the central and southern coasts of Florida on dunes and rock barrens and in mangroves and coastal hardwood forest edges.

The five petals of Scaevola are arranged in a semicircle.

WHAT IT DOES IN THE ECOSYSTEM
Scaevola forms dense thickets in coastal habitats. Fruits float to new locations and stems touching the soil can root. On sand dunes, it displaces plants that are better at erosion control and it covers open dune that is habitat for several threatened and endangered plant species. Fruits are eaten by pigeons and sea birds.

HOW IT CAME TO NORTH AMERICA
Scaevola has been sold in nurseries since the 1960s and was promoted for controlling beach erosion and in coastal landscaping in the 1970s and 1980s.

Native around the Indian and western Pacific oceans, probably introduced from Hawaii.

MANAGEMENT
Plants can be hand pulled, but care should be taken to remove all pieces of underground stem. Stems can also be cut and the cut ends painted with triclopyr.

FOR MORE INFORMATION
Langeland, K. A., H. M. Cherry, C. M. McCormick, and K. A. Craddock Burks et al. 2008. *Identification and Biology of Nonnative Plants in Florida's Natural Areas,* 2nd ed. SP 257. Gainesville, FL: University of Florida IFAS.
Langeland, K. A., J. A. Ferrell, B. Sellers, G. E. MacDonald, and R. K. Stocker. 2011. "Integrated Management of Nonnative Plants in Natural Areas of Florida." SP 242. Gainesville, FL: University of Florida IFAS. http://edis.ifas.ufl.edu/wg209.

Brazilian Pepper Tree   *Schinus terebinthifolius*

NAME AND FAMILY
Brazilian pepper tree, Christmas berry tree, Florida holly (*Schinus terebinthifolius* Raddi); sumac family (Anacardiaceae). A second species, Peruvian peppertree (*Schinus polygamus* (Cav.) Cabrera) is considered invasive in California.

IDENTIFYING CHARACTERISTICS
Brazilian pepper distinguishes itself by the peppery, turpentine-like smell of its crushed leaves and by the clusters of round, peppercorn-size fruits that begin green and turn bright red before drying to a paper-like shell around the pepper seed. The tree often spreads along the ground or grows as a shrub but can grow to 30 ft. (9 m), with a short trunk and tangled branches. Dark green leaves grow alternately on the stem and have 3 to 12 toothed leaflets, 1 to 2 in. (2.5 to 5 cm) long, arranged opposite each other, often along a winged rachis (the stem holding the leaflets). Small, five-petaled, white flowers grow in clusters at the

Fruits turn red as they mature.

Fruit clusters hang along new stems.

leaf axils of new stems. Flowers appear year-round but are more concentrated in the fall. Male and female flowers usually occur on separate plants.

Peruvian peppertree grows as a spiny shrub with abundant, small, greenish-white flowers in spring followed by purple to black peppercorns. It has simple leaves with smooth edges.

HABITAT AND RANGE
This cold-sensitive plant is confined to Florida, Arizona, Texas, Louisiana, and California, though it grows in some areas that experience rare light frosts. Pepper tree is an opportunist that moves in quickly when hurricanes or humans clear wooded or

Brazillian pepper forms dense stands.

brushy areas or abandon farmland. Often found along drainage ditches and canals. It also, however, establishes colonies in relatively undisturbed areas.

WHAT IT DOES IN THE ECOSYSTEM
Brazilian pepper competes aggressively and successfully where the American climate favors it, and forms dense thickets, where its branches cast a heavy

shade that impedes or prevents the growth of native plants. It has become a major competitor in mangrove stands and hardwood hammocks, displacing some rare plants like beach jacquemontia (*Jacquemontia reclinata*) and beach star (*Remirea maritima*). It also produces chemicals that suppress competitors. Pepper tree responds to damaged branches and trunks by sending out new sprouts from either the trunk or roots. Florida authorities estimate it dominates some 700,000 acres in central and southern Florida. While some birds, especially robins, feed on and spread the seeds, the Brazilian pepper colonies eliminate plants that feed many other animals. The sap, which contains alkenyl phenols, causes skin irritation in some people and reports say that people in close proximity to the trees (e.g., sitting in their shade) sometimes experience sneezing, burning eyes, and headaches. Brazilian pepper produces abundant nectar that is the source of several million pounds of Florida honey each year.

HOW IT CAME TO NORTH AMERICA
This native of Argentina, Paraguay, and Brazil appears in American seed catalogs in the 1830s, promoted for its decorative value. In the 1920s when Florida began its first boom as an American winter mecca, Dr. George Stone of Punta Gorda apparently raised and distributed hundreds of plants that became popular ornamentals along city streets. The red fruits are used to make wreaths and dried flower arrangements.

MANAGEMENT
Once Brazilian pepper colonies are established, bulldozers, front-end loaders, and other heavy equipment become the most practical means of removing them. This kind of disturbance, however, opens up the land for new invasions by pepper trees or other exotics. Stumps and roots left in place are likely to resprout unless herbicides follow mechanical removal. Cutting single or scattered trees should also be followed by painting herbicide on the cut trunks. Basal bark treatment is also effective. Systemic herbicides containing triclopyr, glyphosate, hexazinone, or imazapyr can also be sprayed on foliage for an effective kill. Studies indicate that herbicides are most effective when used during active growing seasons—at the end of autumn to March. Avoid contact with the sap, as it can cause skin irritation. Fire can be used to reduce seed germination, but plants over 3 feet tall will resprout readily after fire.

FOR MORE INFORMATION
Dalrymple, G. H., R. F. Doren, N. K. O'Hare, M. R. Norland, and T. V. Armentano. 2003. "Plant colonization after complete and partial removal of disturbed soils for wetland restoration of former agricultural fields in Everglades National Park." *Wetlands* 23:1015–1029.

Hight, S. D., J. P. Cuda, and J. C. Medal. 2002. "Brazilian Pepper." In *Biological Control of Invasive Plants in the Eastern United States,* edited by R. Van Driesche et al. U.S. Department of Agriculture Forest Service Publication FHTET-2002-04. http://wiki.bugwood.org/Archive: BCIPEUS/Brazilian_Peppertree.

Langeland, K. A., J. A. Ferrell, B. Sellers, G. E. MacDonald, and R. K. Stocker. 2011. "Integrated Management of Nonnative Plants in Natural Areas of Florida." SP 242. Gainesville, FL: University of Florida IFAS. http://edis.ifas.ufl.edu/wg209.

Li, Y., and M. Norland. 2001. "The role of soil fertility in invasion of Brazilian pepper (*Schinus terebinthifolius*) in Everglades National Park, Florida." *Soil Science* 166(6):400–405.

Meyer, R. 2011. *"Schinus terebinthifolius."* *Fire Effects Information System*. U.S. Department of Agriculture, Forest Service, Rocky Mountain Research Station, Fire Sciences Laboratory. http://www.fs.fed.us/database/feis/.

Tropical Soda Apple   *Solanum viarum*

NAME AND FAMILY
Tropical soda apple, Sodom's apple (*Solanum viarum* Dunal); nightshade family (Solanaceae). Listed as a federal noxious weed in the United States. Many species of *Solanum* grow in the southeastern United States, both native and nonnative. Wetland nightshade or aquatic soda apple (*S. tampicense* Dunal) is another species of *Solanum* considered particularly problematic in natural areas in Florida.

IDENTIFYING CHARACTERISTICS
A prickly, bushy perennial plant, tropical soda apple is named for its fruits. Fruits are rounded, about 1 in. (2.5 cm) in diameter, and mottled green maturing to yellow. A mature fruit can contain more than 400 red-brown seeds. Hairy, lobed leaves resembling oak leaves with pointed lobes are 6 to 8 in. (15 to 20 cm) long and 2 to 6 in. (5 to 15 cm) wide, alternating along hairy stems.

Tropical soda apple is a sprawling plant bearing mottled green and white fruits.

Leaf veins and stems are covered with 0.5 to 1 in. (1.3 to 2.5 cm) long spines. Clusters of one to five white, five-petaled flowers form below the leaves, flowering year-round with a concentration of flowering and fruiting from September to May. Distinguished from other varieties of *Solanum* by the straight prickles and green fruits with darker green stripes. Wetland nightshade, a vining or trailing plant found in swamps and marshes in southern Florida, is distinguished by its clusters of pea-sized red fruits.

HABITAT AND RANGE
Occurs from North Carolina south through Florida and west into Louisiana, preferring to establish in dry, sunny to partly shady disturbed areas like fields, pastures, and roadsides. Increasingly found along the edges of pine forests and hardwood hammocks.

WHAT IT DOES IN THE ECOSYSTEM
In only six years, from 1990 to 1996, naturalized colonies of tropical soda apple increased from 25,000 acres (10,000 ha) to 500,000 acres (202,000 ha) and may now cover more than 1 million acres (405,000 ha). In natural areas, the plants displace native plants through their aggressive growth and large leaves that shade the ground. Some monoculture stands have covered more than 50 acres (20 ha). The leaves are unpalatable to livestock, diminishing the value of pastures it has invaded. Livestock do eat the fruits, as do wild pigs, deer, raccoons, and birds, among other animals. Movement of cattle and composted cow manure are two of the major pathways for the spread of tropical soda apple. Plants are alternate hosts for many crop pests and diseases.

HOW IT CAME TO NORTH AMERICA
Accidentally introduced to Florida and first noticed in Glades County in 1988. Native to South America. Plants are grown in some parts of the world as a source of steroids (from the solasodine compound), but it is considered a serious weed in many countries.

MANAGEMENT
Small patches of soda apple can be pulled up, although removal of all the roots can be difficult for mature plants. Mow plants before fruits form, then use herbicide containing triclopyr, glyphosate, or aminopyralids on regrowth. Some recommended herbicides additionally contain chemicals to keep seeds from sprouting. If seeds remain in the soil, retreatment will probably be necessary. A new biological herbicide, SolviNix, containing a tobacco mosaic virus, is proving effective at controlling soda apple. A beetle, *Gratiana boliviana,* is available from Florida extension offices as a biological control agent for sparse, remote populations of tropical soda apple.

FOR MORE INFORMATION

Langeland, K. A., H. M. Cherry, C. M. McCormick, and K. A. Craddock Burks et al. 2008. *Identification and Biology of Nonnative Plants in Florida's Natural Areas*, 2nd ed. SP 257. Gainesville, FL: University of Florida IFAS.

Miller, J. H., E. B. Chambliss, N. J. Loewenstein. 2010. *A field guide for the identification of invasive plants in southern forests*. General Technical Report SRS-119. Asheville, NC: U.S. Department of Agriculture, Forest Service, Southern Research Station. http://www.srs.fs.fed.us/pubs/gtr/gtr_srs119.pdf.

Sellers, B., J. Ferrell, J. Mulahey, and P. Hogue. 2010. *Tropical soda apple: biology, ecology and management of a noxious weed in Florida*. SS-AGR-77. Gainesville, FL: University of Florida IFAS. http://edis.ifas.ufl.edu/uw097.

Wunderlin, R. P. et al. 1993. "*Solanum viarum* and *S. tampicense* (Solanaceae): Two weedy species new to Florida and the United States." *SIDA* 15:605–611.

Seaside Mahoe   *Thespesia populnea*

NAME AND FAMILY

Seaside mahoe, portia tree (*Thespesia populnea* [L.] Soland. ex Correa); mallow family (Malvaceae). The genus name Thespesia is derived from the Greek word meaning divine. Daniel Solander, traveling on Captain Cook's expedition to the south seas, named the plant when he found it planted around tem-

Left: *Hibiscus-like flowers open during the day.* **Right:** *Leathery fruits mature from yellow to black.*

Seaside mahoe grows as a shrub or small tree in coastal areas.

ples in Tahiti. May be confused with another introduced species, sea hibiscus (*Hibiscus tiliaceus* L.) or the native endangered wild cotton (*Gossypium hirsutum* L.).

IDENTIFYING CHARACTERISTICS

This small evergreen tree or shrub growing to 40 ft. (12 m) tall is probably most notable for its 3 in. (7.6 cm) wide, hibiscus-like yellow flowers with red centers. The flowers bloom singly from the leaf axils at the upper ends of the branches and last only one day. By nightfall they change to a maroon color. Heart-shaped leaves alternate along the stems. Each leaf is 2 to 8 in. (5.1 to 20 cm) long, 2 to 4 in. (5 to 10 cm) wide, and shiny dark green above with five veins radiating from the base of the leaf. Leaves are held on stalks (petioles) 2 to 4 in. (5 to 10 cm) long. Leathery, ball-shaped fruits mature from yellow to black and contain a few hairy brown seeds. The plant produces fruits and flowers year-round. Sea hibiscus has wider leaves and dense, star-shaped hairs on the leaf underside. The leaves of wild cotton are arranged in pairs instead of alternating along the stems.

HABITAT AND RANGE
Seaside mahoe grows in coastal areas in mangrove swamps, coastal hardwood forests, and along beaches of bays and inlets in southern Florida. Plants are very salt and wind tolerant, but intolerance of frost restricts their range.

WHAT IT DOES IN THE ECOSYSTEM
Trees form dense thickets on hammocks, dunes, and among mangroves, displacing native plants and changing habitat for animals that rely on open beaches. Plants produce prolific fruit crops that float to new locations. Host to the cotton stainer bug (*Dysdercus decussatus*), a pest of cotton.

HOW IT CAME TO NORTH AMERICA
Probably introduced in 1928 to Miami, Florida, as an ornamental plant. It is native to India but is widely naturalized in most tropical coastal areas. Seaside mahoe is a plant of many uses—the wood (known as Pacific rosewood) is used to make small items; the bark is used for making rope; red and yellow dyes are made from other parts of the plant; and it is used in traditional medicines.

MANAGEMENT
Seedlings and saplings can be pulled or dug up. Larger trees can be cut and the stumps treated with herbicide containing triclopyr, or trees can be be treated using a basal bark spray.

FOR MORE INFORMATION
Friday, J. B., and D. Okano. 2005. "*Thespesia populnea* (milo)." In *Species Profiles for Pacific Island Agroforestry*, edited by C. R. Elevitch. Holualoa, HI: Permanent Agriculture Resources (PAR). http://agroforestry.net/tti/Thespesia-milo.pdf.
Langeland, K. A., H. M. Cherry, C. M. McCormick, and K. A. Craddock Burks et al. 2008. *Identification and Biology of Nonnative Plants in Florida's Natural Areas,* 2nd ed. SP 257. Gainesville, FL: University of Florida IFAS.

Terrestrial Plants—Trees and Shrubs— Evergreen—Leaves—Opposite or Whorled—Trees

Schefflera 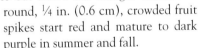 *Schefflera actinophylla*

NAME AND FAMILY
Schefflera, Queensland umbrella tree, octopus tree (*Schefflera actinophylla* [Endl.] Harms); Aralia family (Araliaceae).

IDENTIFYING CHARACTERISTICS
Most people are familiar with this plant because of its popularity as a graceful but tough houseplant as well as a landscaping plant. Outdoors it can grow into a tree 40 ft. (12 m) tall. The evergreen, palmlike leaves are made up of 7 to 16 leaflets that radiate out like the spokes of an umbrella attached to a long leaf stalk. At the tip of each branch, 1 to 2 ft. (0.3 to 0.6 m) long dense clusters of deep red, 1 in. (2.5 cm) flowers radiate out above the foliage. Clusters of round, ¼ in. (0.6 cm), crowded fruit spikes start red and mature to dark purple in summer and fall.

HABITAT AND RANGE
Schefflera grows in a wide variety of habitats in southern Florida, including cypress forests, pine scrublands, dunes, and hardwood hammocks. It can grow in full sun to full shade, but cannot tolerate temperatures below 35°F (1.7°C).

WHAT IT DOES IN THE ECOSYSTEM
Even in undisturbed areas schefflera forms dense stands that shade out native plants. It threatens to shade out populations of several rare plant

Spikes of flowers are held above the foliage.

Top: *This one umbrella-shaped leaf is made up of many leaflets.* Right: *Most often seen as a houseplant, Schefflera can grow into a small tree.*

species in Florida. Often seeds germinate on other plants like cabbage palms or on rocks and can survive until their roots eventually reach the ground. Roots are strongly aggressive and the plant drops large amounts of litter. Birds eat the fruits and spread the seeds.

HOW IT CAME TO NORTH AMERICA
Introduced as a landscaping plant in Florida in 1927. Noted as growing in natural areas in the 1970s. Native to northern Australia, New Guinea, and Java.

MANAGEMENT
Seedlings and saplings can be hand pulled. Larger trees can be cut and the stumps treated with triclopyr herbicide or triclopyr mixed with oil can be applied to the base of trees. Large trees can take up to nine months to kill.

FOR MORE INFORMATION
Hammer, R. 1996. "*Schefflera actinophylla* Queensland umbrella tree." In *Invasive Plants: Weeds of the Global Garden,* edited by J. M. Randall, and J. Marinelli. Brooklyn Botanic Garden.
Langeland, K. A., H. M. Cherry, C. M. McCormick, and K. A. Craddock Burks et al. 2008. *Identification and Biology of Nonnative Plants in Florida's Natural Areas,* 2nd ed. SP 257. Gainesville, FL: University of Florida IFAS.

Terrestrial Plants—Trees and Shrubs—Evergreen—Opposite or Whorled Leaves—Shrubs

Scotch Broom   *Cytisus scoparius*

NAME AND FAMILY
Scotch broom, scots broom (*Cytisus scoparius* [L.] Link); pea family (Fabaceae). Several brooms are invasive in North America. Scotch broom and French broom (*Genista monspessulana* [L.] L. Johnson) are the most common.

IDENTIFYING CHARACTERISTICS
The casual observer often sees broom in stands of tall bushes as high as 15 ft. (4.6 m) but usually 3 to 6 ft. (1 to 1.8 m), its thousands of small yellow or white

blossoms brightening a road embankment, old field, or timbered woodland. Flowers look like sweet pea flowers, usually bright yellow. They occur singly where the leaf meets the stem of the shrub. Scotch broom has small alternate dark green leaves each made up of three leaflets, but often all that is noticeable are the green stems. The flowers form seedpods that explode when mature. Seedpods are fuzzy on the edges, flattish, 1 to 2 in. (2.5 to 5 cm) long. French broom appears similar to scotch broom but has hairy seedpods and lighter yellow flowers in a stalk (raceme) of three to nine flowers.

HABITAT AND RANGE
Broom has spread quickly in the Pacific Northwest and British Columbia coastal areas as far north as Alaska

Yellow, pealike flowers grow from the leaf axils.

Scotch broom's green stems are more noticeable than the tiny leaves.

and lands where the Pacific weather systems make for mild wet winters. This region also has an abundance of well-drained soils. Both conditions mimic broom's native Mediterranean habitat. It grows rapidly and densely in timbered areas. Broom has also claimed large acreages in Alabama, Georgia, Kentucky, Tennessee, Virginia, and West Virginia.

WHAT IT DOES IN THE ECOSYSTEM

Broom often grows in dense stands that outcompete native vegetation. In heavily logged or surface mined areas of the West it takes hold rapidly and can prevent regrowth of native species, including trees. Befitting a plant that withstands long dry summers and poor subsoils, broom has a deep root system, and bacteria in the roots fix nitrogen the plant can use. Through allelopathy and changes it causes in soil nutrients, its presence results in an increase in exotic species and a decrease in native species in some ecosystems. On thousands of acres of Oregon coastal dunes it dominates the plant community almost to the exclusion of native species. When two or three years old, broom begins to produce thousands of seeds in pods that disperse by exploding when mature. The seeds remain viable for up to 30 years.

HOW IT CAME TO NORTH AMERICA

Broom was planted ornamentally in the eastern United States beginning in the early 1800s. Captain Walter Grant brought seeds to Canada's Vancouver Island in the 1850s and planted them on his estate; by 1850 the plant was also sold commercially in Canada. In the Sierra Nevada, broom twigs and seeds arrived as the cushioning between bottles in shipments of scotch for miners. Broom's hearty growth and bright flowers made it useful in gardens and as roadside stabilization. It is still commonly sold as an ornamental plant. It was called broom because the twigs were used to make brooms in Europe. Early European settlers roasted broom seed for a coffee substitute and used sprouts instead of hops in beer production. However, broom contains alkaloids that can damage the heart and nervous system.

MANAGEMENT

Given broom's rapid growth, deep roots, and bountiful and long-lived seeds, the best control is early prevention, eradication, and planting of fast-growing native plants. Once the plant is established, manual or mechanical control of broom is very difficult, time consuming, and expensive. In some cases, seeding fast-growing, tall native plants among the broom can eventually lead to its being shaded out. Experiments in Washington state's prairie areas showed that repeated close mowing was the best way to eliminate broom and restore native species. In areas where grazing is feasible, repeated grazing by goats or llamas can help eliminate stands. Older plants growing in drier sites may be killed by being cut close to the ground at the end of the dry season. Cutting plants in spring and treating cut stems with triclopyr will kill plants. Herbicides with glyphosate or triclopyr and 2,4-D herbicide can be sprayed on plants after the leaves are fully expanded. Fire can also be used in combination with herbicides for effective control.

FOR MORE INFORMATION

Graves, M., J. Mangold and J. Jacobs. 2010. *Biology, ecology, and management of Scotch broom* (Cytisus scoparius *L.*). Invasive Species Technical Note No. MT-29. USDA NRCS. http://www.plant-materials.nrcs.usda.gov/pubs/mtpmstn9888.pdf.

Oneto, S.R., G. B. Kyser, J. DiTomaso. 2010. "Efficacy of mechanical and herbicide control methods for Scotch broom *(Cytisus scoparius)* and cost analysis of chemical control options." *Invasive Plant Science and Management* 3(4):421–428.

Peterson, D., and R. Prasad. 1997. "The biology of Canadian weeds, *Cytisus scoparius* [L.] Link." *Canadian Journal of Plant Science* 78:497–504.

Shaben, J. and J. H. Myers. 2010. "Relationships between Scotch broom *(Cytisus scoparius),* soil nutrients, and plant diversity in the Garry Oak Savannah ecosystem." *Plant Ecology* 207:81–91.

Winter Creeper Euonymus   *Euonymus fortunei*

NAME AND FAMILY
Winter creeper euonymus, emerald'n'gold, gaiety (*Euonymus fortunei* [Turcz.] Hand.-Maz.); bittersweet family (Celastraceae).

IDENTIFYING CHARACTERISTICS
Winter creeper euonymus grows as a climbing vine or as a vining shrub when there is nothing to climb. Its opposite dark green evergreen to semievergreen leaves with finely toothed or wavy edges are egg-shaped, less than 1 in. (2.5 cm) long, and often veined with silvery white. Leaf color and size can be extremely variable among cultivars. Inconspicuous ¼ in. (0.6 cm), five-petaled greenish white flowers bloom in midsummer, maturing to pinkish red capsules in fall that split open to reveal orange-coated seeds. Flowers and fruits tend to form only on plants that have the opportunity to climb. The green stems form aerial roots as English ivy does.

Left: *The bright orange-red-coated seeds attract birds.* **Right:** *Leaves are evergreen to semievergreen.*

Winter creeper grows as a vine or sprawling shrub.

HABITAT AND RANGE
Tolerant of heavy shade, winter creeper grows in forests and forest gaps. It prefers drier soils but thrives in poor or rich, acidic or basic soils. Mostly invasive in the Southeast, but also occurs in southern New England, the Midwest, and Ontario.

WHAT IT DOES IN THE ECOSYSTEM
The dense ground cover restricts native plant establishment. Climbing vines reach 40 to 70 ft. (12 to 21 m) and can kill shrubs and small trees. Birds and other animals consume and disperse the seeds.

HOW IT CAME TO NORTH AMERICA
Introduced in 1907 from China as an evergreen ground cover, this plant is still a popular seller because of its drought tolerance and rapid growth. Native to Japan, Korea, and China.

MANAGEMENT
Plants can be pulled or dug up, but plant parts should be removed since it can resprout from the roots and stems. Glyphosate or triclopyr can be painted on cut stems or sprayed on foliage.

FOR MORE INFORMATION
Dirr, M. A. 1998. *Manual of Woody Landscape Plants.* Champaign, IL: Stipes Publishing.

Miller, J. H., E. B. Chambliss, N. J. Loewenstein. 2010. *A field guide for the identification of invasive plants in southern forests.* General Technical Report SRS-119. Asheville, NC: U.S. Department of Agriculture, Forest Service, Southern Research Station. http://www.srs.fs .fed.us/pubs/gtr/gtr_srs119.pdf.

Zouhar, Kris. 2009. *"Euonymus fortunei." Fire Effects Information System.* U.S. Department of Agriculture, Forest Service, Rocky Mountain Research Station, Fire Sciences Lab. http://www.fs.fed.us/database/feis/ plants/vine/euofor/all.html.

Lantana   *Lantana camara*

NAME AND FAMILY
Lantana, largeleaf lantana, shrub verbena (*Lantana camara* L.); verbena family (Verbenaceae). Recently, some invasive populations of lantana have been identified as a newly named hybrid of unknown origins, *Lantana strigocamara* Sanders. Creeping lantana (*L. montevidensis* [Spreng.] Briq.) is also considered invasive but is not as widespread. Pineland lantana (*L. depressa* Small) is an endangered Florida plant threatened by largeleaf lantana, and there is the more common native relative, Florida sage (*L. involucrata* L.).

IDENTIFYING CHARACTERISTICS
Flat, compact, 1 in. (2.5 cm) clusters of small, colorful flowers range from orange to pink to white, often with different-colored flowers in the same cluster. Flowers bloom almost year-round on spreading shrubs that grow 6 to 15 ft. (1.8 to 4.6 m) in height. Each flower has a tubular shape with four petals, producing a two-seeded, blackish fruit. The leaves are opposite, pointed ovals with teeth along the leaf edges and rough surfaces. Small prickles are sometimes scattered along the square stems. Broken stems and leaves smell of black currants. Shiny green-black fruits form throughout the year and are dispersed by birds. Creeping lantana trails along the ground, seldom reaching more than 3 ft. (1 m) tall. Its

Left: *Flower clusters grow from the leaf axils.* **Right:** *Leaves occur in pairs along the stem.*

Flowers range from orange to pink to white.

Lantana forms large, dense stands.

flowers tend to be pink or purple and its stems are not thorny. Pineland lantana and Florida sage have yellow flowers and the bases of the leaves are tapered rather than being rounded off.

HABITAT AND RANGE

Lantana grows in open, sunny to partly shady moist areas, such as roadsides, pastures, forest edges, dunes, and pinelands. Cannot tolerate heavy frost. Found in natural areas in Georgia and Florida and as far west as Texas. Grown as an ornamental annual or perennial plant throughout much of the United States.

WHAT IT DOES IN THE ECOSYSTEM

Lantana can form dense stands, and the soil under these stands becomes enriched with nitrogen and poisoned with allelochemicals released from the roots and stems. Shoots will root where they touch the ground and seeds are bird dispersed. After cutting, the plant produces a new crop of roots. Lantana reduces the productivity of pastures and forest plantations. The leaves are poisonous to livestock. Flowers attract butterflies, bees, and other insects. Lantana has hybridized with the endangered pineland lantana, changing its gene pool. In some areas of the world it is ranked as one of the 100 worst invaders.

HOW IT CAME TO NORTH AMERICA

Introduced from Central and South America as an ornamental shrub in the early 1800s. There are hundreds of cultivars and hybrids of lantana, mostly varying in flower color. Some newer cultivars are considered to be sterile. Lantana may be native to southern Texas, but invasive populations have been identified as escaped cultivars by their genetic characteristics.

MANAGEMENT

Small plants can be hand pulled or grubbed out. Herbicides applied while shrubs are actively growing can effectively control plants. Cut stump and basal bark treatments with triclopyr are also effective. A combination of a hot fire and herbicide treatment can also be used. Lantana will resprout after light fires.

FOR MORE INFORMATION

Gentle, C. B. and J. A. Duggin. 1997. "Allelopathy as a competitive strategy in persistent thickets of *Lantana camara* L. in three Australian forest communities." *Plant Ecology* 132:85–95.

Langeland, K. A., H. M. Cherry, C. M. McCormick, and K. A. Craddock Burks et al. 2008. *Identification and Biology of Nonnative Plants in Florida's Natural Areas,* 2nd ed. SP 257. Gainesville, FL: University of Florida IFAS.

Maschinski, J., E. Sirkin, and J. Fant. 2010. "Using genetic and morphological analysis to distinguish endangered taxa from their hybrids with the cultivated exotic pest plant *Lantana strigocamara* (syn: *Lantana camara*)." *Conservation Genetics* 11(5):1607–1621.

Privets   *Ligustrum* spp.

NAME AND FAMILY

Privets (*Ligustrum* species); olive family (Oleaceae). At least eight species of introduced privet grow wild in North America and are difficult to tell apart. Chinese privet (*L. sinense* Lour.), common or European privet (*L. vulgare* L.), Japanese privet (*L. japonicum* Thun.), Amur privet (*L. amurense* Carr.), glossy privet (*L. lucidum* Ait.), blunt-leaved or border privet (*L. obtusifolium* Sieb. and Zucc.), waxy-leaved privet (*L. ovalifolium* Hassk.), and wax-leaf privet (*L. quihoui* Carr.). Distinguishing the

Privet leaves grow in pairs along the stems.

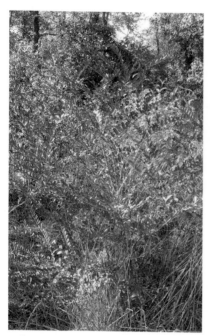

Left: *A privet shrub at the edge of a woodland.* **Right:** *Privet flowers cluster at the ends of the branches in early summer.*

species is difficult and relies on very small differences in some cases. May be confused with native swamp privets (*Forestiera* spp.).

IDENTIFYING CHARACTERISTICS
Privets are shrubs growing to 15 ft. (4.6 m) tall. The leathery leaves are semievergreen to evergreen and arranged opposite along the stems. The leaves are usually less than 2.5 in. (6.2 cm) long and oval with a glossy green surface. In early summer small plumes of tiny white flowers with a fused base breaking up into four petals appear at the ends of the twigs and from the upper leaf axils. By fall, clusters of dark blue-black berries have matured. Native swamp privets have small clusters of flowers held tightly to the stems and few fruits in each cluster.

HABITAT AND RANGE
Japanese privet, glossy privet, and wax-leaf privet all occur principally in the southeastern United States. Amur privet, Chinese privet, and waxy-leaved privet occur from New England south and in some western states. Common privet has the widest range, occurring in Newfoundland, Nova Scotia, Ontario, and British Columbia as well as across much of the United States. Privets grow along woodland edges, in floodplains, old fields, riparian forests, and upland forests. Tolerant of some shade and of occasional drought.

Clusters of dark-blue fruits mature in fall.

WHAT IT DOES IN THE ECOSYSTEM

Privet forms dense stands that outcompete native plants for space, light, and water. Few insects feed on it because chemicals in the leaves inhibit digestion. One study in Georgia showed that butterfly diversity increased when privet was removed. Deer, however, will feed on privet. Birds eat the fruits and disperse the seeds to forest gaps and into fields.

HOW IT CAME TO NORTH AMERICA

Privets were introduced for use in gardens. Gardeners and landscapers love privet hedges because they grow easily and can be pruned, even sculpted into different shapes (topiaries). Privets are native to Europe, Asia, and northern Africa. Common privet was probably the first to be introduced to North America in the 1700s. Blunt-leaved privet was introduced in 1860, Japanese privet in 1945, and Chinese privet in 1952.

MANAGEMENT

Young plants can be hand pulled or pulled with the aid of an uprooting tool like a weed wrench or mattock. Shrubs can also be cut repeatedly but this usually just keeps them from spreading. For large stands, plants can be cut and the resprouts sprayed with a glyphosate or triclopyr in late autumn or early spring when most other plants are dormant. Glyphosate or triclopyr can also be painted on cut stumps or applied to the bark.

FOR MORE INFORMATION

Greene, B. T. and B. Blossey. 2011. "Lost in the weeds: *Ligustrum sinense* reduces native plant growth and survival." *Biological Invasions*. doi: 10.1007/s10530-011-9990-1

Hanula, J. L. and S. Horn. 2011. "Removing an exotic shrub from riparian forests increases but-
    terfly abundance and diversity." *Forest Ecology and Management* 262:674–680.
Maddox, V., J. Byrd Jr., B. Serviss. 2010. "Identification and control of invasive privets (*Ligus-
    trum* spp.) in the middle southern United States." *Invasive Plant Science and Management*
    3(4):482–488.
Miller, J. H., E. B. Chambliss, N. J. Loewenstein. 2010. *A field guide for the identification of
    invasive plants in southern forests.* General Technical Report SRS-119. Asheville, NC: U.S.
    Department of Agriculture, Forest Service, Southern Research Station. http://www.srs.fs
    .fed.us/pubs/gtr/gtr_srs119.pdf.

Nandina   *Nandina domestica*

NAME AND FAMILY

Nandina, sacred bamboo, heavenly bamboo (*Nandina domestica* Thunb.); barberry family (Berberidaceae). Nandina has been cultivated for centuries in
Asia, where it was used in religious observances.

IDENTIFYING CHARACTERISTICS

Nandina is a popular shrub because of
its evergreen leaves and ability to
grow almost anywhere. The large,
alternately arranged leaves are divided
two to three times into many small

Left: *Nandina fruits cluster at the ends
of the branches.* **Bottom:** *Each leaf is
twice or thrice divided and made up
of many leaflets.*

Left: *Nandina's upright branches can resemble stalks of bamboo.* Right: *Flowers can have three to six petals.*

leaflets each 1 to 2 in. (2.5 to 5 cm) long arranged opposite each other and with a leaflet at the tip. Leaflets are oval to lance shaped with a pointed tip. Leaves start out bronzy green and mature to a dark green, sometimes turning reddish in winter. The branched flowering stalk can be up to a foot long, covered with ¼ to ½ in. (0.6 to 1.3 cm) wide, three- to six-petaled pinkish-white flowers. The flowering stalk is held upright until the fruits begin to mature. Hanging clusters of small red fruits mature in fall and winter. Nandina grows 6 to 8 ft. (1.8 to 2.4 m) tall, usually with upright branches that resemble stalks of bamboo.

HABITAT AND RANGE
Most prevalent in the southeastern United States from Virginia to Florida and Texas, nandina tolerates sun to full shade. It prefers moist soils, but once established is drought tolerant. Found on floodplains and in forests. It is more widely planted as an ornamental.

WHAT IT DOES IN THE ECOSYSTEM
Nandina can form dense stands in natural areas, but is only problematic locally so far. In Florida it occupies forests where it is displacing several state rare plant species. Birds disperse the seeds and plants also reproduce from root fragments.

HOW IT CAME TO NORTH AMERICA

Introduced as an ornamental plant in 1804 from China by William Kerr, but first discovered in natural areas in North Carolina in the 1960s. Native to eastern Asia and India. Some cultivars do not produce viable seeds and others produce no fruits and so may be less likely to escape cultivation.

MANAGEMENT

Small plants can be hand pulled or dug up. Herbicides containing glyphosate or triclopyr can be sprayed on foliage or painted on cut stumps to control plants. Seeds will sprout from the seed bank, so retreatment may be necessary.

FOR MORE INFORMATION

Langeland, K. A., H. M. Cherry, C. M. McCormick, and K. A. Craddock Burks et al. 2008. *Identification and Biology of Nonnative Plants in Florida's Natural Areas,* 2nd ed. SP 257. Gainesville, FL: University of Florida IFAS.

Miller, J. H., E. B. Chambliss, N. J. Loewenstein. 2010. *A field guide for the identification of invasive plants in southern forests.* General Technical Report SRS-119. Asheville, NC: U.S. Department of Agriculture, Forest Service, Southern Research Station. http://www.srs.fs .fed.us/pubs/gtr/gtr_srs119.pdf.

Common and Strawberry Guava   *Psidium* spp.

NAME AND FAMILY

Common guava, apple guava (*Psidium guajava* L.) and strawberry guava, cattley guava, pineapple guava (*P. cattleianum* Sabine); myrtle family (Myrtaceae).

IDENTIFYING CHARACTERISTICS

Both of the guavas grow as evergreen small trees or shrubs 30 to 40 ft. (9 to 12 m) and produce edible fruits. The two species can be distinguished by the twigs and leaves. Common guava has twigs with four angles, whereas strawberry

Strawberry guava's purplish fruits.

Four- to five-petaled guava flowers have many stamens.

guava has rounded twigs. The bark of guavas is often reddish and flaking. Common guava has leaves up to 6 in. (15 cm) long and with raised veins on the undersides; whereas the leaves of strawberry guava are up to 3 in. (7.5 cm) long and glossy. Both species' leaves are oblong, with smooth edges, and occur oppositely along the stems. Plants flower and set fruit year-round. White flowers about 1-in. (2.5 cm) wide with four to five petals and a cluster of many stamens are borne singly or in clusters of three at the leaf axils. The fruits of strawberry guava are rounded, 1.2 to 2.4 in. (3 to 6 cm) long, colored purplish on the outside with sweet white flesh when mature. Common guava fruits are yel-

Guavas have reddish, flaky bark.

lowish to pink-tinged and 1 to 4 in. (2.5 to 10 cm) long. Each fruit contains numerous seeds.

HABITAT AND RANGE
Guavas grow in forests in subtropical Florida, but they are also known to naturalize in abandoned fields, pastures, and tree plantations. They are both shade and salt tolerant. Guavas have naturalized readily and are considered a threat to natural vegetation on several southern hemisphere islands such as Mauritius and Fiji and also in Hawaii.

WHAT IT DOES IN THE ECOSYSTEM
By root suckering, guavas can form dense stands that occupy space and shade out native plants. They form dense mats of roots near the soil surface. Guavas may also release allelopathic chemicals that inhibit the establishment or growth of other plants. Seeds are dispersed by birds and mammals. Serves as a host for Caribbean fruit fly, a pest of citrus crops. Guavas yield a fruit that is

Guava shrubs form dense stands.

increasingly popular in the States, high in Vitamins C and antioxidants. The wood is often used for barbecues.

HOW IT CAME TO NORTH AMERICA
Common guava was growing in natural areas in Florida by 1765, introduced from tropical Central or South America as an edible fruit. Strawberry guavas were introduced to Florida in the 1880s as both an ornamental plant and for fruit production. Recorded as occurring in natural areas in Florida in the 1950s.

MANAGEMENT
Seedlings and saplings can be pulled up by hand or with a weed wrench, but root sprouts cannot be easily pulled up. Larger plants can be cut and the stumps painted with glyphosate or triclopyr. These herbicides can also be used for the hack-and-squirt method.

FOR MORE INFORMATION
Huenneke, L. F., and P. M. Vitousek. 1989. "Seedling and clonal recruitment of the invasive tree, *Psidium cattleianum:* Implications for management of native Hawaiian forests." *Biological Conservation* 53:199–211.

Langeland, K. A., H. M. Cherry, C. M. McCormick, and K. A. Craddock Burks et al. 2008. *Identification and Biology of Nonnative Plants in Florida's Natural Areas,* 2nd ed. SP 257. Gainesville, FL: University of Florida IFAS.

Morton, J. 1987. "Cattley guava." In *Fruits of Warm Climates.* 363–364. Miami, FL: Julia F. Morton. http://www.hort.purdue.edu/newcrop/morton/cattley_guava.html.

Terrestrial Plants—Trees and Shrubs—Deciduous—Alternate Leaves—Trees

Tree of Heaven   *Ailanthus altissima*

NAME AND FAMILY
Tree of heaven, ailanthus, Chinese sumac, stinking sumac (*Ailanthus altissima* [Mill.] Swingle); quassia family (Simaroubaceae). Tree of heaven can be confused with sumac, walnut, or pecan trees because the leaves look similar at first glance.

IDENTIFYING CHARACTERISTICS
This fast-growing tree attains a height of 80 ft. or more. The 11 to 41 leaflets on a straight stem are actually part of a single 1 to 4 ft. long compound leaf that appears very late in spring. Each leaflet is lance shaped with a long pointed tip and has just one to five teeth at the base of the leaflet. Sumacs and nut trees have teeth along the entire margin of the leaflets. Crushed foliage has an unpleasant odor often described as burnt peanut butter. The leaves are alternately arranged, but the leaflets are arranged oppositely. The stout twigs are covered with fine hairs when young and have a core of yellowish pith. On older trees the bark is relatively smooth and either light brown or striped grayish

Leaflets have only one to five teeth at the base of each leaflet.

The large leaf scars are shaped like narrow hearts.

brown. In early summer trees produce yellowish flowers held in a large cluster above the leaves at the ends of the branches. Male and female trees are separate. A single large female tree can produce more than 300,000 papery-winged, wind-dispersed, tan seeds in late summer to fall.

HABITAT AND RANGE
Many urban commuters and residents of blighted neighborhoods know tree of heaven as the tree that grows in the garbage and among stones along rail lines, in pavement cracks, and in vacant lots. (In Betty Smith's best-selling novel *A Tree Grows in Brooklyn*, it is a symbol of poor families' endurance and hope.) Tree of heaven grows in southern Canada and throughout most of the United States. It tolerates a wide range of soils, from coastal sandy soils to rocky mountain soils, but is not very shade tolerant. It thrives in disturbed areas and tolerates air pollution and acidic soils. It can be found along highway embankments, field edges, urban pavement cracks, railroad beds, mine spoils, and in disturbed forests. It is often the only tree found in heavily polluted and disturbed city neighborhoods.

WHAT IT DOES IN THE ECOSYSTEM
Once established, tree of heaven sends up many root sprouts, rapidly forming a dense colony. Chemicals released from the roots and from leaf litter hinder the growth of other plants. The aggressive root system can damage sewer pipes and foundations. Because of its tolerance to pollutants and acid soils, it has been used for reclamation of mined lands. Deer and goats occasionally browse the foliage.

Warning: Ailanthus sap contains quassinoids, chemicals that have caused heart problems and debilitating headaches and nausea in people who do not protect themselves from exposure when cutting and handling the trees.

Large seed clusters form on female trees at the ends of the branches.

A papery wing surrounds each seed.

HOW IT CAME TO NORTH AMERICA

Tree of heaven is native to China where it was called tree of heaven because it grew out of the rocks on mountain heights where other trees would not grow. Tree of heaven first reached Philadelphia in 1748, introduced by a gardener named William Hamilton. Nurseries, particularly on the East Coast, sold tree of heaven because it was pest-free, fast growing, and easy to grow in any soil. Chinese immigrants brought seeds to the west coast in the 1850s during the Gold Rush, probably because of traditional medicinal uses.

MANAGEMENT

Cutting trees only encourages a plethora of root sprouts, and even trunks left on the ground can resprout. Small trees can be pulled out by hand or with a weed wrench, but fragments of root can result in more root sprouts. The hack-and-squirt method works well for larger trees, as does cutting trees and applying herbicide to the cut stump in summer. Applying a mixture of herbicide and oil to the base of the tree in late winter/early spring or spraying the foliage of the trees in summer is also effective. A combination of cutting and spraying can also be effective. Several fungal pathogens and insects are under investigation as possible control agents.

FOR MORE INFORMATION

Bisognano, J. D., K. S. McGrody, and A. N. Spence. 2005. "Myocarditis from the Chinese sumac tree." *Annals of Internal Medicine* 143:159–160.

Fryer, J. L. 2010. *"Ailanthus altissima." Fire Effects Information System.* U.S. Department of Agriculture Forest Service, Rocky Mountain Research Station, Fire Sciences Lab. http://www.fs.fed.us/database/feis/plants/tree/ailalt/all.html.

Gomez-Aparicio, L. and C. D. Canham. 2008. "Neighborhood analyses of the allelopathic effects of the invasive tree *Ailanthus altissima* in temperate forests." *Journal of Ecology* 96(3):447–458.

Hu, S. 1979. *"Ailanthus." Arnoldia* 39(2):29–50.

Knapp, L. B., and C. D. Canham. 2000. "Invasion of an old-growth forest in New York by *Ailanthus altissima:* Sapling growth and recruitment in canopy gaps." *Journal of the Torrey Botanical Society* 127(4):307–15.

Plant Conservation Alliance Alien Plant Working Group. *Tree of heaven fact sheet.* http://www.nps.gov/plants/alien/fact/aial1.htm.

Mimosa   *Albizia julibrissin*

NAME AND FAMILY

Mimosa, silk tree (*Albizia julibrissin* Durazz.); pea family (Fabaceae). Woman's tongue (*A. lebbeck* L. Willd.) is a closely related tree species currently considered invasive in Florida. Leaves look similar to those of honey locust (*Gleditsia triacanthos*), but trees are not thorny.

IDENTIFYING CHARACTERISTICS

Mimosas are small trees (20 to 40 ft. [6 to 12 m] tall), often multistemmed with vase-shaped branching. The bark is smooth and tan colored. Trees leaf out late in spring. Their feathery double compound leaves are arranged alternately along the stem, and each leaf (average 20 in. [50 cm] long) is divided

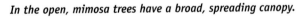

In the open, mimosa trees have a broad, spreading canopy.

Double compound leaves are fern-like in appearance.

Seedpods mature in late summer.

into 10 to 25 "pinnae" (leaflets) (each 5 to 8 in. [12.7 to 20.3 cm] long), which are further divided into 40 to 50 tiny leaflets. Most distinct are the fragrant, pink, powder-puff-looking flowers at the ends of branches that bloom in groups in early summer. They are about 1.5 in. (3.7 cm) across. Clusters of tan to brown 6 in. (15 cm) long flat pods form in late summer, each containing several oval, flattened seeds. Woman's tongue is similar in appearance but has yellow powder-puff flowers and larger (1 to 2 in. [2.5 to 5 cm]) leaflets. Mimosa and woman's tongue are not thorny.

HABITAT AND RANGE

Occurs throughout the United States except in New England and the Pacific Northwest. Grows in open areas, forest edges, and river floodplains in a wide range of soil types. It also tolerates drought, wind, and salt. Woman's tongue grows in central and southern Florida in disturbed woodlands and forest edges.

WHAT IT DOES IN THE ECOSYSTEM
Dense stands can reduce light and water available to other plants. Trees fix nitrogen and their leaf litter is high in nitrogen, resulting in higher soil nitrogen levels. Flowers are visited by many insects and by hummingbirds. Trees are short-lived and have weak wood. They often die due to a fungal disease, mimosa fusarium wilt.

HOW IT CAME TO NORTH AMERICA
Introduced in the late 1700s by botanist André Michaux to Charleston, South Carolina. Jefferson grew mimosa at Monticello in Virginia in the late 1700s. Plantsman Ernest Henry Wilson also brought seeds of mimosa back from eastern Asia to Boston around 1918. Mimosa is native from Iran to Japan. Woman's tongue is native to tropical Asia and was introduced to Florida as an ornamental in the 1880s.

MANAGEMENT
Seedlings can be hand pulled. Trees resprout readily after being cut, and they sucker from roots. Seeds can remain viable for up to 50 years. Trees can be cut while flowering to eliminate seed production, but resprouts will have to be cut or the stumps treated with herbicide. Glyphosate or triclopyr sprayed on leaves

Powder-puff flowers bloom in early summer.

will also kill trees. An accidentally introduced beetle from mimosa's native range, *Bruchidius terrenus*, feeds on the seeds and may make the tree more susceptible to fusarium wilt.

FOR MORE INFORMATION

Chang, S., E. Gonzalez, E. Pardini, J. L. Hamrick. 2011. "Encounters of old foes on a new battleground for an invasive tree, *Albizia julibrissin* Durazz (Fabaceae)." *Biological Invasions* 13(4):1043–1053.

Langeland, K. A., H. M. Cherry, C. M. McCormick, and K. A. Craddock Burks et al. 2008. *Identification and Biology of Nonnative Plants in Florida's Natural Areas,* 2nd ed. SP 257. Gainesville, FL: University of Florida IFAS.

Remaley, T. "Silk tree." Plant Conservation Alliance, Alien Plant Working Group. http://www.nps.gov/plants/alien/fact/alju1.htm.

Spongberg, S. A. 1993. "Exploration and introduction of ornamental and landscape plants from eastern Asia." In *New Crops,* edited by J. Janick and J. E. Simon, 140–147. New York: Wiley. http://www.hort.purdue.edu/newcrop/proceedings1993/v2-140.html.

Black Alder   *Alnus glutinosa*

NAME AND FAMILY
Black alder, European alder (*Alnus glutinosa* [L.] Gaertn.); birch family (Betulaceae). May be confused with native alders including speckled alder (*A. incana*) or smooth alder (*A. serrulata*).

IDENTIFYING CHARACTERISTICS
Black alder grows 40 to 60 ft. (12 to 18 m) tall and can spread its branches to 40 ft. (12 m). It takes a generally pyramidal shape if it grows in the open but will also grow as a multitrunk tree. The *glutinosa* (Latin for "sticky") in its scientific name refers to the sticky young twigs and leaves that distinguish it from native alders like speckled alder or smooth alder. While native alder leaves tend to have pointed ends, black alder has rounded leaves. Leaf edges are toothed. Leaves alternate along the stems. In the fall, clusters of three to four male catkins about 4 in. (10 cm) long form. They wait for the spring appearance of female flowers that appear as green round catkins or "berries" before leaves emerge. Once fertilized, the female catkins harden into ¾ in. (2 cm) long "cones" that shelter winged seeds. The young bark can be greenish to brown, but older bark becomes dark brown to blackish—thus, black alder.

HABITAT AND RANGE
Black alder occurs locally across temperate North America from southern Ontario, Nova Scotia, and Newfoundland across the northern United States to Minnesota and Kansas and as far south as Tennessee. It prefers wet to moist soils and sun.

WHAT IT DOES IN THE ECOSYSTEM

Black alder's limited appearance in natural areas of North America has not caused great problems. In the last 50 years some park and wild land managers have found colonies of black alder, usually along stream- and riverbanks. They can grow in dense stands. They tend to fix in place sand bars and riverbanks, and of course, no two plants can occupy the same space, so some displacement of natives occurs. Alder roots form nodes that harbor a bacterium (*Frankia*) that fixes nitrogen in the soils.

HOW IT CAME TO NORTH AMERICA

In its native range across temperate Europe and Asia, black alder has many uses that also appeal to people in North America. Its wood is durable in water, and Venetians are said to have used it for pilings to support their buildings. It is also the reddish

Female, cone-like catkins hold the seeds of black alder.

wood of European cigar boxes and a common folk remedy for cancer, fevers, and inflammations. Plants and seed began arriving in North America in the late nineteenth century.

MANAGEMENT

Black alder can be killed by repeated cutting or mowing or by using herbicides, glyphosate, or triclopyr, in hack-and-squirt or cut-stump treatments.

FOR MORE INFORMATION

CABI Invasive Species Compendium. http://www.cabi.org/isc/?compid=5&dsid=4574&load module=datasheet&page=481&site=144.

Eckel, P. M. 2003. "Two problems in Betulaceae along the Niagara River: *Alnus glutinosa* and *Betula cordifolia.*" *Clintonia* 18(4):3–4.

Missouri Botanical Garden. http://www.mobot.org/plantscience/ResBot/niag/Misc/Clintonia _Alnus_2003.pdf.

U.S. Department of Agriculture. Germplasm resources information network (GRIN). http://www .ars-grin.gov/cgi-bin/npgs/html/taxon.pl?2448.

Leucaena   *Leucaena leucocephala*

NAME AND FAMILY
Leucaena, white lead tree, koa haole, tamarind tree (*Leucaena leucocephala* [Lam.] de Wit); pea family (Fabaceae). Leaves look similar to those of mimosa (*Albizia julibrissin*, p. 107).

IDENTIFYING CHARACTERISTICS
This small to medium-sized thornless tree or shrub tends to grow in dense stands, resulting in slender trunks with a tuft of leafy branches at the top. Leaves are made up of many leaflets. Each leaflet is a pointed oblong 0.5 to 0.75 in. (1.3 to 1.9 cm) long and less than 0.25 in. (0.6 cm) wide. Each leaf is twice divided, with 4 to 9 pairs of pinnae (first division) and 13 to 21 pairs of leaflets per pinna. Creamy white flowers occur in small rounded clusters, with each flower having 10 stamens and hairy anthers. The hairy anthers (which can be clearly seen with a hand lens) distinguish leucaenas from mimosa trees. Flowers are self-fertile and so do not require pollination to set seed. Clusters of 5 to 20 flattened seed pods, 4 to 7.5 in. (10 to 19 cm) long, each hold 8 to 18 glossy brown seeds. Seeds are dispersed by rodents, seed-eating birds, and cattle.

HABITAT AND RANGE
Leucaena has adapted to tropical and subtropical climates with dry or moist soils and full sun. It grows in Georgia, Florida, Arizona, Texas, and California. It generally grows in disturbed sites and in coastal areas, preferring limey soils.

Creamy flowers bloom in ball-like clusters.

Growing as a shrub or small tree, leucaena prefers basic soils.

WHAT IT DOES IN THE ECOSYSTEM
Once planted for forage, leucaena is now considered a weed in more than 20 countries. It naturalizes rapidly and forms dense thickets that replace native forests. Trees form an association with a bacteria to fix nitrogen, increasing the nitrogen content of the soil.

HOW IT CAME TO NORTH AMERICA
One of the world's most widely seeded nitrogen-fixing plants, used for forage, reforestation, erosion control, and a condiment. The plants are easily coppiced (cut back for rapid regrowth) and the high-energy content wood is used in place of fuel oil. Leaves are high in

Clusters of flattened seedpods hang down from a branch.

protein, although young leaves and seeds can be poisonous to grazers if eaten in large quantities. Latin Americans add unripe seeds to salsas. The Spanish probably introduced leucaena to the southern United States from coastal Mexico and Central America between the late 1500s and early 1800s.

MANAGEMENT
Leucaena resprouts vigorously after cutting and seeds can remain viable in the soil for 10 to 20 years. Repeated browsing by goats can kill plants. Herbicides

containing triclopyr sprayed on foliage or painted on cut stumps can be used to control plants. Basal bark treatment using 2,4-D is also effective.

FOR MORE INFORMATION

Duke, J. A. 1983. *"Leucaena leucocephala* (Lam.) de Wit." *Handbook of energy crops.* http://www .hort.purdue.edu/newcrop/duke_energy/leucaena_leucocephala.html.

Hughes, C. E., and R. J. Jones. 1999. "Environmental hazards of Leucaena." In *Leucaena: Adaptation, Quality, and Farming Systems,* edited by H. M. Shelton, R. C. Gutteridge, B. F. Mullen, and R. A. Bray. Proceedings of a Workshop (February 1998), Hanoi, Vietnam. Australian Centre for International Agricultral Research (ACIAR).

Langeland, K. A., H. M. Cherry, C. M. McCormick, and K. A. Craddock Burks et al. 2008. *Identification and Biology of Nonnative Plants in Florida's Natural Areas,* 2nd ed. SP 257. Gainesville, FL: University of Florida IFAS.

Parrotta, J. A. 1992. Leucaena leucocephala *(Lam.) de Wit, Leucaena, tantan.* SO-ITF-SM-52. New Orleans, LA: U.S. Department of Agriculture, Forest Service, Southern Forest Experiment Sation. http://www.fs.fed.us/global/iitf/Leucaenaleucocephala.pdf.

Chinaberry Tree   *Melia azedarach*

NAME AND FAMILY
Chinaberry tree, umbrella tree, Persian lilac, bead tree, Syringa berry (Melia azedarach L.); mahogany family (Meliaceae).

IDENTIFYING CHARACTERISTICS
The large clusters of round, yellow, cherry-sized fruits that hang from long stalks distinguish this tree in the fall and winter. They begin as green fruits and eventually turn leathery brown in fall, hanging on into winter after the leaves have fallen. The leaves growing alternately on the branches are collections of many leaflets that grow along larger and smaller divisions of the entire leaf (two to three times compound). Leaflets are narrow, sharply pointed, toothed, and blue green in color. The entire leaf grows to 2 ft. (0.6 m) long while leaflets are 1 to 3 in. (2.5 to 7.6 cm) long. The 0.75 in. (1.9 cm) lilac to whitish flowers have five petals around a purple tube and grow in clusters from stems that start at nodes or leaf axils on new growth near the ends of branches. Flowers are fragrant. Wood has distinctly separated rings and an auburn color. Chinaberry can grow as a bush or as a tree up to 50 ft. (15 m) high.

HABITAT AND RANGE
Chinaberry is an opportunist that naturalizes in disturbed ground, along fence lines, around old homesites, and in old pastures, but it is increasingly found in relatively undisturbed areas. It grows from Virginia to Florida and west to Oklahoma, Texas, and Utah. Chinaberry tolerates high temperatures, poor

Chinaberry leaves grow very large but are divided two or three times.

soils, and periods of drought. In Texas it has formed significant natural stands in riparian areas and higher grasslands.

WHAT IT DOES IN THE ECOSYSTEM

This fast-growing tree can form thickets by vegetative reproduction from root suckers, and these thickets reduce plant diversity within the thicket area. Birds readily eat its fruits and disperse the seeds, but the bark, leaves, and seeds are poisonous to most domestic animals and to humans. Crushed seeds have been used to stupefy freshwater fish. Its decaying leaves increase soil nitrogen and pH (making it more alkaline) and have been sometimes used as soil condition-ers. The tree has been honored in Asia both for its medicinal properties and for its appearance. Some sources say that Chinaberry leaves mixed with stored fruits and vegetables repel insects since the leaves have insecticidal properties.

HOW IT CAME TO NORTH AMERICA

Chinaberry was introduced to the Southeast in the 1830s as a shade tree and for its colorful flowers and berries and graceful leaves. Growing in the open, it has a distinct domed umbrella shape that serves as a parasol. Michael Dirr, whose work on landscape plants is well known, describes it as having no land-scape value and being "a genuine weed tree." This member of the mahogany family is native to Asia and northern Australia.

Left: *Large trees produce thousands of fruits.* Right: *Clusters of small, leathery, yellow fruits.*

MANAGEMENT
Hand pulling or digging must be done carefully since roots can send up suckers from broken pieces. Larger trees can be killed by painting cut stumps and basal bark with triclopyr-based herbicides.

FOR MORE INFORMATION
Batcher, M. S. 2000. *Element stewardship abstract for* Melia azedarach *chinaberry, umbrella tree.* http://www.invasive.org/weedcd/pdfs/tncweeds/meliaze.pdf.
Dirr, M. A. 1998. *Manual of Woody Landscape Plants.* Champaign, IL: Stipes Publishing.

Mulberries — *Morus alba* and *Broussonetia papyrifera*

NAME AND FAMILY
White mulberry (*Morus alba* L.) and paper mulberry (*Broussonetia papyrifera* [L.] L'Hér. ex Vent.); mulberry family (Moraceae). The native red mulberry (*M. rubra* L.) is easily confused with white mulberry and hybridizes with it.

IDENTIFYING CHARACTERISTICS
Mulberry trees are easily recognized by their leaves, which have toothed edges and vary in shape from heart shaped to very lobed. White mulberry leaves are smooth on top, usually glossy, and 2 to 8 in. (2.5 to 20.3 cm) long. Paper mulberry leaves are rough and fuzzy on top and velvety underneath. Red mulberry

leaves are rough on top, but not fuzzy like those of paper mulberry. Bark of white mulberry is light gray with ridges or furrows. White mulberry fruits range in color from red to purple to white. Often, mature fruits are white on white mulberry. They look a little like a blackberry and mature in early summer to midsummer. Paper mulberry fruits are round (¾ in. [1.9 cm] diameter) and red to orange. Young paper mulberry tree bark is mottled with tan splotches. White and paper mulberries grow to about 40 ft. (12 m) and have spreading branches. Paper mulberry often forms thickets, sending up suckers from spreading roots. Mulberries have a milky sap that exudes from the leaf stems. Mulberries are deciduous, losing their leaves in winter.

HABITAT AND RANGE
White mulberry occurs in southern Ontario, British Columbia, and Quebec and throughout most of the United States. Paper mulberry grows from southern New England south to Florida and west to Texas. Mulberries grow in floodplains, meadows, field edges, and thickets.

WHAT IT DOES IN THE ECOSYSTEM
Paper mulberry forms dense, shallowly rooted thickets that exclude other plants. Because of their shallow roots they blow over easily. White mulberry is considered a severe threat to the red mulberry, particularly in Canada, where red mulberry is uncommon. White mulberry pollen appears to overwhelm that of red mulberry, causing many hybrid mulberries to form. Because white mulberry is so much more abundant, red mulberry's genes may eventually disap-

Left: *White mulberry fruits can be white or purple when mature.* **Right:** *Mulberry leaves can be heart-shaped or lobed.*

Left: *Leaves alternate along the stems.* **Right:** *The upper surfaces of white mulberry leaves are glossy.*

pear. White mulberry fruits are edible but not nearly as flavorful as those of red mulberry. Birds, fox, raccoon, oppossums, and other animals love to eat the fruits. The Chinese use white mulberry leaves for sheep and cattle fodder.

HOW IT CAME TO NORTH AMERICA
Paper mulberry is an economically important tree in Asia for paper making and for medicinal uses. It was probably introduced in the mid-1700s to the United States, where it is mostly planted as an ornamental shade tree. The British introduced white mulberry to the eastern United States in the 1600s in an attempt to establish a silk industry. The worms failed, but the tree did not. White mulberry originated in China.

MANAGEMENT
Seedlings can be hand pulled, but spreading roots make it difficult to pull up all roots. Repeated, frequent cutting can eventually kill smaller trees. Larger trees are best treated using the cut-stump, hack-and-squirt, or basal bark methods with herbicides containing triclopyr. Foliar spraying with glyphosate will kill young trees.

FOR MORE INFORMATION
Burgess, K. S. and B. C. Husband. 2006. "Habitat differentiation and the ecological costs of hybridization: the effects of introduced mulberry *(Morus alba)* on a native congener *(M. rubra)." Journal of Ecology* 94:1061–1069.

Stone, K. R. 2009. "*Morus alba.*" *Fire Effects Information System.* U.S. Department of Agriculture, Forest Service, Rocky Mountain Research Station, Fire Sciences Laboratory. http://www.fs.fed.us/database/feis.

Swearingen, J. M. 2009. "Paper Mulberry." Fact sheet from Plant Conservation Alliance's Alien Plant Working Group Least Wanted. http://www.nps.gov/plants/alien/fact/brpa1.htm.

White Poplar   *Populus alba*

NAME AND FAMILY
White poplar, silver poplar (*Populus alba* L.); willow family (Salicaceae). Other poplars that are less common, less widespread, and less aggressive, but still considered locally invasive include Lombardy poplar (*P. nigra*), European aspen (*P. tremula*), and gray poplar (*P. x canescens*).

IDENTIFYING CHARACTERISTICS
The silvery undersides of the leaves have given this tree a favored place in many created landscapes. The leaves are 2.5 to 6 in. (6.2 to 15.2 cm) long and 2 to 4 in. (5 to 10 cm) wide, with toothed margins and three to seven lobes, while young leaves on shorter shoots are more oval and the teeth along the edges are irregular. Young leaves are covered top and bottom with dense, very short hairs that give them their silvery look, but as leaves age the upper sides become green and smooth. Leaves alternate along the stems. Yellow to tan catkins (flower clusters) begin forming in early spring before the leaves appear and grow to 2 in. (5 cm) long. Trees have either male or female flowers, male catkins being longer than female ones. The dry fruits on female trees are dark brown, about ⅓ in. in diameter. The small seeds have fluffy hairs that help them ride the winds. The bark of younger trees is whitish and darkens and becomes furrowed with age, while branches are often white barked. The tree grows to 100 ft. (30.5 m) high with a trunk up to 6 ft. (1.8 m) in diameter.

HABITAT AND RANGE
The adaptable white poplar has naturalized in all the lower 48 states as well as in some parts of Alaska and across Canada's southern provinces. It prefers open sunlight and grows in acid and alkaline soils as well as saline soils. It is most often found in grasslands and at forest edges.

White poplar's branches have distinctive silvery bark.

WHAT IT DOES IN THE ECOSYSTEM

White poplar seeds travel on the wind but few are viable. Once established, however, trees reproduce vegetatively, forming large colonies that overwhelm many natives, especially in edge areas. Cutting stimulates the resprouting of multiple stems. A solution containing rooting hormones can be made from soaking white poplar root cuttings.

HOW IT CAME TO NORTH AMERICA

This European and Russian native came to North America in 1784 as a landscape tree. It has also been planted as a windbreak or hedgerow tree.

MANAGEMENT

As with native aspens, cutting, girdling, and herbicides are the most effective management tools. Girdling close to the ground reduces resprouting, but any new sprouts must be cut or girdled. Parallel cuts about 5 inches apart through the bark and cambium, followed by peeling away the bark, is the suggested method. Cutting the entire tree near ground level during summer and mowing or cutting any suckers also works. Burning also works if done for several years at two-year intervals. It is best done in combination with cutting, allowing sprouts to form before repeating a burn. Triclopyr or glyphosate herbicides can be painted on cut sprouts and stumps or sprayed on foliage. Care should be taken to target all branches and sprouts. Basal bark treatment is also effective.

Groves form as plants sucker.

Left: *Leaf undersides are silvery.* **Right:** *Leaves alternate along the stems.*

FOR MORE INFORMATION

Dirr, M. 1990. *Manual of Woody Landscape Plants.* Champaign, IL: Stipes Publishing Company.

Glass, W. 1996. *"Populus alba."* In *Invasive Plants: Weeds of the Global Garden,* edited by J. M. Randall and J. Marinelli, 39. Brooklyn Botanic Garden.

Remaley, T. and J. M. Swearingen, 2009. "White Poplar." Fact sheet from Plant Conservation Alliance's Alien Plant Working Group Least Wanted. http://www.nps.gov/plants/alien/fact/poal1.htm.

Bird Cherry   *Prunus avium*

NAME AND FAMILY

Bird cherry, sweet cherry, Mazzard cherry (*Prunus avium* L.); rose family (Rosaceae). Several cherries are invasive, but we will focus on this more widespread species. Others include: European bird cherry (*P. padus*), blackthorn (*P. spinosa*), Portugal laurel (*P. lusitanica*), sour cherry (*P. cerasus*), and cherry plum (*P. cerasifera*). Several native cherries look similar, including black cherry (*P. serotina*), pin cherry (*P. pensylvanica*), and chokecherry (*P. virginiana*).

IDENTIFYING CHARACTERISTICS

Distinguishing cherries can be difficult, but in the wild, bird cherry may be distinguished by its size (up to 60 ft. [18 m] but more typically near 30 ft. [9 m]) and its small red to black cherries that ripen on spurs of branchlets in early summer. Sweet-smelling white flowers 1.5 in. (3.7 cm) across have long stems, and can be single or in clusters of up to five flowers that appear just before the leaves unfold in spring. Leaves are oval, dark green, up to 6 in. (15.2 cm) long, and yellow in fall. The European bird cherry is similar, but flowers hang down in 3 to 6 in. (7.6 to 15.2 cm) clusters after leaves appear, and fruit is very astringent (puckery) and the size of garden peas, more like that of native chokecherry (*P. virginiana*).

Dark-green leaves have toothed edges.

HABITAT AND RANGE

Bird cherry grows across southern Canada's forested areas south to the Carolinas, and west to the Mississippi River. Also found in Alaska, Washington, Oregon, California, Idaho, and Colorado. Bird cherry grows in many hardwood-dominated forests and in fields, often in acidic soils.

WHAT IT DOES IN THE ECOSYSTEM

Most of the nonnative cherries are not significant threats to native vegetation and do not make large changes in natural areas. Local exceptions may occur where stands of cherry become dense. The Emerald Chapter of the Native Plant Society of Oregon lists bird cherry as a high-impact invasive that is "an enormous, widespread problem." Most of the cherries, native and alien, provide food for birds, chipmunks, and other animals. One case of hybridization with a native cherry has been recorded.

Bird cherry grows as a small tree.

HOW IT CAME TO NORTH AMERICA
Bird cherry is native to Europe and Asia. Early in the European colonization of North America, settlers began planting European cherries, including bird cherry. Bird cherry has been used to create many commercial varieties, including bing cherry.

MANAGEMENT
Small trees can be dug out or hand pulled. Larger trees can be girdled near the base, or trunks cut and painted with glyphosate or triclopyr.

FOR MORE INFORMATION

Jacobson, A. L. and P. F. Zika. 2007. "A new hybrid cherry, *Prunus x pugetensis* (*P. avium x emarginata*, Rosaceae) from the Pacific Northwest." *Madrono* 54:74–85.

University of Connecticut. "The Cherries, Plums and Peaches." http://www.hort.uconn.edu/plants/keys/trees/Prunuskey.htm.

Callery Pear   *Pyrus calleryana*

NAME AND FAMILY
Callery pear, Bradford pear (*Pyrus calleryana* Dcne.); rose family (Rosaceae). Sometimes confused with native shadbush (*Amelanchier canadensis*) in spring. "Bradford" pear is a cultivar of Callery pear, but because it is one of the most commonly sold cultivars, the species is often referred to by the common name Bradford pear.

Clusters of white, five-petaled flowers bloom in early spring.

IDENTIFYING CHARACTERISTICS

Their spring explosion of white flowers has made Callery pear trees very popular for urban and suburban landscaping, along with their rapid growth and low maintenance. Flowers are five-petaled, ½ to ¾ in. (1.3 to 2 cm) across, growing in clusters. Fruits are small greenish-brown balls. Leaves alternate along the stems and are heart-shaped to oval with finely toothed edges. Trees tend to grow in a cone shape, 30 to 50 ft. (9 to 15 m) tall. Bark on young branches is smooth and reddish brown, turning gray brown with shallow furrows with age. Branches are often thorny. Shadbush also has white flowers that bloom in spring, but trees have smooth, gray bark and are often multitrunked.

HABITAT AND RANGE

Most commonly found in disturbed woodlands, along roadsides, in clearcuts (even replanted clearcuts), and in old fields. Trees can grow in a wide range of soil types. They spread most rapidly in temperate climates, requiring some cold for seeds to germinate but not withstanding extremely cold temperatures. The most extensive populations are in the eastern U.S., but callery pear has naturalized in a few central and western states as well.

Callery pear has high visibility along roadsides in spring.

WHAT IT DOES IN THE ECOSYSTEM

Dense, often thorny thickets of callery pear prevent colonization by native species. Trees are vulnerable to storm and ice damage and are short-lived (25 to 30 years), so they are not recommended as street trees or for planting close to sidewalks and buildings. Branches of trees that fall over often form multiple vertical trunks. Trees can produce fruits after only three years. Birds and small mammals eat the fruits and disperse the seeds. Although cultivars are self-sterile—they cannot produce fruit if they receive pollen from the same cultivar—enough other cultivars are now widely planted that viable seeds are frequently produced.

HOW IT CAME TO NORTH AMERICA

Callery pear was first introduced by the Arnold Arboretum in Massachu-

Top: *The seeds in the small green pears are dispersed by animals.* Right: *Leaves have finely toothed edges.*

setts in 1908 but it was also widely propagated by the USDA in Glenn Dale, Maryland, and in Meford, Oregon. 'Bradford' originated in Glenn Dale in the 1950s. Many cultivars are used in ornamental landscape plantings and the plants are so widespread it is difficult to find a town in America without a callery pear planted. The fruit industry has used callery pear as root stock for commercial pears, as a pollen donor in orchards, and in breeding programs for fire blight resistance.

MANAGEMENT
Seedlings can be hand pulled. Saplings can be pulled out using tools like a weed wrench. Larger trees can be girdled or cut down and the cut stumps painted with triclopyr. Resprouts can be sprayed with glyphosate or triclopyr.

FOR MORE INFORMATION
Dirr, M.1998. *Manual of Woody Landscape Plants*. Champaign, IL: Stipes Publishing.

Culley, T. M. and N. A. Hardiman. 2007. "The beginning of a new invasive plant: A history of the ornamental callery pear in the United States." *Bioscience* 57:956–964.

Vincent, M. A. 2005. "On the spread and current distribution of *Pyrus calleryana* in the United States." *Castanea* 70:20–31.

Sawtooth Oak   *Quercus acutissima*

NAME AND FAMILY
Sawtooth oak (*Quercus acutissima* Carruthers); beech family (Fagaceae).

IDENTIFYING CHARACTERISTICS
This oak is among several that have long, pointed oval leaves with toothed edges, very much like chestnut leaves, but the leaves of sawtooth oak are much narrower (1 to 2.25 in. [2.5 to 6.2 cm] wide) than those of others in this group. Leaves are 4 to 8 in. (10 to 20 cm) long, broader near the tip, bright green in spring, and dull yellow to brown in fall. The teeth have small bristles. The trees bear large crops of inch-long acorns, usually every other year. They begin as green nuts and turn brown, and each has a distinctly fringed cap covering a third to half the acorn. The tree grows to 50 feet (15 m) with a sturdy trunk that flares out widely at the base.

HABITAT AND RANGE
This native of China, Japan, Korea, and northern parts of Vietnam and India has adapted to sunlight or light shade, but new trees begin best in full sun. Preferred soils are well drained, somewhat acid soils, but it grows in a wide variety of soil types. Grows in temperate climates across the United States and in southern Canada but not in the upper Midwest or upper New England.

Sawtooth oak acorns have fringed caps.

Left: *Sawtooth oak has furrowed, gray bark.* **Right:** *Long, narrow leaves with saw teeth.*

WHAT IT DOES IN THE ECOSYSTEM

To date no solid evidence indicates that sawtooth oak effects a significant change in the overall natural systems of areas where it has naturalized. Some observers are concerned that it may displace native oaks. Its abundant acorn crops, developing as soon as five years after sprouting, are much earlier than many native oaks, thus making it a popular wildlife planting. A variety called "gobbler oak" has acorns of convenient size for wild turkeys while still providing significant food for squirrels, deer, and other wildlife. Some reports say that sawtooth acorns are not as nutritious for wildlife as native acorns. So far studies indicate that hybridization between sawtooth oak and native oaks is very unlikely.

HOW IT CAME TO NORTH AMERICA

Sawtooth oak came to the United States in 1862 but did not become widespread until wildlife biologists began to recommend it in restoration projects. More recently it has been used as a landscaping tree, especially at the edges of parking lots and in highway medians. It tolerates poor drainage, compact soil, drought, and air pollution.

MANAGEMENT

The heavy drop of acorns is more of a problem than the spread of the tree itself. Trees can be easily removed by pulling or digging when young or by girdling when older.

Sawtooth oak is planted for its abundant acorn production.

FOR MORE INFORMATION
Whittemore, A. T. 2004. "Sawtooth oak (*Quercus acutissima,* Fagaceae) in North America." *SIDA* 21(1):447–454.

Black Locust   *Robinia pseudoacacia*

NAME AND FAMILY
Black locust (*Robinia pseudoacacia* L.); pea family (Fabaceae). Although native to the Appalachian Mountains from Pennsylvania to Georgia and to the Ozark Plateau, black locust is spreading outside of its original range. It can be confused with honey locust (*Gleditsia triacanthos*).

IDENTIFYING CHARACTERISTICS
A fast-growing tree from 40 to 100 ft (12 to 30 m) in height, black locust grows very straight in forests but has spreading, curvy branches in open areas. The bark of young trees is greenish and in older trees the gray-brown bark is deeply furrowed with flat-topped ridges. Seedlings and young sprouts have paired ½ in. (1.3 cm) thorns at the base of the leaves. Leaves are compound with 7 to 21 oval, smooth-edged leaflets, giving the tree a ferny appearance. Leaves turn bright yellow before falling off in autumn. In late spring, fragrant clusters of white flowers hang down from the branches. The uppermost petal of

Left: *Black locust bark is deeply furrowed with flat-topped ridges.* **Right:** *Pea-type flowers hang in long clusters.*

each flower has a yellow blotch on it. The flattened, bean-shaped, 3 to 5 in. (7.6 to 12.7 cm) long pods are black and the uncooked seeds are toxic. Trees tend to form colonies through root suckering. Honey locust has longer, sometimes branched thorns on the stems (although some varieties are thornless) and larger pods, 3 to 18 in. (7.6 to 45 cm) long. Honey locust leaves are sometimes twice compound.

HABITAT AND RANGE
Black locust occurs throughout the United States and also in Nova Scotia, Quebec, and Ontario. It can grow on very poor, dry soils. Intolerant of shade and competition, it tends to grow on disturbed ground and in prairies, meadows, and savannahs.

Trees form colonies by suckering.

Oval-shaped leaflets make up a black locust leaf.

WHAT IT DOES IN THE ECOSYSTEM

Black locust roots have nitrogen-fixing nodules, so it can alter the nitrogen content of soils. It aggressively invades prairies and meadows, where it shades out native plants by forming dense colonies. Its flowers attract bees and other insects. The seeds are eaten by a wide range of animals, but the seeds, leaves, and bark contain toxins poisonous to people and some livestock. Although the blooming period is limited to less than two weeks, honey made from the nectar is highly prized in the U.S. and Europe.

HOW IT CAME TO NORTH AMERICA

Although native to southeastern mountain areas, because it is very resistant to rot, burns well, and grows quickly, black locust spread beyond its original range beginning in the early 1900s. It was used for mine reclamation, reforestation, and erosion-control windbreaks.

MANAGEMENT

Cutting or burning black locust tends to stimulate resprout, although repeated cutting or mowing during the growing season for several years can exhaust the root system. Seedlings may be outcompeted by some grasses and forbs. Grazing by cattle and goats can eventually kill black locust. Herbicide sprayed on foliage or painted on cut stumps can be effective, but effectiveness varies considerably depending on the herbicide used and the concentration and timing of application.

FOR MORE INFORMATION

Converse, C. K. and T. Martin 2001. "Element stewardship abstract for *Robinia pseudoacacia*." http://wiki.bugwood.org/Robinia_pseudoacacia.

Willows   *Salix* spp.

NAME AND FAMILY

Willows (*Salix* spp.); willow family (Salicaceae). There are over 100 species of willows in North America, and at least 10 nonnatives have naturalized in North America, some of which hybridize with each other and with native willows. The nonnatives include white willow (*S. alba* L.), goat willow (*S. caprea* L.), large gray willow (*S. cinerea* L.), crack willow (*S. fragilis* L.), corkscrew willow (*S. matsudana* Koidzumi), bay willow (*S. pentandra* L.), purple willow, purple osier (*S. purpurea* L.), weeping willow (*S.* x *pendulina* Wenderoth and *S.* x *sepulcralis* Simonkai), and basket willow (*S. viminalis* L.).

IDENTIFYING CHARACTERISTICS

The common and most visible traits of willows are their long, narrow leaves that alternate on the twig, buds with a single scale, and flowers borne in catkins. All invasive willows are deciduous and prefer moist to wet locations. The reader will need a key to make good identifications and we suggest two in the list of references. Here are some distinguishing features of three of the more common nonnatives.

White willow bark.

Crack willow (S. fragilis): Very brittle, dusky green to tannish, upright (not bending over) branches that make a cracking sound as they break (usually at the base). Serrated leaves are 1 to 7 in. (2.5 to 17.8 cm) long, shiny on top and blue green beneath, darker on top than those of white willow and with a V-shaped base. The leaves of the very similar native black willow (S. *nigra*) have rounded bases. Catkins (flower clusters) 1 to 2 in. (2.5 to 5 cm) long. Grows to 60 ft. (18.3 m).

White willow (S. alba): The only willow with silky hair beneath and sometimes above the leaves. Dark-green leaves are 1.5 to 6 in. (3.7 to 15.2 cm) long and have finely toothed margins. Branches often droop, but not as completely as those of weeping willow. Short trunks can often reach 3 ft. (1 m) in diameter, with the crown rising as high as 100 ft. (30.5 m).

Goat willow or **French pussy willow (S. caprea):** This smaller willow grows as a shrub or a tree but no higher than 30 ft. (9 m). Gray-green leaves are broadly oval and covered with gray hairs on the undersides of the leaves. Fat catkins that earn it the name "pussy willow" develop in the leaf axils. Grows in lowlands and forest edges and does not depend on proximity to surface water. A drought-tolerant willow. The similar-looking native pussy willow (S. *discolor*) has sparse red or gray hairs on the underside of the leaf.

HABITAT AND RANGE

Willows grow across the entire continent. The most widespread invasive willows are white willow, crack willow, and hybrids of crack willows that grow across most of the United States and Canada. Willows prefer sunny, moist areas along river banks and in wetlands.

Willows often have many branches originating near the base of the tree.

WHAT IT DOES IN THE ECOSYSTEM

Willows often grow in dense stands that shade out other species, many of which are important to wildlife along waterways. Willows, however, also provide food and shelter for many species of animals. While willows can stabilize stream and riverbanks, they can also prevent floods from spreading out, thus increasing scouring action and downstream flooding.

HOW THEY CAME TO NORTH AMERICA

The invasive willows of North America are natives of Europe, Asia, and north Africa. Willows were culti-vated for centuries, different species having different uses that range from hedgerows to making charcoal to bas-ket weaving to specimen plants. The salicylates produced by willows were among the earliest painkillers, con-

White willow leaves have silky hairs on the leaf underside.

taining the same active ingredient as aspirin. The supple new growth was often encouraged by coppicing or cutting back the tops. The willows were brought to North America at various times for various purposes. Because willows are easily planted and grow fast, they are the subject of a number of experiments, including growing willow fodder and biomass for energy.

MANAGEMENT

Many willows will reproduce from even small fragments left on wet ground or in shallow water, so hand pulling or digging must be thorough and often repeated. Young foliage can be sprayed with herbicides containing triclopyr, imazapyr, or glyphosate, and cut stems and trunks can be painted with these herbicides.

FOR MORE INFORMATION

Argus, G. W. 2005. "Interactive Identification of New World *Salix* (Salicaceae) using Intkey."
http://aknhp.uaa.alaska.edu/botany/salix-salicaceae-identification-using-intkey/.
Glenn, S. D. 2010. "*Salix.*" *New York Metropolitan Flora Project* Brooklyn Botanical Garden.
http://nymf.bbg.org/genus/173.

Czarapata, E. J. 2005. *Invasive Plants of the Upper Midwest*. Madison, WI: University of Wisconsin Press.
Newsholme, C. 2003. *Willows: The Genus* Salix. Portland, OR: Timber Press.
Ohio Trees. Extension Bulletin 700-00. Ohio State University.

Chinese Tallow Tree   *Triadica sebifera*

NAME AND FAMILY
Chinese tallow tree, popcorn tree, chicken tree, Florida aspen (*Triadica sebifera* [L.] Small) (also known as *Sapium sebiferum*); spurge family (Euphorbiaceae).

IDENTIFYING CHARACTERISTICS
The names for this tree come from its most easily recognized features. "Popcorn tree" refers to its numerous seeds coated in white wax and resembling popcorn that hang on the tree into winter. "Tallow tree" refers to the wax produced by the tree. Tallow trees grow to 60 ft. (18 m) tall and 3 ft. (1 m) in diameter. The alternating leaves have a distinctive spade shape, wide at the base and pointed at the end, 2 to 3 in. (5 to 7.6 cm) long, dark green, turning red in fall. From April to June flowers appear on drooping spikes about 8 in. (20 cm) long. Flowers have yellow-green sepals and no petals. The female flowers mature as greenish fruits about 0.5 in. (1.3 cm) in diameter. These turn black, then open to reveal the "popcorn," three seeds in white wax. Trees can flower and set fruit by age three.

Chinese tallow tree is often planted for its fall color in the South.

Left: *Spikes of greenish flowers bloom in late spring.* **Right:** *Popcorn-like fruits are seeds covered in a waxy coating.*

HABITAT AND RANGE

Chinese tallow tree has spread across the southeastern states from Florida to Texas and as far north as North Carolina and Arkansas. Also found in California where it has been planted as an ornamental. The trees prefer abandoned fields but also grow in forests. They tolerate saline soils, shade, and sun. Tallow tree is particularly invasive along streams, rivers, and irrigation ditches.

WHAT IT DOES IN THE ECOSYSTEM

In favorable conditions, tallow tree can nearly monopolize a natural area, earning it the name "Terrible Tallow" in northern Florida and a place on The Nature Conservancy's list of "The Dirty Dozen: America's Least Wanted." A single tallow tree produces as many as 100,000 seeds that germinate prolifically and survive for a long time when afloat. Its tannins and leaf drop change soil and water chemistry, adding nitrogen and phosphorus. Since horses and cattle do not eat it, it grows readily in pastures. Leaves, fruit, and sap are toxic to humans and may irritate the skin.

HOW IT CAME TO NORTH AMERICA

For over 1,500 years the Chinese cultivated tallow trees and harvested their oil and wax, a fact that moved Benjamin Franklin to import the tree for soap and

Husks open to reveal three wax-coated seeds.

candle making. It arrived in Charleston, South Carolina, in 1776. In the 1920s and 1930s, the U.S. Department of Agriculture promoted the tallow tree for oil production. During World War II, Chinese and American forces in China used the oil for diesel operation. A plantation can yield "the equivalent of 500 gallons (12 barrels) of fats and oils per acre per year" (Mason 1997). The oils and waxes it produces have stimulated some research into using the tree as a producer of fuels to substitute for petroleum products. The flowers also support honey bees. While some states have banned sales of tallow tree, it is still sold in others and is popular as a fast-growing shade tree and ornamental with fall color.

MANAGEMENT

Tallow trees are difficult to remove once rooted and they can regrow from root fragments. Saplings can be sprayed with glyphosate. Late summer and early fall, before seeds are dispersed and when the tree will transport chemicals to the roots, is the best time to spray. Triclopyr painted on cut stumps or mixed with oil and sprayed on the bark at the base will kill the trees any time of year.

FOR MORE INFORMATION

Langeland, K. A. 2009. *Natural Area Weeds: Chinese Tallow* (Sapium sebiferum, L.). SS-AGR-45. Gainesville, FL: University of Florida IFAS. http://edis.ifas.ufl.edu/pdffiles/AG/AG14800.pdf.

Mason, S. I. 1997. "The Chinese tallow tree: Growing oil on trees." *America's Inventor Online Edition*. http://www.inventionconvention.com/americasinventor/dec97issue/section12.html.

Siberian Elm   *Ulmus pumila*

NAME AND FAMILY
Siberian elm, dwarf elm, littleleaf elm (*Ulmus pumila* L.); elm family (Ulmaceae). Can be confused with Chinese elm (*U. parviflora*), which is also invasive, and with native slippery elm (*U. rubra*) and American elm (*U. americana*).

IDENTIFYING CHARACTERISTICS
Siberian elm is easily confused with native elms because the leaves are almost the same except Siberian elm's are smaller. Siberian elm leaves grow alternately on their twigs. They are broad, heart-shaped at the base and taper to a long point, with small teeth bordering the entire margin; 0.3 to 1 in. (.8 to 2.5 cm) wide and typically twice as long. Flowers are greenish-red to brown without petals, producing clusters of round-winged seed cases with a single seed in the center. In uncrowded conditions the tree grows into a round shape as high as 60 ft. (18 m). Because of brittle branches and frequent sprouting along the trunk, the tree often has a very irregular appearance. Bark of mature trees is light gray brown and furrowed. The native American and slippery elms have leaves that are usually over 2.5 in. (6.2 cm) long and asymmetrical at the base, and instead of single teeth,

Siberian elm branches are brittle, giving the tree an irregular outline.

they have double-toothed margins. The less invasive Chinese elm flowers later in summer or early fall, and its leaves and the teeth on its leaves are not as pointed.

HABITAT AND RANGE
Siberian elm is well adapted to withstand bitter cold winters and dry summers. In North America it grows in most states and well into Canada. It seeds readily even in poor sandy soils, and often grows along roads, in rangeland, in pastures, and along streams.

Leaves alternate along the stems and mature trees have furrowed bark.

Winged seeds of Siberian elm.

WHAT IT DOES IN THE ECOSYSTEM

Siberian elm can occupy significant areas of rangeland or pasture and grow in dense thickets where its shade stifles most native species that need sunlight. Arborists have developed many cultivars for landscaping and windrows, although noted horticulturist Michael Dirr describes it as "[a] poor ornamental tree that does not deserve to be planted anywhere!" It holds some promise for creating hybrids that are resistant to the Dutch Elm Disease that has essentially eliminated America's once most popular shade tree, the American elm. Readily eaten by the elm leaf beetle.

HOW IT CAME TO NORTH AMERICA

Siberian elm came to the United States in the 1860s to be used in the Midwest as a windbreak. Siberian elm is native to northern China, eastern Siberia, Manchuria, and Korea, while a dwarf variety grows even farther north in Siberia.

Leaves have slightly uneven bases and are slightly smaller than those of native elms.

MANAGEMENT

Girdling the trees in mid to late spring to midsummer when the bark will slip off easily will kill them in 1 or 2 years. After cutting parallel lines, being careful not to cut deep, use a mallet or hammer to loosen the bark. Then remove a 4 in. wide band of bark. (The tree reacts to a deep cut as if it had been cut down, sending up shoots from the base). Small trees and shoots can be dug out. Since seeds germinate quickly, any elms left standing nearby will reestablish a stand. For chemical control, glyphosate painted on cut stems or sprayed on new saplings is effective. Controlled burns in fire resistant plant communities will also eliminate most new elm growth.

FOR MORE INFORMATION

Dirr, M. 1998. *Manual of Woody Landscape Plants.* Champaign, IL: Stipes Publishing.
Kennay, J., and G. Fell. 1990. "Vegetation management guideline: Siberian elm (*Ulmus pumila*)."
   Illinois Vegetation Management Guideline 21(1). Springfield: Illinois Nature Preserves
   Commission. http://www.inhs.uiuc.edu/research/VMG/sibelm.html.

Terrestrial Plants—Trees and Shrubs— Deciduous—Alternate Leaves—Shrubs

Camel Thorn   *Alhagi maurorum*

NAME AND FAMILY
Camel thorn (*Alhagi maurorum* Medik.); pea family (Fabaceae).

IDENTIFYING CHARACTERISTICS
This perennial shrub defends itself with 0.25 to 1.75 in. (0.6 to 4 cm) long
yellow-tipped spines on its fine green branches. It grows 1.5 to 4 ft. (0.5 to 1.3
m) tall from a root system that can extend 6 to 7 ft. (1.8 to 2.1 m) below
ground. Rhizomes spread underground from which new shoots can arise up to
25 ft. (7.6 m) from the parent plant. The whole shrub has a greenish color
with small oval to lance-shaped leaves arranged alternately along the stems.
Leaves are yellowish above and gray green underneath. Dark pinkish-purple ⅜
in. (1 cm) wide flowers bloom in midsummer on spine-tipped stems that grow
from the leaf axils. Seed pods are constricted between the seeds and look like
beads on a string.

HABITAT AND RANGE
Camel thorn prefers deep, moist soils
along riverbanks, river bottoms, and
canals, but the deep root system also
allows camel thorn to grow in dry,
rocky, saline soils on rangelands. Grows
from Washington state south to Texas.

WHAT IT DOES IN THE ECOSYSTEM
Shrubs form dense stands along waterways, displacing native vegetation. Most
animals (except camels and goats) do
not prefer to eat camel thorn, so it
reduces the value of rangelands, replacing favored forage plants. Grazing animals will eat the seedpods, though,

*Seedpods look like beads on a string
and long thorns defend the branches.*

These short shrubs can form dense stands.

and become collaborators in the plant's spread since traveling through the grazer's gut sears seeds, enhancing germination. Research shows that camel thorn nourishes goats as well as alfalfa does.

HOW IT CAME TO NORTH AMERICA
The first camel thorn arrived in California in the early 1900s as a contaminant in alfalfa seed from Turkey and in camel dung used as packing

Camel thorn blooms in mid-summer.

material around date palm shoots. Beekeepers considered it a good honey plant and may have helped spread it in the southwest. Native to western Asia, Iran, Anatolia, and parts of the Mediterranean.

MANAGEMENT
Camel thorn is difficult to control once established. The extensive root system makes mowing or cutting generally futile. Herbicides, including picloram and glyphosate, can be painted on cut stems. 2,4-D has been effective as a foliar spray. Even herbicide treatments will probably have to be repeated.

FOR MORE INFORMATION

Parker, K. F. 1972. *An Illustrated Guide to Arizona Weeds.* Tucson: University of Arizona Press.
   http://www.uapress.arizona.edu/onlinebks/weeds/titlweed.htm.
Moser, L. and D. Crisp. "San Francisco peaks weed management area fact sheet on *Alhagi maurorum.*" Coconino National Forest. http://sbsc.wr.usgs.gov/research/projects/swepic/
   factsheets/alma12sf_info.pdf.
Whitson, T. D. et al., eds. 1996. *Weeds of the West.* Western Society of Weed Science in cooperation with Cooperative Extension Services. Laramie: University of Wyoming.

Japanese Barberry   *Berberis thunbergii*

NAME AND FAMILY
Japanese barberry (*Berberis thunbergii* DC.); barberry family (Beberidaceae).
Also invasive is common or European barberry (*B. vulgaris*). The native Amer-
ican barberry (*B. canadensis*) looks similar to common barberry.

IDENTIFYING CHARACTERISTICS
Japanese barberry shrubs are most noticeable in fall when their many branches
shine with bright-red berries. It seldom tops 3 ft. (1 m) but can grow to 6 ft. (2
m). It often grows as dense understory in woodlands. Leafing out early in
spring, it takes advantage of sunlight before the canopy brings shade. Small,
smooth-edged leaves, generally oval to narrowed at one end, cluster close
along the branches, with single spines emerging from the points where leaves
and branches join. Leaves range from dark green to dark red, verging on purple

Barberry's bright-red fruits remain on plants into winter.

Shrubs form impenetrable masses.

in some cultivars. In spring many small, pale-yellow, semiglobular, six-petaled flowers bloom along the branches, singly or in clusters of two to four. Slightly oblong bright-red, dry fruits, 0.25 to 0.4 in. (0.6 to 1 cm) long, mature in late summer and hang on through autumn and into winter if not eaten by birds. Common barberry can be distinguished from Japanese barberry by its spiny, toothed leaves and flowers that grow in a long cluster. American barberry can be distinguished from common barberry by the brown-red color of the young twigs and the flower petals, which are notched at the tip on American barberry.

HABITAT AND RANGE
Japanese barberry prefers partial sunlight but also does well in shade, especially in younger forests. It grows from North Carolina and Tennessee north to Nova Scotia and Ontario and as far west as Montana.

WHAT IT DOES IN THE ECOSYSTEM
Birds, especially turkey and grouse, eat barberries and spread seeds far and wide where the shrubs invade forest edges as well as oak forests and savannas. Barberry can also grow from root creepers or from branches rooting upon contact with the ground. This enables a single bush to form a thicket. Japanese barberry at high densities lowers plant diversity, and the leaf litter causes changes in soil chemistry. It can also be an annoying obstacle to hikers when it grows in thick stands. Japanese barberry stands harbor higher abundances of deer ticks than uninvaded woodlands.

Leaves form a whorl defended by spines.

HOW IT CAME TO NORTH AMERICA

Early settlers imported and used common barberry for dyes and jams, but after farmers discovered it harbored wheat rust, a nationwide eradication campaign began. Japanese barberry seeds were sent from Russia to the Arnold Arboretum in Boston in 1875 as a substitute for common barberry. Other botanical gardens and landscapers soon planted the shrub. Many cultivars of this species are sold as ornamentals because of its bright berries and leaves that provide fall color.

Pale-yellow flowers bloom in spring.

MANAGEMENT

Where numbers are limited, shallow but tough roots allow this shrub to be hand pulled using a hoe or mattock to dig up root systems. Spring is the best time for manual or mechanical control. Controlled burns in fire-resistant plant communities will kill barberry, as will using a propane torch to burn stems in spring just before or after leaves appear, with a followup burn-off in summer. Herbicides labeled for brush control can be painted on stumps after cutting.

FOR MORE INFORMATION

Manning, W. H. 1913. "*Berberis thunbergii* naturalized in New Hampshire." *Rhodora* 15:225–26.

Silander, J. A., and D. M. Klepeis. 1999. "The invasion ecology of Japanese Barberry (*Berberis thunbergii*) in the New England landscape." *Biological Invasions* 1:189–201.

Ward, J. S., T. E. Worthley, and S. C. Williams. 2009. "Controlling Japanese barberry (*Berberis thunbergii* DC) in Southern New England U.S.A." *Forest Ecology and Management* 257:561–566.

Zouhar, K. 2008. "*Berberis thunbergii*." *Fire Effects Information System*. U.S. Department of Agriculture, Forest Service, Rocky Mountain Research Station, Fire Sciences Laboratory. http://www.fs.fed.us/database/feis/.

Autumn and Russian Olive   *Elaeagnus* spp.

NAME AND FAMILY

Autumn olive, silverberry (*Elaeagnus umbellata* Thunb.) and Russian olive (*E. angustifolia* L.); autumn olive family (Elaeagnaceae). Silverthorn (*E. pungens*) is an evergreen planted in the southeastern United States and considered a more restricted invader geographically.

IDENTIFYING CHARACTERISTICS

Autumn and Russian olive grow as shrubs to small trees that can reach 30 ft. (9 m) and are covered with silvery leaves. The leaves are arranged alternately along the stem, with Russian olive having narrower leaves (Russian olive leaves are 0.5 in. [1.3 cm] wide versus 1 in. [2.5 cm] wide for autumn olive). Both have fragrant, creamy yellow flowers in spring. They produce large numbers of small, less than 0.5 in. (1.3 cm) fruits, each with one seed, eaten mainly by birds. Russian olive has dry, mealy fruits, whereas autumn olive has juicy red fruits with silver netting. Russian olive tends to be thornier than autumn olive. Silverthorn keeps its leaves all winter and tends to grow as a dense shrub with long shoots coming out the top.

Russian olive's yellow fruits and slender leaves.

A thicket of Russian olive, common in the central and western United States.

HABITAT AND RANGE

Autumn olive grows from southern and eastern Ontario south to Florida and west from Wisconsin to Kansas, Arkansas, and Louisiana. It will grow in open forests, prairies, and along floodplains, preferring well-drained soils and sun to part shade. Russian olive is most abundant in the central and western United States but does occur from Virginia north into southern Canada. Western states report dense stands along streams, rivers, and drainage ditches as well as in fields and open areas.

WHAT IT DOES IN THE ECOSYSTEM

Autumn olive has red, juicy fruits.

The roots form an association with bacteria to fix nitrogen, which may result in soils becoming more nitrogen rich, changing the composition of the plant community growing on that site. The leaf litter of Russian olive can also affect nutrient dynamics in streams. Shrubs can grow so densely that they out-

Autumn olive's fragrant flowers bloom in spring.

compete other species. Autumn and Russian olive provide cover for wildlife but may reduce the diversity of cover types over the long term. Fruits tend to remain on the plants into winter and may be an important food source for wildlife in winter. Autumn olive fruits make a good thirst quencher for hikers and a jam rich in vitamin C.

HOW IT CAME TO NORTH AMERICA
Native to China, Korea, Pakistan, and Japan, autumn olive was brought to North America from China and Japan in 1830 for cultivation. Russian olive is native to southeastern Europe and western Asia and arrived in the United States in the late 1800s as an ornamental plant and for wildlife planting and mine reclamation. In addition to being sold as ornamental plants, autumn and Russian olives were extensively promoted by wildlife managers as cover and food plants and by agricultural extensions as fast-growing windbreaks.

MANAGEMENT
Young seedlings can be hand pulled, especially when the soil is moist. Plants readily resprout after cutting, but applying glyphosate or triclopyr to cut stumps can kill plants. This treatment is most effective in late summer. Other herbicide treatments include foliar spray and applying herbicides to the base of the trunk. Burning the root crown with a flame weeder can kill plants as well.

FOR MORE INFORMATION
Hussain, Imtiaz. 2011. "Physiochemical and sensory characteristics of *Elaeagnus umbellata* (Thunb) fruit from Rawalakot (Azad Kashmir) Pakistan." *African Journal of Food Science and Technology* (ISSN: 2141-5455) Vol. 2(7): 151–156,

Miller, J. H., E. B. Chambliss, N. J. Loewenstein. 2010. *A field guide for the identification of invasive plants in southern forests.* General Technical Report SRS-119. Asheville, NC: U.S.

Department of Agriculture, Forest Service, Southern Research Station. http://www.srs.fs
.fed.us/pubs/gtr/gtr_srs119.pdf.

Mineau, M. M., C.V. Baxter, and A. M. Marcarelli. 2011. "A non-native riparian tree *(Elaeagnus
angustifolia)* changes nutrient dynamics in streams." *Ecosystems* 14:353–365.

Zouhar, Kris. 2005. "*Elaeagnus angustifolia.*" *Fire Effects Information System.* U.S. Department
of Agriculture, Forest Service, Rocky Mountain Research Station, Fire Sciences Laboratory.
http://www.fs.fed.us/database/feis/.

Sea Buckthorn   *Hippophae rhamnoides*

NAME AND FAMILY
Sea buckthorn, sea berry (*Hippophae rhamnoides* L.); autumn olive family
(*Elaeagnaceae*).

IDENTIFYING CHARACTERISTICS
Sea buckthorn grows as a thorny shrub 6 to 20 ft. (1.8 to 6.1 m) tall. The leaves
are silvery on top, narrow, lance-shaped, and 1.5 to 2.25 in. (3.7 to 5.1 m) long.
Sea buckthorn's yellowish flowers bloom on separate male and female plants
before the leaves appear in spring. The flowers are inconspicuous and wind pollinated. The ⅛ in. (0.3 cm) globular fruits line the branches and turn from yellow to orange when mature in early fall. Each fruit has a single, very hard seed.

HABITAT AND RANGE
Despite its name, sea buckthorn is currently most widely established in the
Canadian prairies, particularly along river floodplains. Sea buckthorn prefers
lighter soils but can grow in a variety of moist or wet soils, including clays and
nutrient-poor land. It tolerates salty maritime environments very well.

Edible fruits line the branches of sea buckthorn.

Left: *Narrow leaves are silvery on top.* **Right:** *Sea buckthorn shrubs can grow tall and form thickets.*

WHAT IT DOES IN THE ECOSYSTEM

Sea buckthorn creates dense thickets that displace native plants and alter the food and shelter available to birds and other animals. The shrubs form an association with the microorganism actinomycete *Frankia* that fixes nitrogen, increasing nitrogen levels in the soil. Birds occasionally eat the fruits and disperse seeds, and plants also spread via rhizomes.

HOW IT CAME TO NORTH AMERICA

Introduced from Siberia to Canada in the 1930s. Sea buckthorn has a multitude of uses, some documented for more than a thousand years, including as food (edible fruits), medicine (oils), soil enhancement, erosion control in saline soils, and as a shelterbelt. The fruits are high in vitamin C and E and in protein. It is widely planted for shelterbelts and is promoted as an orchard crop in Canada. It was not reported as invasive in Canada until 1997. Native to northern Europe and Asia, where it is often grown as a fruit producer and for forage and, in Russia, for wine production.

MANAGEMENT

Young seedlings can be hand pulled, especially when the soil is moist. Plants readily resprout after cutting, but applying glyphosate to cut stumps can kill plants. This treatment is most effective in late summer. Other herbicide treatments include foliar spray and applying triclopyr mixed with oil to the base of the trunk.

FOR MORE INFORMATION

Catling, P. M. 2005. "New 'top of the list' invasive plants of Canada." *Botanical Electronic News* 345:March 25. http://www.ou.edu/cas/botany-micro/ben/ben345.html.

Li, T. S. C. 1999. "Sea buckthorn: New crop opportunity." In *Perspectives on new crops and new uses,* edited by J. Janick, 335–337. Alexandria, VA: ASHS Press.

Lian, Y. 1988. "New discoveries of the genus Hippophaë L. (Elaeagnaceae)." *Acta Phytotaxonomica Sinica* 26(3):235–37.

Small, E., and P. M. Catling. 2002. "Blossoming treasures of biodiversity. 5. Sea buckthorn (*Hippophae rhamnoides*): An ancient crop with modern virtues." *Biodiversity* 3(2):25–27.

Multiflora Rose   *Rosa multiflora*

NAME AND FAMILY
Multiflora rose (*Rosa multiflora* Thunb.); rose family (Rosaceae). There are many species of native and nonnative roses that may grow alongside or among multiflora roses.

IDENTIFYING CHARACTERISTICS
Multiflora rose has long, arching canes that can clamber into trees or whose tips root in the ground. Canes are covered with stout thorns. Canes are green when young, and brownish gray when older. In spring, 0.5 to 1 in. (1.3 to 2.5 cm)

White, five-petaled flowers cover multiflora rose branches in spring.

Left: *Small red rose hips persist on plants into winter.* **Right:** *Fringed petioles distinguish multiflora rose from native roses.*

wide, white to pinkish, five-petaled, fragrant flowers cover the plants. Flowers occur in clusters at the ends of small twigs and mature into clusters of small 0.25 in. (0.8 cm) red rose hips that often stay on the plants well into winter. The flower color and size and fruit size distinguish it from many rose species. Leaves are divided into 5 to 11 oblong leaflets, 1 to 1.5 in. (2.5 to 3.7 cm) long, with serrated edges. The petiole (the stalk of the leaf where it attaches to the branch) of multiflora rose looks fringed, unlike those of most other rose species.

HABITAT AND RANGE

Colonizes gaps in woodlands, forest edges, prairies, and old fields. Flowers most prolifically in full sun but can survive in shade. Occurs throughout most of the United States except in the Rocky Mountains, deserts, and subtropical areas, and has naturalized in southwestern Ontario in Canada.

WHAT IT DOES IN THE ECOSYSTEM

The arching canes and thorns form impenetrable thickets. Canes that climb into trees can add weight to the branches, making them vulnerable to breaking in windstorms. Many species of birds and mammals eat the fruits and disperse the seeds. A single plant can produce a million seeds a year that can survive in the soil for up to 20 years. In many states multiflora rose is considered a noxious weed because of its tendency to colonize pastures and fields. The dense shelter of the multiflora rose is a preferred nesting place for many birds, and a

Rose canes can climb up into trees.

haven for rodents, including rabbits. Predation of nests can be higher for some birds species nesting in multiflora rose.

HOW IT CAME TO NORTH AMERICA

The rose industry introduced multiflora rose to the eastern United States from Japan in 1866 as a rootstock for ornamental roses. In the 1930s the U.S. Soil Conservation Service promoted it for erosion control and as a living fence. Many states promoted it as a plant for food and cover for wildlife. More recently, highway departments often planted it as a snow fence and soft but effective crash barrier in highway medians.

MANAGEMENT

To eliminate small patches, have one or more persons hold the canes back with pitchforks while another person cuts the canes toward the base of the plant. Resprouts will need to be cut repeatedly or the rootstock dug out. Cut canes can also be painted with glyphosate or triclopyr. In fields, multiflora rose can be mowed repeatedly during the growing season with a brush mower for two to four years. Glyphosate can be sprayed on foliage after flower buds form or fosamine can be sprayed during summer. Rose rosette disease, a virus native to the western United States and spread by mites, has moved eastward and could kill dense patches of multiflora rose, but it also infects other rose species. Goats love to eat multiflora rose.

FOR MORE INFORMATION

Epstein, A. H., J. H. Hill, and F. W. Nutter. "Augmentation of rose rosette disease for biocontrol of multiflora rose (*Rosa multiflora*)." *Weed Science* 45(1):172–178.

Munger, Gregory T. 2002. "*Rosa multiflora.*" *Fire Effects Information System*. U.S. Department of Agriculture, Forest Service, Rocky Mountain Research Station, Fire Sciences Laboratory. http://www.fs.fed.us/database/feis/.

Szafone, R. 1991. "Vegetation management guidelines: Multiflora rose (*Rosa multiflora* Thunb.)" *Natural Areas Journal* 11(4):215–216.

Rugosa Rose   *Rosa rugosa*

NAME AND FAMILY
Rugosa rose, Japanese rose, beach rose (*Rosa rugosa* Thunb.); rose family (Rosaceae).

IDENTIFYING CHARACTERISTICS
Rugosa rose is a tough rose covered with 2 in. (5.1 cm) wide, dark-pink to white flowers in summer. Flowers occur singly or in small clusters in summer, have a whorl of five petals or multiple whorls of petals, and yellow stamens. In fall, large, red, rounded but slightly flattened rose hips mature to 1 in. (2.5 cm) in diameter. Dark-green leaves with serrated edges have a heavily veined surface (giving it the name rugosa, for rugose or wrinkled) and are made up of five to nine leaflets. Each leaflet is 1 to 2 in. (2.5 to 5 cm) long and leaves are 3 to 6 in. (7.6 to 15 cm) long. Shrubs grow 4 to 7 ft. (1.2 to 2 m) tall with stems densely covered with straight gray prickles. Young stems may be hairy.

HABITAT AND RANGE
Rugosa rose is often found on sandy dunes on the East Coast and around the Great Lakes but also grows in fields. It ranges from southern Canada south to North Carolina and west to Wisconsin and Missouri. Also occurs in Washington State and British Columbia.

WHAT IT DOES IN THE ECOSYSTEM
Plants seem to spread slowly into new places, but once established, dense stands form from root suckers. On sand dunes they provide some erosion control but dis-

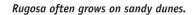

Rugosa often grows on sandy dunes.

place native dune vegetation. Rugosa rose hybridizes with the native *Rosa blanda*. Plants provide some cover for birds and deer, and fruits may be eaten by animals. The plant's tendency to grow densely has made it popular as a "living hedge."

HOW IT CAME TO NORTH AMERICA
Rugosa rose was brought from Japan to Europe as an ornamental rose in the early 1770s and probably arrived in North America soon after. It is native to Japan, China, and Korea. In China it is listed as an endangered species because of habitat loss and disturbance of coastal dunes. It is often hybridized with other roses because it is more resistant to common rose diseases. Rugosa rose is used for erosion control along beaches due to its tolerance to drought and salt spray. The rose hips are high in Vitamin C and can be used in cooking and for tea.

MANAGEMENT
Small plants can be pulled out, but all roots must be removed to prevent resprouting. Cut canes can also be painted with glyphosate or triclopyr. In fields, rugosa rose can be mowed repeatedly during the growing season with a brush mower. Glyphosate can be sprayed on foliage after flower buds form for effective control.

FOR MORE INFORMATION
Dickerson, J., and C. Miller. 2002. *Rugosa rose.* U.S. Department of Agriculture Natural Resources Conservation Service. http://plants.usda.gov/factsheet/pdf/fs_roru.pdf.
Mercure, M. and A. Bruneau. 2008. "Hybridization between the escaped *Rosa rugosa* (Rosaceae) and native *R. blanda* in Eastern North America." *American Journal of Botany* 95:597–607.
Verrier, S. 1999. Rosa rugosa. Buffalo, NY: Firefly Books.

Blackberries   *Rubus* spp.

NAME AND FAMILY
Blackberry, Himalayan blackberry, Asian blackberry (*Rubus armeniacus* Focke), and cutleaf or evergreen blackberry (*R. laciniatus* Willd.); rose family (Rosaceae). There are many species of both native and nonnative *Rubus* (blackberry, raspberry, and dewberry) species. Himalayan and cutleaf blackberry are the two most widespread invasive blackberry species. In the confusing taxonomy of introduced *Rubus*, *R. ulmifolius*, *R. bifrons*, and *R. discolor* are sometimes referred to as different species or the same as *R. armeniacus*. Take care in identifying *Rubus* to distinguish native and nonnative species in your region.

IDENTIFYING CHARACTERISTICS
Himalayan and cutleaf blackberry are distinguished from native blackberries by their leaves, stems, and denser thickets. Himalayan blackberry first-year canes have leaves with five leaflets; second-year side shoots have leaves with three to

Left: *Nonnative blackberries often have five rounded leaflets.* **Right:** *Native trailing blackberry,* **Rubus argutus,** *with three leaflets.*

five leaflets. The leaflets are round to oblong in shape. Many natives have narrower leaves with three leaflets and deeper serrations on the edges. Himalayan blackberry leaves are dark green on top, and a hairy grayish green on the bottom. Stems are five-angled and armed with curved, wide-based spines. Flowers are white and clustered, sometimes pinkish, and the berries produced from midsummer to late summer mature as shiny black fruits. Berries are usually round, from 0.5 to 1 in. (1.3 to 2.5 cm) in diameter. Berries are on average larger than native blackberries, often ripening at a different time. Canes often reach 10 ft. (3 m) and sometimes 20 ft. (6 m) or more, and drape themselves over nearby tree branches. Cutleaf blackberry is easily distinguished from other Rubus species by its five very deeply lobed leaflets.

HABITAT AND RANGE
Himalayan and cutleaf blackberry have spread to most subarctic and subalpine regions with more than 10 in. (25.4 cm) annual rainfall and in a wide range of soils from acid to alkaline. They occur most commonly in the Pacific northwest but may be overlooked in the East, where several native species are similar in appearance. They grow most thickly in wet areas along streambanks, near ponds, along drainage ditches, and so on.

WHAT IT DOES IN THE ECOSYSTEM
Himalayan blackberry has become so common in North America that most people simply call it blackberry, not distinguishing it from native blackberry species. Himalayan and cutleaf blackberry generally grow much more rapidly and densely and much taller than native species. Researchers demonstrated that Himalayan and cutleaf blackberries are able to photosynthesize more efficiently compared with two native blackberry species, giving them an edge in growth and reproduction. Blackberries often block access to water sources, overgrow fences, and grow in such dense thickets that native plants, including

sun-dependent tree shoots, cannot compete. In replanted clearcuts, their dense thickets grow so much faster than fir or pine that they shade out the new trees. A thicket expands through root suckers and as first-year canes root when their tips stay in contact with the ground. Thickets can reach a density of over 500 canes per square yard ($600/m^2$), with many dead canes, since canes last no longer than three years. The thickets, however, do provide good shelter for some birds, small mammals, and insects, and they provide a good source of nectar for bees, followed by abundant and useful berries. More than 150 animal species, ranging from mice to bears, consume blackberries.

HOW IT CAME TO NORTH AMERICA

Himalayan blackberry probably invaded the Himalayas from its native region of western Europe. Given its abundant berries, agriculturalists probably introduced it to Asia, then to North America in about 1885 when famous plant breeder Luther Burbank introduced it as "Himalayan Giant." By the 1950s many nurseries and agricultural experiment stations were using it. Cutleaf blackberry was also introduced from Europe for both its ornamental leaves and its edible fruits.

Fruits form on second-year canes of Himalayan blackberry.

Pink to white, five-petaled flowers bloom in spring.

MANAGEMENT

Seedlings can be hand pulled if wearing gloves. Goats can be used to clear dense patches. Heavy blackberry infestations can be burned or bush hogged, then sprayed, and the new shoots sprayed, mowed, or grazed by sheep or goats for several years in succession. It is also very vulnerable to systemic herbicides, especially when sprayed on first-year green-stemmed growth after canes are at least 3 ft. long, or after fruiting when canes begin to carry sugars down to the rhizomes instead of up, as they do early in the season. Do not mow for two weeks after spraying. Blackberry also fails to reproduce well in heavy shade. Fast-growing trees will eventually rob blackberry stands of needed sunlight.

FOR MORE INFORMATION

Hoshovsky, M. *"Rubus discolor." Invasive Plants of California's Wildlland.* California Invasive Plant Council. http://www.cal-ipc.org/ip/management/ipcw/online.php.

McDowell, S. C. L. 2002. "Photosynthetic characteristics of invasive and noninvasive species of *Rubus* (Rosaceae)." *American Journal of Botany* 89:1431–1438.

McHenry, W. B. 1990. *Wild blackberry control.* Oakland: University of California Agriculture and Natural Resources OSA 7186.

Soll, J. 2004. *Controlling Himalayan blackberry* (Rubus armeniacus [R. discolor, R. procerus]) *in the Pacific Northwest.* The Nature Conservancy. http://www.invasive.org/gist/moredocs/rubarm01.pdf.

Wineberry   *Rubus phoenicolasius*

NAME AND FAMILY
Wineberry, wine raspberry, Japanese wineberry (*Rubus phoenicolasius* Maxim.);
rose family (Rosaceae).

IDENTIFYING CHARACTERISTICS
As the Latin name suggests, this wineberry is a berry like the raspberry and
the blackberry but with purple hairs. The long canes of wineberry distinguish
themselves by reddish hairs and spines. Canes may grow to 10 ft. (3 m) and
root where the tip contacts the ground. The leaf twigs end in a three-part
leaf, each part somewhat heart-shaped, toothed, sharply pointed at the tip.
The underside of the leaf is silvery white. Late in spring or early summer,
small flowers with white petals and greenish centers appear, followed by a red

to yellow-red fruit that slips off the
core when mature like a raspberry.
Fruits are smaller than and not as flavorful as raspberries.

HABITAT AND RANGE
Wineberry grows from eastern Canada
south to North Carolina and west to
Michigan and states east of the Mis-

Left: *Ripe, red wineberry fruits.*
Bottom: *Red, spiny hairs cover
wineberry flower buds.*

Reddish hairs and spines cover wineberry stems.

sissippi. It prefers moist soils and sunlight. It grows in open woods, fields, and along waterways and in some savannahs and prairies.

WHAT IT DOES IN THE ECOSYSTEM

Dense wineberry thickets can cover and dominate land once used by native plants. The thickets provide shelter for a variety of birds and small animals and the berries are food for a variety of wildlife. Wineberry reproduces from seeds as well as from contact points between canes and the ground.

Wineberry often grows along woodland edges in moist soils.

HOW IT CAME TO NORTH AMERICA

Agronomists introduced this native of eastern Asia and Japan to the United States in 1890 to improve the breeding stock of several commercial berries. The same qualities (growing quickly, hardiness) that make it a favorite for creating hybrids also allow it to compete successfully in the wild against native plants.

MANAGEMENT

In moist, loose soils, hand pulling or using a spade will remove wineberry. Live pieces of cane should be cleared off the site and any berries should be eaten or disposed of in bags to prevent seeds from establishing new plants. Mowing or cutting wineberry just stimulates resprouting but can be used in combination with herbicides. Herbicides containing glyphosate or triclopyr can be sprayed on foliage before seeds have matured or can be painted or sprayed on cut canes.

FOR MORE INFORMATION

Innes, Robin J. 2009. "*Rubus phoenicolasius*." *Fire Effects Information System*. U.S. Department of Agriculture, Forest Service, Rocky Mountain Research Station, Fire Sciences Laboratory. http://www.fs.fed.us/database/feis/.

Rattlebox   *Sesbania punicea*

NAME AND FAMILY

Rattlebox, Spanish gold, purple or red sesban, scarlet wisteria, Brazilian rattlebox, scarlet locust tree (*Sesbania punicea* (Cav.) Benth.); bean family (Fabaceae).

IDENTIFYING CHARACTERISTICS

A fast-growing, leguminous shrub, rattlebox is named for its distinctive four-winged, 3 in. (8 cm) long pods that hold up to nine seeds. The pods develop from hanging clusters of 5 to 30 bright red-orange flowers that bloom from late spring to early summer. Each flower is about 1 in. (2.5 cm) long. Compound leaves up to 12 in. (30 cm) long alternate along the stems. Each leaf is made up of 7 to 16 oblong, 1 in. (3 cm) long, oppositely arranged leaflets. Shrubs can grow to 15 ft. (5 m) and have spreading, open branches.

HABITAT AND RANGE

Grows in wetlands and along riverbanks in the southeastern United States from North Carolina to Texas and in California. Tolerant of full sun or part shade and of brackish water. Also considered a major threat in South Africa and parts of Australia.

WHAT IT DOES IN THE ECOSYSTEM

Plants form dense thickets that exclude other plants and alter wildlife habitat. Fixes nitrogen that may change soil fertility levels, and may release chemicals that hinder the growth of other plants. Along streams the shrubs can alter stream flow and cause flooding and streambank erosion and impede boating. Plants can begin to produce seeds in just two years, and each plant can pro-

Left: *Inflated seed pods are winged.* Right: *Bright red-orange flowers bloom in spring and early summer.*

duce thousands of seeds. Flowing water disperses the seeds. All parts of the plant contain toxic pyrrolizidine alkaloids, and few, if any, mammals and birds are immune to them.

HOW IT CAME TO NORTH AMERICA
Native to South America, rattlebox was introduced as an ornamental plant and for its ability to fix soil nitrogen. It was collected in Florida as early as 1927.

MANAGEMENT
Small plants can be hand pulled or pulled out using a weed wrench. Larger shrubs can be cut and the stumps treated with an herbicide containing glyphosate and triclopyr and formulated for use near water. Sprouting seeds can be killed using a flame weeder.

FOR MORE INFORMATION
Global Invasive Species Database. 2010. *"Sesbania punicea."* http://www.issg.org/database/species/ecology.asp?si=1673&fr=1&sts=&lang=EN.

Langeland, K. A., H. M. Cherry, C. M. McCormick, and K. A. Craddock Burks et al. 2008. *Identification and Biology of Nonnative Plants in Florida's Natural Areas,* 2nd ed. SP 257. Gainesville, FL: University of Florida IFAS.

Japanese Spiraea
Spiraea japonica

NAME AND FAMILY
Japanese spiraea, Japanese meadowsweet, fortune's meadowsweet, little princess (*Spiraea japonica* L. f.); rose family (Rosaceae). Baby's breath spiraea, Thunberg's meadowsweet (*S. thunbergii* Sieb. ex Blume) is also considered invasive but is not as widespread.

IDENTIFYING CHARACTERISTICS
Japanese spiraea grows as a small shrub to 6.5 ft. (2 m) tall with upright, slender stems. In early summer, flat-topped clusters of small pink flowers (0.2 in. [0.5 cm] wide) bloom on short stems at the ends of the branches. Seeds mature inside hard, small capsules. Alternate, lance-shaped leaves, 3 to 4.5 in. (7.6 to 11.2 cm) long and 1 to 1.5 in. (2.5 to 3.7 cm) wide, have finely toothed edges. Leaves drop in fall, revealing brown to reddish-brown stems. Young twigs are densely hairy. Japanese spiraea can be distinguished from native spiraeas by its

pink flowers and softly hairy young twigs. Thunberg's meadowsweet has single or small clusters of white, five-petaled flowers all along the stems in spring. Flower buds may be pink.

HABITAT AND RANGE
Found along stream- and riverbanks and in canopy gaps and old fields. More likely to establish on disturbed, well-watered ground in sun to part shade. Reported as invasive from Pennsylvania south to North Carolina and west into Kentucky and Tennessee, but grows from southern Canada south to Georgia and west to Illinois.

WHAT IT DOES IN THE ECOSYSTEM
Forms dense thickets that shade out existing plants and prevent establishment of new plants. Can occupy the same habitats as native spiraeas. Seeds often spread along waterways and with movement of fill dirt.

The pink flowers and lance-shaped, toothed leaves of Japanese spiraea.

Japanese spiraea blooms in early summer.

HOW IT CAME TO NORTH AMERICA

Introduced from Asia around 1870 to the northeastern United States as an ornamental plant. It continues to be popular among gardeners as an easy perennial that grows in many soil types.

MANAGEMENT

Cutting will control the spread of spiraea but will not kill it. Glyphosate or triclopyr can be used as a foliar spray or painted on cut stumps to kill plants.

FOR MORE INFORMATION

Remaley, T. "Japanese spiraea." Plant Conservation Alliance Alien Plant Working Group. http://www.nps.gov/plants/alien/fact/spja1.htm.

Terrestrial Plants—Trees and Shrubs— Deciduous—Opposite or Whorled leaves—Trees

Norway Maple   *Acer platanoides*

NAME AND FAMILY
Norway maple (*Acer platanoides* L.); maple family (Aceraceae). Many species of maples grow in the United States and Canada. Sycamore maple (*Acer pseudoplatanus*) and Amur maple (*A. ginnala*) are also reported as invasive in parts of the Northeast and Midwest.

IDENTIFYING CHARACTERISTICS
A spreading shade tree, Norway maple generally grows to be 40 to 50 ft. (12 to 15 m) tall. The dark green leaves, 4 to 7 in. (10 to 18 cm) long, positioned opposite each other, have five sharply pointed lobes. Leaves turn yellow in fall. The trees produce copious quantities of winged seeds in a broad V shape with one seed in each wing. Seeds mature in late summer but often remain on the tree into winter. The bark of mature trees is gray brown, forming long, narrow, interlacing ridges. Sycamore maple is a tall tree with leaves that appear to have three toothed lobes (they actually have five lobes, but the lowermost

Bottom: *Prolific seeds have a broad V shape.* **Right:** *Typical long, interlacing ridges of mature bark.*

Left: *Norway maple leaves have five sharply pointed lobes.* Right: *Norway maple trees are often planted as shade trees.*

lobes are very small). Amur maple generally has small three-lobed leaves with the middle lobe much longer than the side lobes. It grows as a small tree or shrub. Norway maple can be distinguished from native maples by its leaf shape and milky sap that exudes from broken leaves or stems.

HABITAT AND RANGE
Norway maple seeds itself in Ontario, Newfoundland, and Quebec, and from Maine to North Carolina and west to Wisconsin. It also occurs in Washington and Idaho. Norway maple seedlings quickly establish in deciduous forests. It is most often found in urban and suburban natural areas and urban woodlots.

WHAT IT DOES IN THE ECOSYSTEM
Norway maple casts heavy shade and its shallow, dense root system makes it difficult for other plants to establish. Generally only Norway maple seedlings can grow under mature Norway maple trees. Forests with Norway maple in the canopy have much lower plant species diversity. Norway maple, because of its shallow roots, is also subject to blowdowns in strong storms. Since it tolerates poor soils and air pollution, it is often the dominant urban tree, either by choice or chance. The wood has been used in musical instruments, including the backs of Stradivarius violins.

HOW IT CAME TO NORTH AMERICA
Norway maple is native from southern Scandinavia to northern Iran. Botanist John Bartram introduced it into his famous botanical garden in Philadelphia in 1776 and sold many for ornamental shade trees. Norway maple quickly became

popular for its fast growth and deep shade and is still widely sold throughout the United States. In the 1930s and 1940s, when the avenues of American towns lost their towering aisles of elms to Dutch elm disease, Norway maples were often planted as replacements.

MANAGEMENT

Norway maple seedlings can be pulled up by hand and saplings removed with a weed wrench. The soil disturbance can result in new seedling establishment, however. Adult trees can be cut down. Stumps may resprout unless painted with triclopyr. Small trees and resprouts can be sprayed with a solution of glyphosate in summer. Basal bark treatment with triclopyr is effective in summer or winter.

FOR MORE INFORMATION

Galbraith-Kent, S. L. and S. N. Handel. 2008. "Invasive *Acer platanoides* inhibits native sapling growth in forest understorey communities." *Journal of Ecology* 96:293–302.

Kloeppel, B. D., M. and D. Abrams. 1995. "Ecophysiological attributes of the native *Acer saccharum* and the exotic *Acer platanoides* in urban oak forests in Pennsylvania, U.S.A." *Tree Physiology* 15:739–746.

Nowak, D. J., R. and A. Rowntree. 1990. "History and range of Norway maple." *Journal of Arboriculture.* 16(11):291–296.

Webb, S. L., T. H. Pendergast, IV, M. and E. Dwyer. 2001. "Response of native and exotic maple seedling banks to removal of the exotic, invasive Norway maple (*Acer platanoides*)." *Journal of the Torrey Botanical Society* 128:141–149.

Royal Paulownia   *Paulownia tomentosa*

NAME AND FAMILY

Royal paulownia, princess tree, empress tree (*Paulownia tomentosa* [Thunb.] Sieb. & Zucc. ex Steud.); figwort family (Scrophulariaceae). Leaves look similar to catalpa (*Catalpa bignoniodes* and *C. speciosa*).

Mature capsules contain thousands of seeds.

Violet flowers bloom in upright clusters in spring.

IDENTIFYING CHARACTERISTICS

Paulownia reveals itself in its large, heart-shaped, velvety leaves and the pale, violet spring flowers that blossom in 8 to 12 in. (20 to 30 cm) long, upright clusters. Flowers are 1.5 to 2 in. (3.7 to 5 cm) long, tube-like. Leaves are opposite pairs on the stems, 5 to 12 in. (12.7 to 30 cm) long. Flowers yield a four-segment capsule that shelters thousands of small, winged seeds. Capsules stay attached during the winter; smaller rounded flower buds are also visible in winter. New branches and stems are greenish to brown, and flattened where stems and branches join. Bark is thin with shallow creases. Second-year branches are pithy to hollow inside. Trees grow very rapidly, taking root even in sidewalk

Royal paulownia grows along forest edges and in disturbed areas.

Paulownia has large, velvety leaves.

crevices, old chimney cracks, and rocky soils. Trees can grow to 50 ft. (15 m) tall and 2 ft. (0.6 m) in diameter. Catalpa tree leaves are sparsely hairy on the upper surface and rough and hairy on the undersides, and they form long cigar-shaped pods that hang down in late summer.

HABITAT AND RANGE
Paulownia grows from Maine and southern Canadian provinces to Texas and also in the Pacific Northwest. Its copious seeds spread on the wind and new trees spring up readily in any available soil, especially in disturbed areas, clearcuts, burns, roadsides, and storm blowdowns.

WHAT IT DOES IN THE ECOSYSTEM
Paulownia can grow more than 15 ft. (4.6 m) a year and sends up new shoots from root sprouts. Millions of seeds from a single tree give paulownia wide reproductive options. The seeds germinate quickly. In small riparian areas and in fire-dependent exposed sites that harbor rare native plants, paulownia can be serious competition. It is estimated to cover more than 23,000 acres in the southeast.

HOW IT CAME TO NORTH AMERICA
The Dutch East India Company brought paulownia to Europe in the 1830s from its native China or Japan. Despite its origin, it is named for Russia's Czar Paul I's daughter Anna Paulovna. The easily dispersed seeds quickly spread the

tree into compatible niches. The wood has great value in Japan and some farmers and foresters cultivate stands for international trade. It is also sold in the nursery trade as an extremely fast-growing shade tree.

MANAGEMENT
Hand pulling of young plants should be done when soil is wet so the entire root system can be removed. Trees can also be cut or mowed before flowering and then any resprouts mowed again or sprayed. Larger trees can be killed by a girdling cut that extends deeper than the bark, but stumps are likely to produce sprouts that must be cut or sprayed. Foliage can be sprayed with herbicides containing glyphosate or triclopyr, or cut stumps can be painted with triclopyr.

FOR MORE INFORMATION
Innes, Robin J. 2009. "*Paulownia tomentosa.*" *Fire Effects Information System.* U.S. Department of Agriculture, Forest Service, Rocky Mountain Research Station, Fire Sciences Laboratory. http://www.fs.fed.us/database/feis/.

Niemeier, J. 1984. "I had to kill the empress." *Arbor Bulletin* 47(2):21–23. Arbor Foundation. Seattle: University Washington.

Rehder, M. A. 1927. *Manual of Cultivated Trees and Shrubs.* New York: MacMillan Co.; reprint, Portland, OR: Dioscorides Press, 1983.

Williams, C. E. 1983. "The exotic empress tree, *Paulownia tomentosa:* An invasive pest of forests." *Natural Areas Journal* 13(3):221–222.

Amur Cork Tree   *Phellodendron amurense*

NAME AND FAMILY
Amur cork tree (*Phellodendron amurense* Rupr.); citrus family (Rutaceae). Several other species of cork tree have also been introduced to the United States and all look very similar.

IDENTIFYING CHARACTERISTICS
Cork trees are named for their corky, furrowed, gray bark. Amur cork tree is a deciduous tree with a short trunk and wide spreading branches growing to 35 to 45 ft. (10 to 14 m) tall. The leaves occur opposite along the stems and are divided into 5 to 11 leaflets that smell of turpentine when crushed. The elliptical, pointed leaflets are 2.5 to 4.5 in. (6 to 11 cm) long, tapering to very pointed tips. The

Corky bark of Amur cork tree.

whole leaf is 10 to 15 in. (25 to 38 cm) long. Male and female flowers occur on separate trees. Bunches of small greenish flowers bloom in late spring, followed on female plants by clusters of green pea-sized fruits that mature to black. Fruits remain on the trees into winter.

HABITAT AND RANGE
The Amur cork tree prefers full sun and rich soils, and it is tolerant of drought and flooding. Found in urban and suburban forests and along roadsides from Massachusetts to Virginia and in Illinois and Ohio. Trees are not tolerant of very cold winters.

WHAT IT DOES IN THE ECOSYSTEM
Trees have mostly been noted as aggressively invading disturbed forests near urban and suburban areas around New York and Philadelphia. There they displace native plants. Birds, squirrels, and other animals eat the fruits.

Left: *Pea-sized fruits found on female trees can persist into winter.* **Right:** *Leaves are divided into five to eleven leaflets.*

HOW IT CAME TO NORTH AMERICA

Introduced in 1856 from eastern Asia and used throughout the Southeast and Mid-Atlantic states. Fruitless male cultivars are available for landscaping, although they can still pollinate females. Amur cork tree is a native of the northern China Amur valley in the Manchuria region as well as Japan. Its chemical compounds made it popular for treating a variety of ailments, from diarrhea to prostate enlargement.

Trees have short trunks and spreading branches.

MANAGEMENT

Seedlings can be hand pulled. Control can focus on removing female plants to reduce seed production. Large trees can be cut and stumps treated with a systemic herbicide.

FOR MORE INFORMATION

Glaeser, W., and D. Kincaid. 2005. "The non-native invasive *Phellodendron amurense* Rupr. in a New York City woodland." *The Arboricultural Journal* 28(3):149–150.

Martin, T. 2003. "Weed alert! *Phellodendron amurense* Rupr." The Nature Conservancy. http://www.invasive.org/gist/alert/alrtphel.html.

Simons, D. 2009. "Amur Corktree." Plant Conservation Alliance's Alien Plant Working Group Least Wanted. http://www.nps.gov/plants/alien/fact/pham1.htm.

Terrestrial Plants—Trees and Shrubs—Deciduous—Opposite or Whorled Leaves—Shrubs

Butterfly Bush   *Buddleja davidii*

NAME AND FAMILY
Butterfly bush, buddleia, summer lilac (*Buddleja davidii* Franchet); butterfly bush family (Buddlejaceae). Several other buddleias, including smokebush (*B. madagascariensis*) in Florida and Lindley's butterfly bush (*B. lindleyana*) in the southeastern United States, are considered invasive locally. The name *Buddleja davidii* refers to two of Europe's greatest naturalists, both clerics. The first species of butterfly bush came to Europe from French missionary and naturalist Father Armand David in 1869. Father David is most famous for describing an unusual Asian deer, now called Père David's deer. The name *buddleja* was

Butterfly bush flowers resemble those of lilacs, but bloom in summer.

Leaves grow in pairs along the stems.

bestowed on the species *globosa*, another plant in this group, by Linnaeus in honor of the English naturalist Reverend Adam Buddle (1665–1715).

IDENTIFYING CHARACTERISTICS
Many people recognize this bush from gardens, where it is called summer lilac because the flower clusters resemble those of lilacs but bloom in summer. This deciduous woody shrub grows 3 to 15 ft. (1 to 5 m) tall with arching branches. Dense clusters of four-petaled flowers, tubular at the base, can be white to pink to many shades of purple. They are usually purple with orange centers. Fragrant flowers produce lots of nectar attractive to butterflies and other insects. Each flower produces many small seeds held in a capsule. Velvety, often gray-green lance-shaped leaves with toothed or wavy edges are usually arranged opposite along the stem. Smokebush has orange flowers and fleshy white fruits that mature to dark blue. Lindley's butterfly bush has purple flowers without an orange center and dark-green leaves that are shiny on the upper sides.

HABITAT AND RANGE
Colonizes riparian areas and disturbed sites such as in urban areas. It prefers dry, sunny sites or well-drained soils. Ranges from British Columbia south into California and in the eastern United States from southern New England and Michigan south into Georgia.

Plants escape into natural areas, preferring sunny sites with well-drained soils.

WHAT IT DOES IN THE ECOSYSTEM

Butterfly bush can form dense stands, particularly along streams or wetland edges, blocking light to understory plants and new growth. In Oregon it is displacing willows along suburban Lake Oswego. It has a fast growth rate and can produce seeds after only one year. It is short lived, however, and densities decline after 10 to 20 years. Wind-and water-dispersed seeds spread the shrub to new sites. Also considered invasive in western Europe and a major problem in New Zealand. It does provide copious nectar for butterflies and other pollinators.

HOW IT CAME TO NORTH AMERICA

Buddleja davidii came to North America around 1900 and is native to southwestern China. Most garden varieties are from seeds collected by E. H. Wilson between 1900 and 1908 for the Arnold Arboretum in Massachusetts.

MANAGEMENT

Plants do not spread vegetatively, so they can be pulled or dug out. Cut plants will resprout, but cut stumps can be painted with glyphosate or triclopyr or the resprouts can be sprayed. In gardens, cutting off the seed heads before seeds are dispersed will limit its spread.

FOR MORE INFORMATION

Richard, S. 1996. "*Buddleia davidii*." In *Invasive Plants: Weeds of the Global Garden,* edited by J. M. Randall and J. Marinelli, 48. Brooklyn, NY: Brooklyn Botanic Garden.
Tallent-Halsell, M. G. and M. S. Watt. 2009. "The invasive *Buddleja davidii* (butterfly bush)." *Botanical Review* 75:292–325.

Surinam Cherry   *Eugenia uniflora*

NAME AND FAMILY
Surinam cherry, pitanga, cayenne cherry (*Eugenia uniflora* L.); myrtle family (Myrtaceae). Several native species of *Eugenia* look similar to Surinam cherry.

IDENTIFYING CHARACTERISTICS
The edible fruits of this shrub or small tree (grows to 30 ft. [9 m]) turn red and juicy when ripe. Each round fruit has eight deep longitudinal grooves and contains one to two light-brown, rounded seeds. Fruits are about 1.5 in. (3.7 cm) wide. Opposite, smooth-edged leaves are shiny, dark green above and paler below, and 1 to 2.5 in. (2.5 to 6.2 cm) long and 1.5 in. (3.7 cm) wide. Young leaves are often a coppery pink color. Fragrant white, four-petaled flowers with many stamens bloom in clusters of one to three from the leaf axils. It flowers and fruits primarily in spring, but sometimes again in fall. Native Eugenias will always have clusters of more than one flower and the fruits are smaller.

HABITAT AND RANGE
Grows in central and southern Florida. Prefers the fertile, moist soils and partial shade typical of hardwood islands.

Left: *Edible fruits have deep grooves like an indented beach ball.* Bottom: *Four-petaled white flowers of Surinam cherry bloom mainly in spring.*

Glossy leaves occur in pairs along the stem.

Surinam cherry grows aggressively in southern Florida.

WHAT IT DOES IN THE ECOSYSTEM

Surinam cherry forms thickets in the hardwood hammocks of southern Florida, where it keeps native vegetation from growing. Fruits are relished by birds and probably also eaten by small mammals. Is a host for Mediterranean fruit fly, a serious agricultural pest of fruit crops.

HOW IT CAME TO NORTH AMERICA

The attractive and edible fruits of this native of South American tropical areas appealed to Portuguese voyagers traveling around the world in the sixteenth century and the plant quickly found homes in India, the Phillipines, and the Caribbean Islands. Surinam cherry is cultivated in many tropical countries. Israeli farmers started planting it for its spring fruits in 1922. Introduced to Florida before 1930, it became a common hedge plant and fruits sometimes appeared in Miami markets.

MANAGEMENT
Small plants can be pulled out. Larger plants can be cut and the stumps treated with a systemic herbicide.

FOR MORE INFORMATION
Langeland, K. A., H. M. Cherry, C. M. McCormick, and K. A. Craddock Burks et al. 2008. *Identification and Biology of Nonnative Plants in Florida's Natural Areas,* 2nd ed. SP 257. Gainesville, FL: University of Florida IFAS.

Burning Bush   *Euonymus alatus*

NAME AND FAMILY
Burning bush, winged euonymus (*Euonymus alatus* [Thunb.] Siebold); staff-tree family (Celastraceae). Resembles the endangered native burning bush, *E. atropurpureus*.

IDENTIFYING CHARACTERISTICS
Called "burning bush" because leaves turn a brilliant red in fall, and "winged" euonymus because four corky wings often line the green stems. Leaves, usually less than 2 in. (5 cm) long, grow opposite along the stems and are elliptical with narrowed points on both ends. Leaf edges are finely toothed. Inconspicu-

Left: *Bright-red fruits attract birds.* **Right:** *Burning bush gets its name from its red fall leaves.*

Left: *Inconspicuous flowers lie flat against leaves in spring.* **Right:** *Branches are often winged.*

ous flowers bloom in spring, each one greenish yellow with four petals. Smooth, red-purple fruits mature in late summer containing four red to orange seeds, which are eaten by birds. Bushes grow to 12 ft. (3.6 m) with broad spreading branches. Native burning bush leaves have hairy undersides and longer leaf stems (petioles) compared to the hairless leaves of the nonnative.

HABITAT AND RANGE
Occurs from southern Ontario and New England south to northern Florida and west to Illinois. Grows well in many soil types and is shade tolerant. Often found in open woods, disturbed lands, early stages of forest growth, and floodplains.

WHAT IT DOES IN THE ECOSYSTEM
Winged euonymus naturalizes in woodlands where it can establish dense stands, outcompeting other understory plants for light and space. Its most competitive colonies are found from New England south to Virginia. Birds feed on the fruits and disperse the seeds.

HOW IT CAME TO NORTH AMERICA
Introduced to the United States in the 1860s as an ornamental plant from northeastern Asia. A very popular landscape plant, burning bush has many cultivars that are widely planted. Horticulturists are working on developing sterile cultivars.

MANAGEMENT
Seedlings are easily hand pulled and larger shrubs can be pulled out with a weed wrench or dug out. For larger shrubs or infestations, cut stems back and paint stumps with glyphosate, or spray foliage with glyphosate in early summer.

FOR MORE INFORMATION

Ebinger, J. E. 1983. "Exotic shrubs a potential problem in natural area management in Illinois." *Natural Areas Journal* 3:3–6.

Fryer, Janet L. 2009. *"Euonymus alatus." Fire Effects Information System.* U.S. Department of Agriculture, Forest Service, Rocky Mountain Research Station, Fire Sciences Laboratory. http://www.fs.fed.us/database/feis/.

Martin, T. 2000. "Weed alert *Euonymus alatus* (Thunb.) Siebold." The Nature Conservancy. http://www.invasive.org/gist/alert/alrteuon.html.

Shrub Honeysuckle   *Lonicera* spp.

NAME AND FAMILY

Amur honeysuckle (*Lonicera maackii* [Rupr.] Herder); honeysuckle family (Caprifoliaceae). Several species of shrubby honeysuckles have invaded North America. In addition to Amur honeysuckle, these include Morrow's honeysuckle (*L. morrowii*), Tatarian honeysuckle (*L. tatarica*), and Bell's honeysuckle (*L.* x *bella*), a cross between Morrow's and Tatarian honeysuckles. Several native shrub honeysuckles can be distinguished from the exotics by their blue or black berries or their flower characteristics.

IDENTIFYING CHARACTERISTICS

All of these exotic honeysuckles have opposite leaves and flower in spring. Amur honeysuckle can be distinguished because its leaves end in a long point. White to pink flowers fade to yellow and produce red fruits. It tends to grow taller (up to 20 ft. [6 m]) than the other species, which average 10 ft. (3 m). Morrow's honeysuckle has oblong, gray-green leaves that are covered with fine

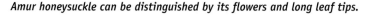

Amur honeysuckle can be distinguished by its flowers and long leaf tips.

hairs on the underside. The flowers are white fading to yellow and are also covered with fine hairs. The fruits are red. Tatarian honeysuckle has leaves similar in shape to those of Morrow's honeysuckle, but they are smooth, not hairy. The flowers are white-pink fading to yellow and the fruits are red or sometimes yellow. The hybrid Bell's honeysuckle has characteristics in between those of its parents—slightly hairy leaves and flowers.

HABITAT AND RANGE
Amur honeysuckle occurs in 24 states in the eastern and central United States and in Ontario, Canada. The other honeysuckles have a similar range. They occupy old fields, forest edges, thickets, floodplains, and maritime forests. Amur honeysuckle is particularly aggressive on calcareous soils typical of serpentine grasslands, shrub fens, and shale barrens. They are tolerant of shade and sun, wet and dry sites.

WHAT IT DOES IN THE ECOSYSTEM
The exotic honeysuckles tend to leaf out earlier in spring than many native shrubs and hold their leaves later into the fall, which probably aids in their competitive ability. The shrubby honeysuckles severely limit the ability of forests to regrow, reducing seedling establishment and herb species richness in forests. Robins experienced higher nest predation in Amur honeysuckle shrubs in Chicago, perhaps because the architecture of the shrub caused the birds to build their nests closer to the ground. Amphibians declined as honeysuckle increased in a study in Missouri woodlands. Deer browse honeysuckle, sometimes eliminat-

Red fruits of Amur honeysuckle.

Amur honeysuckle quickly grows into a tall thicket.

ing it. In places where Amur honeysuckle is the only shrub remaining, its fruits appear to be an important winter food source for birds and deer mice.

HOW IT CAME TO NORTH AMERICA
Because of its attractive flowers, Amur honeysuckle arrived at the Dominion Arboretum in Ottawa, Canada, in 1896 and at the New York Botanical Garden in 1898. It is native to central and northeastern China, Manchuria, and Korea. The U.S. Department of Agriculture promoted its use for wildlife and as a shelterbelt and commercial nurseries sell it is an ornamental.

MANAGEMENT
Shrubs can be pulled or dug out of the ground or repeatedly cut in spring and fall. Using a flame weeder on the root crown two or more times can kill shrubs. Control using mechanical means may take three to five years, especially for large shrubs in sunny areas. Herbicides sprayed on foliage late in the growing season or painted on cut stumps from summer through winter will kill the shrubs. Goats will browse honeysuckle, eventually killing it. Heavy deer populations also severely limit regrowth.

FOR MORE INFORMATION
"Bush Honeysuckles Control." Missouri Department of Conservation. http://mdc.mo.gov/landwater-care/invasive-species-management/invasive-plant-management/bush-honeysuckles-control.

"Bush honeysuckles." *Southeast Exotic Pest Plant Council Invasive Plant Manual*. http://www.se-eppc.org/manual/bushhoney.html.

Luken, J. O., and J. W. Thieret. 1996. "Amur honeysuckle: Its fall from grace." *BioScience* 46 (1):18–24.

Watling, J. I., C.R. Hickman, J. L. Orrock. 2011. "Invasive shrub alters native forest amphibian communities." *Biological Conservation* 144: 2597–2601.

Fragrant Honeysuckle   *Lonicera fragrantissima*

NAME AND FAMILY
Fragrant honeysuckle, sweet breath of spring, January jasmine, winter honey-suckle (*Lonicera fragrantissima* Lindl. & Paxton); honeysuckle family (Caprifo-liaceae). Another introduced shrub honeysuckle, Standish's honeysuckle (*L. standishii* Jacques) also flowers early in spring.

IDENTIFYING CHARACTERISTICS
Flowering as early as January to February before the leaves come out, fragrant honeysuckle's small (0.5 in. [1.3 cm]), white, tubular, five-petaled white flowers are indeed sweetly fragrant, variously described as lemony or fruit loop–scented. Red fruits mature in late spring or early summer and are held out of sight under the leaves. Opposite, bluish-green leaves are rounded (1 to 3 in. [2.5 to 7.6 cm] long and wide) and have hairy leaf edges and bristles on the underside of the leaf along the midrib. Leaves may stay green well into winter. Grows 6 to 10 ft. (1.8 to 3 m) tall with arching branches. Young stems often have a purplish color. Standish's honeysuckle also flowers in winter but has longer, more pointed leaves (2 to 4.5 in. [5 to 11.2 cm] long) with hairs on the undersides of the leaves.

HABITAT AND RANGE
Fragrant honeysuckle grows from New York south to Georgia and Louisiana. Grows in full sun to partial shade, naturalizing in open woodlands, thickets, forest edges, old fields, and logged areas.

Left: *Red fruits hang below the leaves in fall.* **Right:** *Leaves have hairy edges and can stay green long into winter.*

Left: *Fragrant honeysuckle can be found in open woodlands.* Right: *The very fragrant flowers bloom early in spring.*

WHAT IT DOES IN THE ECOSYSTEM

This very cold-resistant plant can form dense thickets that keep native species from establishing. The fruits are eaten by many bird species, which disperse the seeds.

HOW IT CAME TO NORTH AMERICA

Introduced to Europe from China by Scottish explorer Robert Fortune in 1845 and soon after found in North American gardens.

MANAGEMENT

Shrubs can be pulled or dug out of the ground or repeatedly cut in spring and fall. Control using mechanical means may take three to five years, especially for large shrubs in sunny areas. Herbicides sprayed on foliage late in the growing season or painted on cut stumps from summer through winter will kill the shrubs.

FOR MORE INFORMATION

Dirr, M. A. 1998. *Manual of Woody Landscape Plants.* Champaign, IL: Stipes Publishing.
Glenn, S. D. 2005. "*Lonicera fragrantissima* fragrant honeysuckle." *New York Metropolitan Flora.*
   Brooklyn Botanic Garden. http://nymf.bbg.org/profile_species_tech.asp?id=337# description.

Common, Glossy, and Dahurian Buckthorn   *Rhamnus* spp.

NAME AND FAMILY
The three invasive buckthorns are common buckthorn (*Rhamnus cathartica* L.), glossy buckthorn (*Frangula alnus* P. Mill. or *R. frangula* L.), and dahurian buckthorn (*R. davurica* Pallas); buckthorn family (Rhamnaceae). Common and glossy buckthorns occur most frequently. Several native buckthorns also live in North America and can be confused with the exotics.

IDENTIFYING CHARACTERISTICS
Buckthorns grow as shrubs or small trees, and the leaves have distinctly up-curved veins. Clusters of small, four-petaled yellow-green flowers appear in spring, with male and female flowers on separate plants. Common buckthorn can be distinguished from other buckthorns by the spines it often has at the tips of the twigs. It has small black fruits and the leaves are smooth on both the upper and lower surfaces. The leaves are finely toothed and arranged alternately or nearly oppositely. Dahurian buckthorn is similar in appearance, but the leaves are more lance shaped and the leaves turn brown in fall. Common and glossy buckthorns keep their green leaves late into fall. Glossy buckthorn has a shiny upper leaf surface and a hairy lower leaf surface. The leaves are not toothed and are alternate and more rounded at the tips than those of common buckthorn. The fruits turn from red to black. The nonnative buckthorns grow

Left: *Glossy buckthorn fruits mature in fall.* Right: *Common buckthorn in a forest understory.*

Glossy buckthorn flowers bloom in the leaf axils.

20 to 25 ft. (6 to 8 m) tall. The native Carolina buckthorn (*R. caroliniana*) looks similar to glossy buckthorn but has toothed leaves that are smooth on both sides. Alder buckthorn (*R. alnifolia*) seldom grows more than 3 ft. (1 m) tall, and lance-leaved buckthorn (*R. lanceolata*) has narrow leaves.

HABITAT AND RANGE

Common buckthorn grows from Nova Scotia to Saskatchewan south to Missouri and Virginia. It typically grows in woodlands and abandoned fields, and is particularly a problem in open woodlands and woodland edges and in thickets within prairies. Glossy buckthorn's range is from Nova Scotia to Manitoba and south to Illinois and New Jersey. It prefers wetland communities like marshes, fens, and bogs but does invade upland sites with moderate moisture, including fields and roadsides. Dahurian buckthorn has a more limited distribution, occurring from southern New England into the Midwest and potentially as far south as North Carolina. It grows in communities similar to those favored by common buckthorn.

WHAT IT DOES IN THE ECOSYSTEM

Buckthorns form dense thickets under which few other plants can grow. Birds and mammals eat the fruits, but a study in Ohio showed European starlings to

Common buckthorn leaves are edged with shallow teeth.

be a primary disperser. In prairies, the shrubs may reduce or eliminate fires by discouraging undergrowth. Buckthorn leaf litter's high nitrogen content stimulates nonnative earthworms to break down all leaf litter, changing the quantity of leaf litter left on the ground and increasing levels of nitrogen in the soil. These changes may favor further establishment of buckthorn and discourage establishment of native species adapted to prior soil conditions. Common buckthorn is an alternate host plant for oat rust, a fungus damaging to oat crops, and the soybean aphid overwinters on it.

HOW IT CAME TO NORTH AMERICA

Buckthorns were introduced as ornamental plants in the early 1800s and were widely planted as fencerows and for wildlife habitat.

MANAGEMENT

In prairies maintained by fire, fire can be used to kill seedlings and eventually kill mature plants. Repeated cutting or mowing will also reduce growth.

The soil disturbance caused by digging or pulling trees out may cause seeds in the soil to sprout. Systemic herbicides like triclopyr can be sprayed on foliage in late summer and fall or painted on cut stumps.

FOR MORE INFORMATION

Catling, P. M., and Z. S. Porebski. 1994. "The history of invasion and current status of glossy buckthorn, *Rhamnus frangula*, in southern Ontario." *Canadian Field-Naturalist* 108:305–310.

Converse, C. 2007. "Element stewardship abstract for *Rhamnus cathartica, Rhamnus frangula* (syn. *Frangula alnus*)." The Nature Conservancy. http://wiki.bugwood.org/Rhamnus_cathartica (accessed December 7, 2011).

Heidorn, R. 1991. "Vegetation management guideline: Exotic buckthorns—common buckthorn (*Rhamnus cathartica* L.), glossy buckthorn (*Rhamnus frangula* L.), Dahurian buckthorn (*Rhamnus davurica* Pall.)." *Natural Areas Journal* 11:216–217.

Klionsky, S. M., K. L. Amatangelo and D. M. Waller. 2010. "Above- and below-ground impacts of common buckthorn (*Rhamnus cathartica*) on four native forbs." *Restoration Ecology* 19:728–737.

Knight, K. S., J. S. Kurylo, A. G. Endress, J. R. Stewart and P. B. Reich. 2007. "Ecology and ecosystem impacts of common buckthorn (*Rhamnus cathartica*): a review." *Biological Invasions* 9:925–937.

Jetbead   *Rhodotypos scandens*

NAME AND FAMILY

Jetbead, black jetbead, white kerria, jetberry bush (*Rhodotypos scandens* [Thunb.] Makino); rose family (Rosaceae).

IDENTIFYING CHARACTERISTICS

Jetbead became a favorite of landscapers in the early twentieth century for its distinctive four-petaled single white flowers blooming in spring that are followed by blood-red berries that turn black, hanging usually in fours and lasting well into winter. The 2 in. (5 cm) wide flowers open in spring against the back-

ground of dense green leaves. Opposite leaves are generally oval, with a long pointed tip, and have distinct toothed edges and ribbed veins. Leaves grow 2.5 to 4 in. (6.2 to 10 cm) long and half as wide. Jetbead grows as a multistemmed bush up to 6 ft. (1.8 cm) high and equal in width. Stems are green when young, turning reddish brown with gray streaks and orange lenticels when older.

Right: *Jetbead's black, shiny fruits give the plant its name.* Bottom: *Jetbead leaves have prominent veins.*

Leaves grow in pairs along the stems.

White, four-petaled flowers bloom in spring.

HABITAT AND RANGE

Jetbead grows from Massachusetts and Connecticut west to Michigan and Illinois and south to Tennessee. Jetbead is a hardy shrub that grows in full sun or partial shade and a wide range of soils, including poor and heavily compacted soils.

WHAT IT DOES IN THE ECOSYSTEM

Jetbead's adaptability gives it the potential to be a troublesome invasive, but it is not a major competitor in any large areas. It can form dense cover that inhibits native plants from growing. It spreads by seeds and vegetative means. The attractive berries might tempt some people to eat them, but eating the berries risks death since human intestinal secretions turn natural amygdalin from the seeds into hydrogen cyanide. The result can be spasms, convulsions, coma, and finally respiratory failure. Fruits are, however, eaten by birds.

HOW IT CAME TO NORTH AMERICA
Jetbead's native territory is temperate climates of eastern Asia from central China to Japan. Because of its distinct flowers and durable berries, horticulturalists bought it to the United States in 1866. It became very popular among gardeners and landscapers in the early twentieth century.

MANAGEMENT
Small plants can be hand pulled or dug out to prevent reproduction from root pieces. More mature plants and colonies can be killed by cutting close to the ground in fall and using herbicides containing glyphosate to kill new growth in the spring. Triclopyr herbicides can be painted on the cut stumps.

FOR MORE INFORMATION
Albrecht, L. A. 2001. "Jetbead: a new invasive threat." *Northeastern Weed Science Society Newsletter,* April 2001:7. http://www.newss.org/docs/newsletter/2001_nl_apr.pdf.
Dirr, M. A. 1998. *Manual of Woody Landscape Plants.* Champaign, IL: Stipes Publishing.

Linden Viburnum   *Viburnum dilatatum*

NAME AND FAMILY
Linden viburnum, arrowwood linden (*Viburnum dilatatum* Thunb.); honeysuckle family (Caprifoliaceae). Other invasive viburnums include European cranberrybush or Guelder rose (*V. opulus* L.), Japanese snowball (*V. plicatum*), siebold viburnum (*V. sieboldii*), and wayfaring-tree (*V. lantana*).

IDENTIFYING CHARACTERISTICS
Flat, 3 to 5 in. (7.6 to 12.7 cm) wide, disklike clusters of very small white flowers cover this viburnum in late spring and/or early summer. Cherry-red fruits of late summer shrivel into raisins that last through winter unless consumed by birds. Linden viburnum leaves are oval, 3 to 5 in. (7.6 to 12.7 cm) long, with toothed edges, and surfaces corrugated by prominent veins and covered by fine hairs above and below the leaf. Leaves grow opposite on twigs. In fall leaves turn a rich maroon color. The many-stemmed bush can reach 10 ft. (3 m) high and almost as wide. The densely hairy stems distinguish it from other viburnums. European cranberrybush has three-lobed leaves very similar to those of native American cranberrybush (*V. trilobum*), but the lobes of European cranberrybush leaves are more deeply indented.

The fruits of linden viburnum attract birds.

HABITAT AND RANGE

Linden viburnum grows in natural areas from New York to Virginia. Viburnums are found in wetlands, along streams, in open woods, and at forest edges.

WHAT IT DOES IN THE ECOSYSTEM

Viburnum can grow into thickets that exclude many native plants, but they are not exceptionally aggressive colonizers or usurpers. They spread rapidly into natural areas through seed dispersal by birds. Several species of birds feed on the berries, which are particularly attractive to cedar waxwings. Shrubs provide shelter for small birds.

HOW IT CAME TO NORTH AMERICA

Linden viburnum is a native of eastern Asia, particularly Korea, and was

Linden viburnum grows as a tall shrub in the eastern United States.

Clusters of five-petaled flowers bloom in spring.

introduced to the United States before 1845. European cranberrybush is sometimes listed as native, although it has also been imported from Eurasia and North Africa. Some viburnums are favored by horticulturalists and landscapers who have developed a variety of cultivars. *V. opulus* berries have been used to make jellies but are very acidic and can be toxic. In Turkey they are used to treat kidney stones. There are many native species of viburnums that can be planted instead of the nonnative species.

MANAGEMENT

Some viburnums are increasingly attacked by the nonnative viburnum leaf beetle, but linden viburnum is relatively resistant. Young plants can be dug or pulled out. Small shrubs can be sprayed with a systemic herbicide during the growing season. Viburnum can be cut near ground level and the stumps painted with glyphosate.

FOR MORE INFORMATION

Cornell University, Department of Horticulture. "Viburnum leaf beetle." http://www.hort.cornell.edu/vlb/.

Dirr, M. A. 1998. *Manual of Woody Landscape Plants.* Champaign, IL: Stipes Publishing.

Swearingen, J., B. Slattery, K. Reshetiloff, and S. Zwicker. 2010. *Plant Invaders of Mid-Atlantic Natural Areas,* 4th ed. Washington, DC: National Park Service and U.S. Fish and Wildlife Service. http://www.nps.gov/plants/alien/pubs/midatlantic/vidi.htm.

Beach Vitex

Vitex rotundifolia

NAME AND FAMILY
Beach vitex, chasteberry, Monk's pepper, or kolokolo kahakai (*Vitex rotundifolia* L. f.); verbena family (Verbenaceae). Nicknamed "beach kudzu." Two other species of vitex, lilac chaste-tree (*V. agnus-castus* L.) and simpleleaf chastetree (*V. trifolia* L.) are sometimes considered local invaders in the southern United States. Both grow as shrubs or small trees.

IDENTIFYING CHARACTERISTICS
This coastal deciduous shrub or woody vine with runners can sprawl more than 60 ft. (18 m) across the ground. Typically plants are 1 to 2 ft. (0.3 to 1 m) tall and form mounds 12 ft. (3.6 m) in diameter. Brittle stems root at the nodes but tend to break off during high tides, floating off to colonize new beaches. Oval, 1 to 2 in. (2.5 to 5 cm) long, silvery-gray leaves grow oppositely along the branches and have a spicy odor when crushed. Small clusters of violet-blue flowers bloom at the ends of the branches. Each flower is 1 in. (2.5 cm) wide. Fruits mature to a dark purple-black color in fall.

Clusters of violet-purple flowers bloom at the ends of the branches of beach vitex.

HABITAT AND RANGE
Grows in full sun and sandy soils, such as on sand dunes, in North and South Carolina, Virginia, and Maryland.

WHAT IT DOES IN THE ECOSYSTEM
Beach vitex forms dense mounds, outcompeting native dune plants, including some endangered species. Although the foliage traps sand, the roots do not hold sand in place well because they are not very fibrous. Beach vitex also threatens to overgrow loggerhead sea turtle and shorebird nesting areas. Plants spread by seeds and by stem fragments carried in water. North Carolina listed beach vitex as a noxious weed in 2009.

Left: *Silvery leaves grow in pairs along the branches.* **Right:** *Beach vitex stems sprawl over the ground.*

HOW IT CAME TO NORTH AMERICA

Beach vitex arrived in the U.S. as early as 1955, but was not introduced to the nursery trade until around 1985. It was used for dune restoration in the Carolinas after Hurricane Hugo in 1989 due to a shortage of native planting stock. It is used as a drought- and salt-tolerant landscape plant. Native to the Pacific Rim, including Hawaii. Fruits have been used since the Middle Ages as spice and to relieve premenstrual symptoms, but its effectiveness is questionable. The name chasteberry derives from the use of garlands of vitex to signify chastity. Legend says monks chewed it as a kind of medieval anti-Viagra to reduce sex drive. An oil named rotundial, extracted from the leaves, has shown promise as a mosquito repellent.

MANAGEMENT

Plants can be hand pulled or dug out. Herbicides are currently being tested to control beach vitex.

FOR MORE INFORMATION

Beach Vitex Task Force. http://www.beachvitex.org.

Cousins, M. M., J. Briggs, C. Gresham, J. Whetstone, and T. Whitwell. 2010. "Beach vitex (*Vitex rotundifolia*): an invasive coastal species." *Invasive Plant Science and Management* 3:340–345. http://www.gri.msstate.edu/lwa/invspec/beach_vitex_fs.pdf.

Terrestrial Plants—Vines— Evergreen—Alternate Leaves

Cape Ivy   *Delairea odorata*

NAME AND FAMILY
Cape ivy, German ivy, climbing groundsel (*Delairea odorata* Lem.); aster family (Asteraceae).

IDENTIFYING CHARACTERISTICS
This fleshy, perennial vine has leaves similar in shape to those of English ivy, but with five to nine lobes or points and a yellowish-green color. Leaves alternate along the stem and at the base of each leaf there are usually a pair of tiny ears, or stipules. Stems are green or purple and run both above and below the ground. From winter into early spring, the vines are covered with clusters of 20 or more yellow flower heads. Each dime-sized flower head tightly holds many florets that in turn form wind-dispersed seeds, but seeds in California are mostly sterile. Because a Cape ivy plant has to cross with a different individual to produce viable seeds, it is possible that all of the plants on the West Coast are from a single clone. Most of the growth of the vine is from February to June, and in areas with little water vines die back in summer.

Cape ivy's yellow flowers bloom in winter, but most seeds are sterile.

Leaves resemble those of English ivy, but with more points.

HABITAT AND RANGE

Occurs in coastal California and southern Oregon and Montana, and probably also in Washington state. Although it mainly grows in areas with moist soils such as coastal forests and streambanks, recently it has also been spreading to grasslands and shrublands in drier areas.

WHAT IT DOES IN THE ECOSYSTEM

Vines climb over other plants, creating mats 30 in. (75 cm) deep in some places that prevent sunlight from reaching the plants below. Native plant diversity declines by 30 to 50 percent in some areas. Because Cape ivy has shallow roots, it contributes to soil erosion on hillsides where it has displaced more deeply rooted species. Alkaloids in the stems and leaves make it unpalatable to herbivores, and the alkaloids may decrease fish survival in streams. Vines can root at nodes, and seeds are wind dispersed.

Cape ivy covers the ground in coastal California and southern Oregon.

HOW IT CAME TO NORTH AMERICA

Native to South Africa, Cape ivy was first introduced to the eastern United States in 1850 as an ornamental plant, and to California in 1950. So far it has not naturalized outside the West Coast, where it was first found in natural areas in the 1960s.

MANAGEMENT

Cape ivy is difficult to control because even small fragments can root. In some soils, mats of Cape ivy can be rolled up and bagged or the runners pulled out using a hand rake. Pieces of ivy left on the ground will resprout. A mixture of glyphosate, triclopyr, and a surfactant sprayed on foliage in late spring after flowering can be used to control infestations. Clopyralid has been successfully used in Australia.

FOR MORE INFORMATION

Alvarez, M. E. and J. H. Cushman. 2002. "Community-level consequences of a plant invasion: effects on three habitats in coastal California." *Ecological Applications* 12:1434–1444.

Bossard, C. "*Delairea odorata.*" In *Invasive Plants of California's Wildlands,* edited by C. C. Bossard, J. M. Randall, and M. C. Hoshousky. Berkley, CA: University of California Press. http://www.cal-ipc.org/ip/management/ipcw/.

Fagg, P. C. 1989. "Control of *Delairea odorata* (Cape ivy) in native forest with the herbicide clopyralid." *Plant Protection Quarterly.* 4:107–110.

English Ivy   *Hedera helix*

NAME AND FAMILY

English ivy (*Hedera helix* L.); ginseng family (Araliaceae). The invasives of this family come in several species and many cultivars. What many call English ivy can be a near look-alike and close relative, Irish ivy (*H. hibernica* [Kirchn.] Bean). English and Irish ivy are difficult to distinguish and taxonomists sometimes lump them as the same species. Persian ivy (*H. colchica*) is similar but has a larger leaf. Other plants called "ivy" are not in the same genus. These include the look-alike Boston ivy (*Parthenocissus tricuspidata,* Asian origin), which has a leaf made up of leaflets and loses its leaves in the winter. English ivy is not related to poison ivy (*Toxicodendron radicans,* North American) either.

IDENTIFYING CHARACTERISTICS

Most people know English ivy from buildings and landscaping, and of course, from the "Ivy League." The evergreen climbing vine has dark waxy leaves placed alternately along the stem. Leaves can take several forms but are generally three-lobed with a heart-shaped base. Plants can remain immature indefinitely, but when vines climb trees or buildings, sunlight triggers maturity.

English ivy produces small black fruits when it has a chance to climb.

Mature plants branch out and have unlobed, rhomboid-shaped leaves. Sunlight triggers the flowering of clusters of small, greenish-white flowers that in late summer form blackish fleshy fruits around one to several stonelike seeds. Ivy spreads by runners and also by bird-dispersed seeds.

HABITAT AND RANGE IN NORTH AMERICA

This native of Europe, western Asia, and northern Africa has spread across 26 states in the United States and occurs in southern British Columbia and southwestern Ontario, and is scattered in many other areas. It grows wherever it finds a temperate to

English ivy can harm trees if left to grow up the trunks.

subtropical environment with enough water. It is particularly abundant in the Pacific Northwest and along the east coast of the United States. In the Northwest, Irish ivy tends to be more common. Ivy prefers shady or semishady moist areas but is drought tolerant. It thrives in disturbed forests.

WHAT IT DOES IN THE ECOSYSTEM

English ivy grows along the ground and climbs on any plant or object in its path. The dense blanket blocks light and germination of other plants beneath it, reducing local plant diversity. It will grow several stories up the sides of buildings or into the canopy of a forest, adding weight to trees. Covering tree trunks, ivy can loosen bark and hold moisture against the tree trunk, encouraging fungus and decay. In some cases it provides shelter and food for birds. The flowers in late fall and through winter attract many varieties of bees, wasps, and other pollen gatherers. The berries are mildly toxic, though, so few bird species feed on them.

Leaves on climbing vines are more often spade shaped.

HOW IT CAME TO NORTH AMERICA

European colonists introduced ivy early in their settlement of North America, mainly as a decorative plant. The earliest record of English ivy in North America is from 1727. In cultivation it is fast growing and low maintenance. Ivy provides an effective, but monopolistic, ground cover, and can cover trellis and climb walls. On buildings, the rootlets eventually dig into masonry and on wooden siding the plant retains moisture, pries open joints, and encourages rot.

It also had some uses in medicine and tanning.

MANAGEMENT

Vines can be pulled up by hand or dug up. Vines on the ground can also be smothered by covering them with plastic or a thick layer of newspaper and mulch during the summer. Goats will also browse and kill English ivy. Climbing vines can be contained by cutting stems at a convenient height.

On shady ground, leaves are usually lobed.

Clear a 1- to 2-foot area of vines off of tree trunks and do not try to remove vines from high branches. Broadleaf herbicides will also kill ivy, but because of the waxy coating on the leaves, application to cut stems or spraying herbicide when new leaves are expanding in spring may be more effective. Since ivy can reproduce from cuttings or vines in contact with earth, do not leave cut vines on the ground.

FOR MORE INFORMATION

Clarke, M. M., S. H. Reichard, and C. W. Hamilton. 2006. "Prevalence of different horticultural taxa of ivy (*Hedera* spp., Araliaceae) in invading populations." *Biological Invasions* 8:149–157.

No ivy league, The. http://www.noivyleague.com/index.html.

Soll, J. 2005. *Controlling English ivy* (Hedera helix) *in the Pacific Northwest.* The Nature Conservancy. http://www.invasive.org/gist/moredocs/hedhel02.pdf.

Swearingen, J., and S. Diedrich. "English ivy." *Weeds gone wild.* Plant Conservation Alliance Alien Plant Working group. http://www.nps.gov/plants/alien/fact/hehe1.htm.

Terrestrial Plants—Vines— Evergreen—Opposite Leaves

Jasmines   *Jasminum* spp.

NAME AND FAMILY
Gold coast jasmine (*Jasminum dichotomum* Vahl.) and Brazilian jasmine (*Jasminum fluminense* Vell.); olive family (Oleaceae). Several other species of jasmine have naturalized in the United States but do not spread as rapidly as gold coast and Brazilian jasmine.

IDENTIFYING CHARACTERISTICS
The wonderful smell of jasmine emanates from white tubular flowers that grow in clusters from the leaf axils. The 1 in. (2.5 cm) long flowers open at night and have five to nine petals. Jasmine flowers most heavily in spring. Gold coast jasmine grows as an evergreen vine or spreading shrub. The oblong (2 to 4 in.

Gold coast jasmine flowers open at night against a backdrop of glossy green leaves.

[5 to 10 cm] long), glossy, leathery, pointed leaves grow opposite along the stems. Brazilian jasmine has compound leaves made of three oblong leaflets (each 1 to 1.5 in. [2.5 to 3.7 cm] long and 0.75 to 1 in. [1.9 to 2.5 cm] wide) and the young stems are hairy. The fruit is a fleshy, black, two-lobed, rounded berry.

HABITAT AND RANGE
Occurs in southern Florida where average minimum temperatures are above 35° F (1.7°C). Found in hardwood hammocks (tree islands) and on disturbed ground.

WHAT IT DOES IN THE ECOSYSTEM
Jasmines climb into trees, completely covering canopies and blocking light to plants below. Fruits are eaten and seeds dispersed by birds and raccoons.

HOW IT CAME TO NORTH AMERICA
Dr. David Fairchild, a renowned plant explorer, introduced more than 100,000 nonnative plants to the U.S., including both jasmine species, which he

planted in 1920 in Florida as ornamentals. (Fairchild also brought Japanese cherry to Washington, DC.) Most likely, Brazilian jasmine also had an earlier introduction around 1916. These jasmines are native to tropical Africa, but Brazilian jasmine was first described in Brazil, where it had been introduced by Portuguese explorers.

MANAGEMENT
Young plants can be hand pulled. Older vines can be cut at ground level and the cut stems treated with triclopyr herbicide. A solution of triclopyr in oil can be applied as a basal bark treatment to young vines.

FOR MORE INFORMATION
Hammer, R. L. 1996. "*Jasminum dichotomum, J. fluminense.*" In *Invasive Plants: Weeds of the Global Garden,* edited by J. M. Randall and J. Marinelli. Brooklyn Botanic Garden.

Langeland, K. A., H. M. Cherry, C. M. McCormick, and K. A. Craddock Burks et al. 2008. *Identification and Biology of Nonnative Plants in Florida's Natural Areas,* 2nd ed. SP 257. Gainesville, FL: University of Florida IFAS.

Japanese Honeysuckle   *Lonicera japonica*

NAME AND FAMILY
Japanese honeysuckle (*Lonicera japonica*); honeysuckle family (Caprifoliaceae). The nursery industry commonly sells an aggressive variety of Japanese honeysuckle called Hall's honeysuckle. There are several native honeysuckle vines that might at first glance resemble Japanese honeysuckle.

Left: *Leaves at the base of the vines are often lobed.*
Right: *Fragrant flowers begin to bloom in late spring.*

Japanese honeysuckle forms a carpet over a meadow.

Japanese honeysuckle has black fruits; native coral honeysuckle has orange-red fruits.

IDENTIFYING CHARACTERISTICS

The sweetly scented, tubular, white to pink flowers that fade to yellow make Japanese honeysuckle most recognizable in late spring. Flowering can continue throughout the summer and into fall. The twining vines ramble across the ground and over trees and shrubs. Most leaves are simple, oval, and opposite, 1.5 to 3.5 in. (4 to 8 cm) long. Some lowermost leaves are lobed like oak leaves. It spreads by seeds, underground rhizomes, and aboveground runners. The black fruits mature in fall. Leaves remain on the vines year-round in the southern part of its range, but fall off in midwinter in the northern part of its range. Japanese honeysuckle is easily distinguished from native vining honeysuckles by its black fruits (natives have red to orange fruits) and because the uppermost pair of leaves is distinctly separate as opposed to the fused leaves of the native honeysuckle.

HABITAT AND RANGE
Japanese honeysuckle frequents roadsides, field edges, floodplains, and disturbed woods and forest openings. It occurs across the southern United States north to New England. Its range is limited by severe winter temperatures and low precipitation.

WHAT IT DOES IN THE ECOSYSTEM
Japanese honeysuckle overgrows small trees and shrubs and can girdle trees as vines thicken with age, killing hosts and eventually changing forest structure. It forms a dense ground cover in sunny areas, outcompeting native vegetation through above- and below-ground competition. Deer and rabbits eat the leaves, which are relatively nutritious, particularly in spring. Birds feed on the fruits, and the vines provide cover for a variety of animals. Where Japanese honeysuckle forms a monoculture, however, the variety of food and shelter available to animals declines.

HOW IT CAME TO NORTH AMERICA
An east Asian native, Japanese honeysuckle came to Long Island, New York, in 1806, and the variety Hall's honeysuckle appeared in 1862 in Flushing, New York. Japanese honeysuckle spread through the nursery trade and further through promotion for wildlife habitat and erosion control. Some states still promote its use for deer and other wildlife.

MANAGEMENT
In loose soil, plants can be pulled up, but root fragments will resprout. Because Japanese honeysuckle holds its leaves late into the year, it can be sprayed with herbicide, such as glyphosate, after most other plants are dormant. Retreatment may be necessary as plants often resprout. In mature forests, honeysuckle can be significantly reduced by cutting vines off of trees to prevent them from climbing to the light necessary for stronger photosynthesis.

FOR MORE INFORMATION
Bravo, M. A. "Japanese honeysuckle." Plant Conservation Alliance, Alien Plant Working Group.
    http://www.nps.gov/plants/alien/fact/loja1.htm.
Evans, J. E. 1982. "Japanese honeysuckle (*Lonicera japonica*): A literature review of management
    practices." *Natural Areas Journal* 4:4–10.
Munger, Gregory T. 2002. "*Lonicera japonica*." *Fire Effects Information System*. U.S. Department
    of Agriculture, Forest Service, Rocky Mountain Research Station, Fire Sciences Laboratory.
    http://www.fs.fed.us/database/feis/.
Nuzzo, V. 1997. "Element stewardship abstract for *Lonicera japonica*." The Nature Conservancy.
    http://wiki.bugwood.org/Lonicera_japonica.

Cat's Claw Vine   *Macfadyena unguis-cati*

NAME AND FAMILY
Cat's claw vine, yellow trumpet vine (*Macfadyena unguis-cati* [L.] Gentry); bignonia family (Bignoniaceae). Resembles the native cross-vine (*Bignonia capreolata*).

IDENTIFYING CHARACTERISTICS
The tubular yellow flowers of this woody vine lead to its common use as an ornamental plant. The flowers are 3 in. (7.6 cm) long, borne in clusters from the axils of the leaves in spring. The leaves grow opposite along the stems and are divided into two leaflets and a tendril in the middle. The tendril is forked with stiff hooks on each of the three tines that allow it to climb trees. Aerial roots also help the vine cling to surfaces. Each leaflet is elliptical, 1.5 to 3 in. (3.7 to 7.6 cm) long. Long (6 to 20 in. [15 to 50 cm]), flattened capsules hold winged seeds. Older plants develop underground tubers that can be more than 1 ft. (0.3 m) long. Cross-vine has orange flowers and the tendrils are not forked.

HABITAT AND RANGE
Occurs throughout most of Florida around human habitation and in hammocks and hardwood forest islands. Also naturalized in Texas and cultivated in Alabama and South Carolina. Prefers full sun to part shade and moist soils.

Left: *Beautiful yellow flowers bloom in spring.* **Right:** *The claws of cat's claw vine, forked tendrils, grasp a tree trunk.*

Stems of a mature cat's claw vine.

Long, flattened capsules hold winged seeds.

WHAT IT DOES IN THE ECOSYSTEM
After spending several years developing a root system, cat's claw vines begin to grow rapidly, forming a ground cover in open forests and climbing into trees. The weight and shading of vines can eventually kill canopy trees. Vine seeds disperse quickly via wind and water.

HOW IT CAME TO NORTH AMERICA
Introduced to Florida as an ornamental plant in the first half of the twentieth century and noted in the wild by 1947. Native to Central or South America and the West Indies.

MANAGEMENT
It is difficult to pull or dig up plants because fragments of the roots will resprout and vines can root again from stem sections. Vines can be sprayed with a solution of triclopyr or cut and the stumps treated with glyphosate or triclopyr herbicide.

FOR MORE INFORMATION
Langeland, K. A., H. M. Cherry, C. M. McCormick, and K. A. Craddock Burks et al. 2008. *Identification and Biology of Nonnative Plants in Florida's Natural Areas*, 2nd ed. SP 257. Gainesville, FL: University of Florida IFAS.

Weber, E. 2005. *Invasive Plant Species of the World*. Cambridge: CABI Publishing.

Skunk Vine   *Paederia foetida*

NAME AND FAMILY
Skunk vine (*Paederia foetida* L.); madder family (Rubiaceae). The closely related sewer vine (*P. cruddassiana* Prain) currently occurs only in Dade County, Florida.

IDENTIFYING CHARACTERISTICS
The sulfur compounds in these woody vines give off a skunk-like smell when leaves or stems are crushed. Skunk vine grows to 30 ft. (9 m), twining to the right. The leaves vary considerably in shape and size, but most will have a heart-shaped base and long point. Leaves are smooth on the upper side and hairy below and have smooth edges. The leaves occur opposite or sometimes in whorls of three along the stems. Leaves and flowers are held on long stems (petioles) up to 2.5 in. (6.2 cm) long. A leaflike stipule 0.08 to 0.1 in. (2 to 3 mm) long occurs at the base of each petiole. The small, five-petaled, tubular flowers are pink to pale lavender with red centers. They grow in clusters from the leaf axils or ends of the branches. Shiny, brown, globe-shaped capsules 0.3 in. (0.7 cm) contain two seeds. Most flowering and fruiting occurs in summer and fall. The fruits of the related sewer vine are more flattened and the seeds are winged. The leaves of sewer vine are usually larger than those of skunk vine.

Skunk vine's showy flowers bloom in summer.

Seed capsules grow in clusters from the ends of the branches and the leaf axils.

HABITAT AND RANGE

Climbing into trees or sprawling along the ground, skunk vine will grow in a wide variety of habitats, including forest gaps, sand hill communities, and floodplains. Although it has been found as far north as North Carolina and west to Texas, it is mainly a problem in northern and central Florida.

WHAT IT DOES IN THE ECOSYSTEM

The vine can form a ground cover, rooting at the nodes, or it can climb into trees. The dense growth shades species below it. It grows in several habitats where it threatens rare or endangered native plants such as Brooksville bell flower (*Campanula robinsiae*) and Cooley's water willow (*Justicia cooleyi*). The seeds are eaten by some bird species, and plants are also dispersed from fragments.

HOW IT CAME TO NORTH AMERICA

Skunk vine was introduced to Florida from Eastern Asia before 1897 by the U.S. Department of Agriculture as a potential fiber crop. It was first noticed as a weed in Brooksville, Florida, in 1917. It is also sometimes planted as an ornamental vine.

MANAGEMENT

Vines can be sprayed with glyphosate, or stems can be cut and treated with glyphosate or triclopyr. Retreatment will probably be necessary. Where native plants are fire resistant, burning has produced effective control. Biological controls are also being tested on skunk vine.

FOR MORE INFORMATION

Flores, A. 2003. "Scouring the world for a skunk vine control." *Agricultural Research Magazine* 51:16–17. http://www.ars.usda.gov/is/AR/archive/oct03/skunk1003.htm.

Gann, G., and D. R. Gordon. 1998. "*Paederia foetida* (skunk vine) and *P. cruddasiana* (sewer vine): Threats and management strategies." *Natural Areas Journal.* 18(2):169–174.

Langeland, K. A., H. M. Cherry, C. M. McCormick, and K. A. Craddock Burks et al. 2008. *Identification and Biology of Nonnative Plants in Florida's Natural Areas,* 2nd ed. SP 257. Gainesville, FL: University of Florida IFAS.

Common Periwinkle   *Vinca minor*

NAME AND FAMILY

Common periwinkle, periwinkle, running myrtle (Vinca minor L.); dogbane family (Apocynaceae). Sometimes confused with greater periwinkle (V. major L.), which can also be invasive.

IDENTIFYING CHARACTERISTICS

Common periwinkle is familiar to most gardeners as an evergreen ground cover that grows dense and low and bears violet to blue-violet flowers. It grows to 6 in. (15 cm) tall and a single plant can spread over a 3 ft. (1 m) wide area. Trailing stems bear elliptical, pointed leaves that are dark green, shiny, and about 1 in. (2.5 cm) long, and grow opposite on the thin stems. The five petals of the flowers are widest at the outer margins. Flowers are about 1 in. (2.5

Five-petaled blue or purple flowers bloom in spring.

Vinca forms a dense groundcover in a forest.

cm) across, first appearing in spring and continuing in lesser quantities through summer. The winter buds are very small and brown seeds are also tiny. Its close relative greater periwinkle has larger flowers (1 to 1.5 in. [2.5 to 3.8 cm] across) and leaves (1.5 to 3 in. [3.8 to 7.6 cm]).

HABITAT AND RANGE
Periwinkle grows best in rich, moist soils and partial shade and does not tolerate full sun in summer. It is often found around old homesites and in

Vinca's leaves are glossy and evergreen.

open forests. It grows in most states and regions of the United States and southern Canada except high, semiarid plains and mountains and deserts.

WHAT IT DOES IN THE ECOSYSTEM
Periwinkle spreads mainly from underground runners and from rootlets formed at leaf nodes. It seldom reproduces from seeds. Once established, it can form a

carpet that entirely shades the ground, excluding other plants. Thus it has been popular as a garden and landscaping ground cover. One key to its success is that, as an evergreen, it can make use of spring and fall sunlight when the forest canopy is bare. Leaves are toxic to most or all grazers, and seeds are too small for birds, so when it displaces native plants, it also displaces food sources for wildlife.

HOW IT CAME TO NORTH AMERICA
Authorities disagree whether it came to North America as an ornamental or an herbal medicine and aphrodisiac or both, but it had begun to naturalize in the 1700s. Europeans used it against hemorrhage, diarrhea, headache, piles, and even diabetes since some of its alkaloids lower blood sugar. In medieval Britain criminals on the way to the gallows or the executioner's axe were sometimes bedecked with garlands or crowns of this "immortality flower" to mock their fate.

MANAGEMENT
Small populations can be pulled by hand, but often root fragments will produce new growth. Herbicides with glyphosate are the most effective control. Triclopyr also works but not as effectively.

FOR MORE INFORMATION
Miller, J. H., E. B. Chambliss, N. J. Loewenstein. 2010. *A field guide for the identification of invasive plants in southern forests.* General Technical Report SRS-119. Asheville, NC: U.S. Department of Agriculture, Forest Service, Southern Research Station. http://www.srs.fs.fed.us/pubs/gtr/gtr_srs119.pdf.

Terrestrial Plants—Vines— Deciduous—Alternate Leaves

Rosary Pea   *Abrus precatorius*

NAME AND FAMILY
Rosary pea, crab's eyes, jequiriti bean (*Abrus precatorius* L.); pea family (Fabaceae). Rosary pea leaves look similar to some other legumes, but the seeds are distinctive.

IDENTIFYING CHARACTERISTICS
Puffy green to black seed pods 1.5 to 2 in. (3.7 to 5 cm) long (they look similar to sugar snap pea pods) curl back to reveal six to eight bright-red seeds with a black dot at the base. Seeds are less than ½ in. (1.3 cm) in diameter. Rosary pea grows as a woody twining or trailing perennial vine with alternate leaves along the stems. The delicate, feathery leaves are made up of 5 to 15 pairs of leaflets less than 1 in. (2.5 cm) long, oblong in shape with smooth edges. Clusters of small, pale-violet to pink, pealike flowers hanging from the leaf axils bloom in summer.

HABITAT AND RANGE
Rosary pea grows in Alabama, Arkansas, Georgia, and Florida. It prefers full sun and well-drained soils and cannot tolerate frost. In Florida the plants grow in hardwood hammocks and pine rocklands.

WHAT IT DOES IN THE ECOSYSTEM
Rosary pea vines climb over trees and shrubs, blocking light to the vegetation below. It invades Florida's imperiled pine rocklands, adding to the stresses on that community. As a

Summer clusters of rosary pea flowers hang from leaf axils.

legume, it fixes its own nitrogen and could increase nitrogen levels in the soil. The seeds mimic those with fleshy coverings, so birds eat and disperse the seeds but get no nutritional benefit. Dry seeds have been used in some jewelry and rosaries; but if eaten, prayer may not cure the sometimes fatal consequences. The seeds contain abrin, a chemical so toxic that ingestion of a single broken seed can cause blindness or death in people, cattle, and horses.

HOW IT CAME TO NORTH AMERICA

Indian artisans used the pretty red seeds to make necklaces and rosaries. It arrived in Florida sometime before 1932. Rosary pea is native to India or possibly to other parts of Asia and Africa. It

occurs throughout the tropics now, where it is widely used medicinally and for ornaments. Extracts of the plant are being tested as possible insecticides, drugs, and artificial sweeteners (glycyrrhizin, found in the roots and leaves, is the principle flavor component in licorice).

Left: *The bright-red seeds are used to make necklaces and rosaries.*
Bottom: *Each leaf is made up of many pairs of leaflets.*

The fruits persist on the plant even after the leaves have fallen off.

MANAGEMENT

This deep-rooted vine is difficult to eradicate. Small plants can be hand pulled. Larger stands can be sprayed with systemic herbicide. Heavy grazing may also control infestations.

FOR MORE INFORMATION

Francis, J. K. Abrus precatorius *L., crab's eye*. General Technical Report IITF-WB-1. U.S. Department of Agriculture, Forest Service. http://www.fs.fed.us/global/iitf/pdf/shrubs/Abrus%20 precatorius.pdf.

Langeland, K. A., H. M. Cherry, C. M. McCormick, and K. A. Craddock Burks et al. 2008. *Identification and Biology of Nonnative Plants in Florida's Natural Areas,* 2nd ed. SP 257. Gainesville, FL: University of Florida IFAS.

Morton, J. F. 1976. "Pestiferous spread of many ornamental and fruit species in South Florida." *Proceedings of the Florida State Horticultural Society* 89:348–353.

Five Leaf Akebia   *Akebia quinata*

NAME AND FAMILY

Five leaf akebia or chocolate vine (*Akebia quinata* [Houtt.] Dcne.); the mostly tropical lardizabala family (Lardizabalaceae).

IDENTIFYING CHARACTERISTICS

The leaves divided into five broadly oval leaflets arranged in a palm or fan shape and the chocolate-purple flowers they often hide are the most distinctive features of akebia. The leaves are arranged alternately along the stem, emerging with a reddish-purple color but becoming blue green when mature. The name chocolate vine comes from the fragrant, brownish-purple flowers, which are 1 in. (2. 5 cm) wide, with three petals. Individual flowers are either male or female. The plant sometimes produces 2 to 4 in. (5 to 10 cm) seedpods in the fall containing a white pulp with many small black seeds. The outside of the pod is purple. Fruits are seldom produced, and cross-pollination with another plant is necessary for fruits to set. The woody vine either sprawls like a ground cover or twines its way up other plants. This fast-growing (20 to 40 ft. [6 to 12 m] a year) vine loses its leaves in cooler climates but can be evergreen in the deep south.

HABITAT AND RANGE

Five leaf akebia grows in the eastern United Sates from Michigan to Connecticut and south to Georgia and Louisiana. It is often planted in other parts of the United States. Being shade and drought tolerant, it can invade many types of habitats.

Left: *Purple flowers give off a chocolaty fragrance.* **Bottom:** *Akebia leaves are divided into five leaflets arranged in a circle.*

WHAT IT DOES IN THE ECOSYSTEM

This perennial grows along the ground, outcompeting native plants and preventing seed germination. It also climbs over understory trees and shrubs, blocking light to the vegetation below.

HOW IT CAME TO NORTH AMERICA

The nursery industry introduced five leaf akebia to the United States in 1845 from Asia (China, Japan, and Korea). It spreads mostly vegetatively, and because it grows fast and has a life of 3 to 10 years, landscapers like to use it as a wall-climbing decorative plant. In Asia the fruits and young shoots of the plants are eaten and the plant has medicinal uses.

MANAGEMENT

Vines can be cut repeatedly or dug up. Systemic herbicides like glyphosate or triclopyr sprayed on the foliage or painted on cut stumps also kill vines.

Occasionally akebia produces seedpods.

FOR MORE INFORMATION

Swearingen, J., B. Slattery, K. Reshetiloff and S. Zwicker. 2010. *Plant Invaders of Mid-Atlantic Natural Areas, 4th ed.* National Park Service and U.S. Fish and Wildlife Service. Washington, DC. http://www.invasive.org/eastern/midatlantic/akqu.html.

Porcelainberry   *Ampelopsis brevipedunculata*

NAME AND FAMILY

Porcelainberry, Amur peppervine (*Ampelopsis brevipedunculata*); grape family (Vitaceae). Porcelainberry vines look similar to the native *Ampelopsis cordata* and to grapes (*Vitis* spp.).

IDENTIFYING CHARACTERISTICS

The multicolored fruits of porcelainberry do look like little porcelain balls. The beautiful fruits vary in color from white to yellow to pastel shades of green, blue, and purple, with different colored fruits held on the plant at the

Left: *Porcelainberry resembles a grape vine in its climbing habit.* **Right:** *Multi-colored fruits look like porcelain balls.*

same time. Sprawling over fields and trees, porcelainberry is a woody vine that climbs 15 to 20 ft. (4.6 to 6 m) by tendrils growing opposite the leaves on the stems. The leaves are alternate, ranging from slightly three- to five-lobed to deeply dissected. The undersides of the leaves are shiny with delicate hairs along the veins. The vines bear clusters of small greenish-white flowers in mid-summer, and colorful fruits appear in fall. Besides the difference in fruits, the pith inside a grape stem is brown while porcelainberry's is white, and porcelainberry bark does not peel as grape bark does.

HABITAT AND RANGE
Porcelainberry favors the moist edges of streams, ponds, and thickets. It prefers sun to part shade. It grows from New England south to North Carolina and west to Michigan.

WHAT IT DOES IN THE ECOSYSTEM
The fruits attract birds and small animals that disperse the seeds, and some fruits may float down streams. Its vigorous growth shades out and outcompetes small plants. It grows over shrubs and up trees, increasing top weight and "sail area," making them vulnerable to wind damage.

HOW IT CAME TO NORTH AMERICA

Native to eastern Russia, China, Korea, and Japan, porcelainberry came to North America near the end of the nineteenth century as an ornamental plant. It is still sold commonly in the nursery trade.

MANAGEMENT

The vines' flowers come from current season's growth, so pruning vines in fall or spring will prevent flower buds from forming the following season. Pull vines up before fruiting to reduce the spread of seeds. Herbicide containing triclopyr or glyphosate can be sprayed on the foliage or applied to individual leaves for a small infestation. Cut vines back in summer, then spray new sprouts in early autumn before leaves drop off.

FOR MORE INFORMATION

Virginia Native Plant Society. "Invasive alien plant species of Virginia: Porcelainberry (*Ampelopsis brevipedunculata* [Maxim.] Trautv.)." http://www.vnps.org/invasive/invamp.htm.

Young, J. "Porcelainberry: *Ampelopsis brevipedunculata* (Maxim.) Trautv." Alien plant invaders of natural areas fact sheet. Plant Conservation Alliance, Alien Plant Working Group. http://www.nps.gov/plants/alien/fact/ambr1.htm.

Bushkiller   *Cayratia japonica*

NAME AND FAMILY

Bushkiller, sorrel vine (*Cayratia japonica* [Thunb.] Gagnepain); grape family (Vitaceae). Looks similar to native Virginia creeper (*Parthenocissus quinquefolia*).

IDENTIFYING CHARACTERISTICS

The leaves of this perennial vine are divided into three- to five-toothed, pointed leaflets. The lowermost leaflets occur in pairs, unlike Virginia creeper leaflets, which are all attached to a central point. A tendril grows opposite each leaf. Flat-topped clusters of tiny, cup-shaped, salmon-colored flowers bloom in spring. Plants in North America do not seem to set fruit.

Bushkiller vines sprawl on the ground or climb.

Tendrils grow opposite each divided leaf.

HABITAT AND RANGE
Currently occurs in North Carolina, Louisiana, Alabama, and Texas in hardwood river bottoms and near homesites. Could potentially grow throughout most of the United States.

WHAT IT DOES IN THE ECOSYSTEM
As its name implies, this vine grows over top of shrubs and trees, blocking light to the canopy and weighing down branches. It spreads via its aggressive root system. One land manager described it as potentially becoming the next kudzu.

HOW IT CAME TO NORTH AMERICA
Originally introduced as an ornamental plant, bushkiller comes from temperate-subtropical regions in Asia.

MANAGEMENT
Hand pull or dig out plants, including all root fragments, for smaller plants. Larger infestations can be sprayed with imazapyr or another systemic herbicide. However, these herbicides can end up killing the tree or shrub growing under the vine.

FOR MORE INFORMATION
Iverson, R. *Bushkiller* (Cayratia japonica) *a new noxious weed in North Carolina.* North Carolina Dept. of Agriculture and Consumer Services. http://www.invasiveplantcontrol.com/dod/Bushkiller_DODWorkshop.pdf

Tu, M. 2001. "Weed Alert! *Cayratia japonica* (Thunb. x Murray) Gagnep." The Nature Conservancy. http://www.invasive.org/gist/alert/alrtcayr.html.

Oriental Bittersweet   *Celastrus orbiculatus*

NAME AND FAMILY
Oriental bittersweet, Asiatic bittersweet, round-leaved bittersweet (*Celastrus orbiculatus* Thunb.); staff-tree family (Celastraceae). Looks very similar to native American bittersweet (*C. scandens* L.).

IDENTIFYING CHARACTERISTICS
Yellow fruits split open in fall, revealing bright red-orange-coated seeds that stand out as the leaves fall off the vines. Vines can climb more than 60 feet (18 m) into trees, twining around trunks and branches. Twigs and branches are brown with tan lenticels like little dashes on the twigs. Leaves, arranged alternately along the stems, are rounded and 2 to 5 in. (5 to 13 cm) long, with wavy, slightly toothed edges. Greenish-yellow, five-petaled flowers bloom in spring in small clusters (3 to 7 flowers) in the leaf axils. Male and female flowers grow on separate plants. Green, round fruits form, turning yellow in fall. American bittersweet is distinguishable from Oriental bittersweet because its flowers and fruits occur only at the ends of the branches and generally have more than seven flowers and fruits in a cluster.

HABITAT AND RANGE
Often found along forest edges and in forest gaps, as well as in fencerows, old fields, and coastal areas. Prefers sun to part shade. Grows from southern Ontario south to Louisiana and Georgia and west to Iowa.

WHAT IT DOES IN THE ECOSYSTEM
Bittersweet seedlings can establish in dense shade, then take advantage of gaps created after storms, fire, or human disturbance to climb into the light. The vine grows rapidly and it is sometimes called "the kudzu of the North," as it overtops trees, cutting off light to the plants below. The leaf litter is high in calcium and increases soil pH. Seeds may be an important winter food source for birds and are dispersed by birds to new areas. Oriental bittersweet can hybridize with

Bright yellow and orange fruits attract people and birds in the fall.

Twigs are brown with tan lenticels.

Inconspicuous flowers bloom in the leaf axils in the spring.

the relatively rare American bittersweet, potentially threatening the genetic identity of American bittersweet. American bittersweet cannot photosynthesize as well as oriental bittersweet at low light levels, perhaps giving oriental bittersweet a competitive advantage.

HOW IT CAME TO NORTH AMERICA
Native to temperate Japan, China, and Korea, oriental bittersweet arrived from China as an ornamental vine around 1860. The fruits are often used in wreaths and dried flower arrangements.

MANAGEMENT
Small plants are easily hand pulled in loose soil. Cut plants will send up suckers from the roots, but resprouts can then be sprayed with triclopyr or glyphosate. If vines have climbed high into trees, cut plants at 4 to 5 ft. and again at ground level to prevent resprouts from climbing old vines and to prevent harm to the tree from pulling off vines. Weekly mowing will also discourage plants.

Rounded leaves have wavy, toothed edges.

FOR MORE INFORMATION

Fryer, Janet L. 2011. "*Celastrus orbiculatus.*" *Fire Effects Information System.* U.S. Department of Agriculture, Forest Service, Rocky Mountain Research Station, Fire Sciences Laboratory. http://www.fs.fed.us/database/feis/.

Greenberg, C. H., L. H. Smith, and D. J. Levey. 2001. "Fruit fate, seed germination, and growth of an invasive vine: An experimental test of 'sit and wait' strategy." *Biological Invasions* 3:363–372.

Pooler, M. R., R. L. Dix, and J. Feely. 2002. "Interspecific hybridizations between the native bittersweet, *Celastrus scandens*, and the introduced invasive species, *C. orbiculatus.*" *Southeastern Naturalist* 1:69–76.

Field Bindweed   *Convolvulus arvensis*

NAME AND FAMILY

Field bindweed, creeping Jennie, field morning glory, possession vine (*Convolvulus arvensis* L.); morning glory family (Convolvulaceae). Similar in appearance to the nonnative black bindweed (*Polygonum convolvulus* L.) and to the eastern U.S. native hedge bindweed (*Calystegia sepium* [L.] R. Br.).

IDENTIFYING CHARACTERISTICS

This twining perennial vine has typical morning glory funnel-shaped flowers with white to pale-pink petals. Flowers are usually 0.75 to 1 in. (1.9 to 2.5 cm) across. The leaf shape varies from a typical arrow shape with pointed lobes at the base to being round or oblong. Leaves are 1 to 2.25 in. (2.5 to 5.7 cm) long and occur alternately along the stems. Rounded light-brown fruits about 0.13 in. (0.3 cm) wide usually contain two seeds each. The roots can extend up to 18 ft. (5.4 m).

Black bindweed is an annual vine whose leaves are always arrow shaped. It can also be distinguished because it has inconspicuous greenish clusters of flowers and a papery sheath around each node. Hedge bindweed has larger flowers and the bases of the leaves are squared off rather than being pointed.

HABITAT AND RANGE
Field bindweed can be found throughout the United States and in western Canada. It typically grows along streams and in shrub- and grasslands as well as in more disturbed locations such as roadsides and field edges.

WHAT IT DOES IN THE ECOSYSTEM
Primarily considered an agricultural weed, field bindweed can outcompete native plants on disturbed sites. In croplands, it competes by using up moisture in the soil and by exuding chemicals that inhibit seed germination. Seeds can remain in the seed bank for decades. A number of bird species eat the seeds.

HOW IT CAME TO NORTH AMERICA
Native to Eurasia, field bindweed probably arrived as a contaminant in crop seeds. It may have arrived in Virginia as early as 1739, and was certainly estab-

White funnel-shaped flowers adorn bindweed vines.

Black bindweed covering the ground in a nature preserve.

lished in the eastern United States by the early 1800s. It spread west via the railroad lines and possibly through other contaminated seed shipments.

MANAGEMENT
Seedlings can be hand pulled, but repeated pulling may be necessary for older plants. Repeated deep cultivation can be successful. Bindweed can also be shaded out eventually. Herbicides glyphosate and 2,4-D provide best control of field bindweed if sprayed just before or during first flowering. Several insect species have also been released in the United States as biological control agents.

FOR MORE INFORMATION
Lyons, K. E. 1998. "Element stewardship abstract for *Convolvulus arvensis*." The Nature Conservancy. http://www.invasive.org/weedcd/pdfs/tncweeds/convarv.pdf.

Uva, R. H., J. C. Neal, and J. M. DiTomaso. 1997. *Weeds of the Northeast*. Ithaca, NY: Cornell University Press.

Air Potato · *Dioscorea bulbifera*

NAME AND FAMILY
Air potato, bitter yam (*Dioscorea bulbifera* L.); yam family (Dioscoreaceae). The invasive winged yam or white yam (*D. alata* L.) and several native species of *Dioscorea* occur in the same areas as air potato.

IDENTIFYING CHARACTERISTICS
The twining vines of this herbaceous perennial grow 60 ft. (18 m) into the canopy. The stems grow from an underground tuber, usually 4 to 6 in. (10 to 15 cm) in diameter, but in June and July the plant also produces "bulbils," round aerial tubers from which new plants can grow. The leaves are heart shaped, arranged alternately along the stem, and have prominent veins that fan out from the leaf stalk. Leaves are usually 8 in. (20 cm) long or longer. Although plants rarely flower, they can have fragrant green to white flowers in spikes up to 4 in. (10 cm) long hanging down from the leaf axils, with male and female flowers on separate plants (dioecious), flowering in late summer to early fall. Stems die back in winter. Winged yam has square stems with winged corners, the leaves are generally arranged oppositely, and the tuber can be quite substantial. Winged yam produces fewer bulbils than air potato. Winged yam vines twine to the right, whereas air potato vines twine to the left. Native wild yams have smaller leaves and do not produce aerial tubers.

The air "potato" is actually an aerial tuber from which a new plant can grow.

HABITAT AND RANGE

Air potato invades hardwood tree islands (hammocks), woodland edges, wetland edges, and waste places in 17 Florida counties, and it grows along the Gulf Coast as far west as Texas.

WHAT IT DOES IN THE ECOSYSTEM

Vines climb into trees and over small trees and shrubs, blocking light to the leaves and vegetation below. They become particularly aggressive in full sun. The plant spreads rapidly via the bulbils, which—if they drop into water—will float and colonize new places. All parts of the plant are poisonous to humans when raw.

Heart-shaped leaves alternate along the stems.

HOW IT CAME TO NORTH AMERICA

Native to Asia and Africa, air potato first arrived in the Americas with the slave trade from Africa, where the aerial tubers were sometimes eaten. Genetic studies indicate that Florida air potato vines are more genetically similar to ones from China than to ones from Africa, however. The U.S. Department of Agriculture sent a sample to horticulturist Henry Nehrling in Florida in 1905, probably as a potential ornamental plant. Nehrling likened its aggressive growth to that of kudzu, and recommended against cultivating it. The plant had already escaped his garden, however, and began to be sold as a fast growing ornamental vine. The chemical diosgenin in the genus *Dioscorea* is the principal component of many birth control pills.

MANAGEMENT

For small infestations, rake or pick the bulbils and burn them, then cut vines back to the ground frequently during the growing season. Glyphosate or triclopyr plus a surfactant can be sprayed on vines from late spring into summer before bulbils form, but vines sometimes resprout and require re-treatment.

FOR MORE INFORMATION

Croxton, M. D., M. A. Andreu, D. A. Williams, W. A. Overholt, and J. A. Smith. 2011. "Geographic origins and genetic diversity of air potato *(Dioscorea bulbifera)* in Florida." *Invasive Plant Science and Management* 4(1):22–30.

Demers, C., A. Long and R. Williams. 2008. *Controlling Invasive Exotic Plants in North Florida Forests.* SS-FOR19. University of Florida IFAS. http://edis.ifas.ufl.edu/pdffiles/FR/FR13300.pdf.

Langeland, K. A., H. M. Cherry, C. M. McCormick, and K. A. Craddock Burks et al. 2008. *Identification and Biology of Nonnative Plants in Florida's Natural Areas,* 2nd ed. SP 257. Gainesville, FL: University of Florida IFAS.

Cinnamon Vine — *Dioscorea oppositifolia*

NAME AND FAMILY

Cinnamon vine, Chinese yam (*Dioscorea polystachya* Turcz., formerly *D. oppositifolia* L.); yam family (Dioscoreaceae). Leaves look similar to some *Smilax* species and to the native yam (*D. villosa* L.). Cinnamon vine has a more northern distribution than air potato (*D. bulbifera* L.).

IDENTIFYING CHARACTERISTICS

Stems of this fast-growing, counterclockwise-twining vine grow from an underground tuber that can be 3 ft. (1 m) long. Leaves are heart shaped with a long tip and, although usually arranged opposite along the stem, can be alternate or in whorls of three leaves. Each leaf has seven to nine distinct veins that fan out from its base. New leaves are often bronze in color, and older leaves can be reddish along the edges and down the stems. The name cinnamon vine comes from the fragrance of the white to greenish flowers that grow in spikes hanging from the leaf axils. Male and female flowers grow on separate plants. Seeds are held in a papery, three-sided capsule. Plants also reproduce vegetatively from "bulbils," aerial tubers. The native yam does not have bulbils, the leaves are slightly hairy on the upper side, and vines twine clockwise. Smilax has similar-shaped leaves, but most species have thorns and none have bulbils.

Cinnamon vine's aerial tubers.

Heart-shaped leaves have seven to nine veins that fan out from the base of the leaf.

HABITAT AND RANGE
Most common in rich soils along streambanks and in floodplain forests. Also grows along fencerows and roadsides. It can tolerate full sun to full shade but prefers nutrient-rich soils. Occurs from Vermont south to northern Florida and west to Kansas and Texas.

WHAT IT DOES IN THE ECOSYSTEM
Vines quickly overgrow shrubs and small trees, blocking light to the ground. Plant species diversity declines under heavy cinnamon vine cover. Vines can grow so thickly that branches break under their weight. Rodents will chew on and move bulbils. Bulbils are also dispersed by water.

HOW IT CAME TO NORTH AMERICA
This native of China and India was introduced to North America as an ornamental or food plant in the 1800s; it wasn't noticed in the wild until the 1980s. Tubers and bulbils are edible, tasting something like a sweet potato. Parts of the plant are also used medicinally for treatment of snakebite and scorpion stings, as a contraceptive, and as an herbal tonic.

MANAGEMENT
Vines can be repeatedly mowed or clipped, but this may take several growing seasons depending on the size of the tuber. Bulbils can be hand picked to keep vines from spreading. Glyphosate or triclopyr can be sprayed on the foliage.

Spraying leaves late in the growing season will lead to better control, but spraying before bulbils are formed may prevent their formation.

FOR MORE INFORMATION

Miller, J. H., E. B. Chambliss, N. J. Loewenstein. 2010. *A field guide for the identification of invasive plants in southern forests.* General Technical Report SRS-119. Asheville, NC: U.S. Department of Agriculture, Forest Service, Southern Research Station. http://www.srs.fs .fed.us/pubs/gtr/gtr_srs119.pdf

Thomas, J. R., B. Middleton, and D. J. Gibson. 2006. "A landscape perspective of the stream corridor invasion and habitat characteristics of an exotic *(Dioscorea oppositifolia)* in a pristine watershed in Illinois." *Biological Invasions* 8(5):1103–1113.

Tu, M. 2002. "Element stewardship abstract for *Dioscorea oppositifolia.*" The Nature Conservancy. http://wiki.bugwood.org/Dioscorea_oppositifolia.

Mile-A-Minute Vine   *Persicaria perfoliata*

NAME AND FAMILY

Mile-a-minute vine, Asiatic tearthumb, Devil's-tail tearthumb (*Persicaria perfoliata* L., formerly *Polygonum perfoliatum* L.); knotweed family (Polygonaceae). Somewhat similar to native arrow-vine (*P. sagittatum* L.) and halberdleaved tearthumb (*P. arifolium* L.).

IDENTIFYING CHARACTERISTICS

The first key identifier the observer is likely to encounter are the spines on leaves and stems that earn it the name "tearthumb." A quick look at the distinctive equilateral triangles of its leaves, 1 to 3 in. (2.5 to 7.6 cm) wide, might

be a good warning to touch carefully. This annual trailing vine has slender reddish stems on which the leaves alternate. Spaced along the stem grow cuplike leafy structures called ocreas. From these ocreas flower buds emerge, blossoming in small, white flowers that produce metallic blue, pea-sized fruits inside of which are dark seeds. Vines can grow to 15 ft. (4.6 m) long. Mile-a-minute is somewhat similar to the native arrow-vine and the halberdleaved tearthumb, but neither of these has the distinctive ocreas.

Mile-a-minute flowers emerge from the rounded ocrea.

Left: *Metallic blue fruits mature in summer.* Right: *Leaves are shaped like equilateral triangles.*

HABITAT AND RANGE

This fast-spreading species is found mainly in the Mid-Atlantic and northeastern United States but is already naturalizing in other places that have a temperate to subtropical climate. It is also found in Ohio, Oregon, and British Columbia, Canada. It requires about two months of winter with temperatures below 46° F (10° C) for seeds to germinate. It commonly grows on road embankments and in fallow fields, young forests, streams, moist meadows, and recently cut timberland. Where debris or fallen leaves maintain soil moisture, seeds germinate. Tolerates partial shade but needs some sunlight.

WHAT IT DOES IN THE ECOSYSTEM

Mile-a-minute, growing up to 6 in. (15 cm) a day and climbing over other plants by means of downward-pointing spines, sometimes shades out native species. Where it grows in wet areas it can reduce or eliminate local populations of rare native plants. It can also prevent tree seedlings from developing and has been detrimental to reforestation efforts. In warmer climates it produces seeds throughout the summer until first frost. Both birds and ants disperse the seeds since they eat the fruits. Chipmunk, deer, and squirrels also feed on the fruits. Vines that grow up streamsides drop fruits into the water for dispersal.

HOW IT CAME TO NORTH AMERICA

This native of India, China, Japan, and east Asia was brought to Portland, Oregon, in 1890 in ship ballast and to Beltsville, Maryland, in 1937, but

This annual vine grows extremely quickly over surrounding plants.

apparently did not escape or establish naturalized populations until the late 1930s, when seeds probably came into a Pennsylvania nursery with a shipment of rhododendrons. It also seems to have come in again with later rhododendron shipments.

MANAGEMENT

Maintaining good shade from bushes, brush, or trees in invasion prone areas will deny the vine the sunlight it needs. Pull plants in spring or early summer before they form seeds. In early spring, the spines are still soft, but gloves are always adviseable when working near these vines. In areas where mowing is possible, a close mowing removes flowers and thus prevents fruiting. Glyphosate herbicides are effective killers if used before seed formation. An introduced weevil, *Rhinoncomimus latipes*, has been successful at controlling the spread of populations in the Mid-Atlantic.

FOR MORE INFORMATION

Gerlach, J. A., J. Hough-Goldstein, and J. Swearingen. 2010. "Mile-a-Minute Fact Sheet." Plant Conservation Alliance Alien Plant Working Group. http://www.nps.gov/plants/alien/fact/pepe1.htm.

Hill, Rovert J., G. Springer, and L. B. Forer. 1981. "Mile-a-minute, *Polygonum perfoliatum* L. (Polygonaceae): A new potential orchard and nursery weed." Pennsylvania Department of Agriculture. *Regulatory Horticulture* 7(1).

Stone, Katharine R. 2010. "*Polygonum perfoliatum*." *Fire Effects Information System*. U.S. Department of Agriculture, Forest Service, Rocky Mountain Research Station, Fire Sciences Laboratory. http://www.fs.fed.us/database/feis/.

Kudzu   *Pueraria montana* var. *lobata*

NAME AND FAMILY
Kudzu (*Pueraria montana* [Lour.] Merr. var. *lobata* [Willd.] Maesen & S. Almeida); pea family (Fabaceae).

IDENTIFYING CHARACTERISTICS
Sometimes called "the vine that ate the South," these deciduous vines scramble over trees, buildings, and power lines. Kudzu leaves are made up of three leaflets, each of which can be unlobed or have two to three lobes, with the leaflet up to 4 in. (10 cm) wide. Leaves have hairy margins and grow alternately along the stems. Individual flowers, about 0.5 in. (1.3 cm) long, are purple, highly fragrant, and borne in long hanging clusters in midsummer. Brown, hairy, flat seed pods follow quickly with 2 to 10 hard seeds. Kudzu has massive taproots up to 6 ft. (1.8 m) in length.

HABITAT AND RANGE
Kudzu grows from Pennsylvania south to Florida and west to the Mississippi River and in Texas and Arizona. Kudzu adapts to a variety of soils but thrives in areas of mild winters and hot, humid summers with annual average rainfall above 40 in. (100 cm). In the year 2000 Oregonians were startled by the discovery of a kudzu stand southeast of Portland.

Left: *Kudzu leaves are divided into three leaflets, which may be lobed or unlobed.*
Right: *Kudzu rapidly grows over trees and buildings.*

WHAT IT DOES IN THE ECOSYSTEM

Kudzu's massive leaf surface area collects nitrogen and sunlight from the air. Its release of nitric oxide from soils can increase local ozone levels. It is almost entirely an open field and forest edge invader, but it now affects up to 7 million acres (3 million ha) of land by some estimates. While frost kills most of the vines, shoots resprout every spring from the roots and vines can grow up to 1 ft. (0.3 m) a day and send out anchor roots as they travel. In the six- to seven-month frostless periods of the Southeast one root can send out 30 vines and a vine can grow over 100 ft. (30 m) of ground or grow over the tops of trees and onto power lines. Roots can penetrate 10 ft. (3 m) into the ground and weigh over 100 lbs. (45 kg). Few kudzu seeds are viable and those that are may take several years to germinate, so kudzu spreads mostly vegetatively.

HOW IT CAME TO NORTH AMERICA

The Japanese government's garden of native plants at the 1876 Centennial Exposition in Philadelphia, Pennsylvania, included kudzu. Americans began

using it in the Southeast to shade porches with its broad leaves and sweet blossoms, earning it the name "porch vine." Botanist David Fairchild planted it around his home in Wahington, DC, in 1902 after observing its extensive use in Japan as a forage crop. By 1938 he considered it a "tangled nuisance." In the 1930s, as small farmers in the South abandoned unprofitable red clay soils, government work crews planted thousands of acres of kudzu for erosion control. In the 1940s government helped spread kudzu by paying farmers $8 an acre to plant kudzu in old fields. Georgia farm writer and radio pundit Channing Cope often promoted "the miracle vine" or "King Kudzu" on his show and traveled through the South forming Kudzu Clubs. The U.S. Department of Agriculture removed kudzu from its approved cover crops in 1953.

Kudzu, sometimes cut and baled for hay, is also good forage for goats and pigs. Vines are used for basket

Clusters of purple flowers hang from the stems.

Kudzu flowers in mid-summer, but the leaves often hide the flowers.

making, leaves for paper, and its large roots can be processed for food. Tea, jellies, and syrups also can be made from kudzu. Over 2,000 years ago, kudzu was an ingredient in Chinese herbal medicines, as it still is today.

MANAGEMENT

The key action is root destruction. Agriculturalists have found goats will control kudzu and even kill it if an area is grazed for at least three years. Pigs are more effective since they eat the roots as well as the vines. Monthly close-to-the ground mowing for at least two years to deplete carbohydrate storage can eliminate some stands. Systemic herbicides such as glyphosate applied late in the season to cut vine stumps will also control kudzu by killing the roots. Recently, an insect accidentally introduced from Asia called the bean plataspid (*Megacopta cribraria*) has been feeding on kudzu in the southeast, but it also feeds extensively on soybean.

FOR MORE INFORMATION

Hickman, J. E., S. Wu, L. J. Mickley and M. T. Lerdau. 2010. "Kudzu *(Pueraria lobata)* Invasion doubles emissions of nitric oxide and increases ozone pollution." *Proceedings of the National Academy of Sciences* 107:10115–10119.

Miller, J. H., and B. Edwards. 1982. "Kudzu: Where did it come from? And how can we stop it?" *Southern Journal of Applied Forestry:* 165–169.

Pappert, R. A., J. L. Hamrick, and L. A. Donovan. 2000. "Genetic variation in *Pueraria lobata* (Fabaceae): An introduced, clonal, invasive plant of the southeastern United States." *American Journal of Botany* 87(9):1240–1245.

Shurtleff, W., and A. Aoyagi. 1977. *The Book of Kudzu: A Culinary and Healing Guide.* Brookline, MA: Autumn Press.

Stewart, D. 2000. "Kudzu: Love it—or run." *Smithsonian* 31:65–70.

Chinese and Japanese Wisteria   *Wisteria* spp.

NAME AND FAMILY
Chinese wisteria (*Wisteria sinensis* [Sims] DC.); pea family (Fabaceae). A similar exotic is Japanese wisteria (*W. floribunda* [Willd.] DC.). American wisteria (*W. frutescens* [L.] Poir.) also looks similar and is native to the eastern and midwestern United States. Botanist Thomas Nuttal chose the name in honor of American anatomy professor and amateur botanist Caspar Wistar, 1761–1818.

IDENTIFYING CHARACTERISTICS
Most people recognize wisteria in landscaping as a winding, climbing, muscular vine with large, drooping clusters of violet to purple spring flowers, and later, 2 to 6 in. (5 to 15.2 cm) long, fuzzy seedpods that children often use as clubs. Cul-

tivars may have white, pink, or lavender flowers. Wisteria grows rapidly, seeking sunlight, and can reach 70 ft. (21 m) with vines up to 15 in. (38 cm) in diameter. Leaves about 1 ft. (30 cm) long alternate on the stem and are made up of 7 to 13 leaflets (Chinese) or 13 to 19 leaflets (Japanese). Each flower is pealike in shape with a hoodlike petal over the other petals, 0.5 to 1 in. (1.3 to 2.5 cm) long, blooming in drooping 8 to 20 in. (20 to 50 cm) clusters along the vines. Chinese vines twist clockwise around trees, shrubs, or posts, while Japanese wisteria twists counterclockwise. The native American wisteria does not have hairy pods and bears 4 to 6 in. long (10.2 to 15.2 cm) clusters of purple to white flowers in early summer.

HABITAT AND RANGE
Wisteria prefers the sunny South but grows as far north as Oregon and Massachusetts. It tolerates shade and a variety of soils but prefers moist ground. In natural areas it principally occurs in the southeastern United States.

Wisteria is best known for its fragrant clusters of flowers.

Vines twist around trees or each other and thicken with age.

WHAT IT DOES IN THE ECOSYSTEM

The wisterias of China and Japan are most aggressive in warm climates, but in any climate where it grows, wisteria will eventually reach rooftops and treetops since it can live more than 50 years. So far it has done more damage to homes than to the natural environment, although it can kill individual shrubs and trees by robbing them of light or strangling them. When it reproduces by runners, it can also form wisteria thickets.

HOW IT CAME TO NORTH AMERICA

Chinese and Japanese wisteria came to the United States as ornamentals in 1816 and 1830 respectively. Because the vines grow fast and produce heavy shade, they were favored for porches, trellises, and gazebos.

Long, fuzzy seedpods hang on the branches by late summer.

Wisteria's leaves begin to emerge during flowering in spring.

MANAGEMENT

The only way to eliminate an established wisteria stand mechanically is to cut all vines back as close to the ground as possible, then keep cutting all new shoots back every week or two until the cutting effectively starves the root. Single plants can be dug out, but any roots left in the ground might sprout. Cutting close to the ground and applying herbicide like glyphosate or triclopyr to the stems or to regrowth will kill root networks.

FOR MORE INFORMATION

Dirr, M. A. 1998. *Manual of Woody Landscape Plants.* Champaign, IL: Stipes Publishing.

Swearingen, J. and T. Remaley. 2010. "Chinese Wisteria Fact Sheet." Plant Conservation Alliance Alien Plant Working Group. http://www.nps.gov/plants/alien/fact/wisi1.htm.

Swearingen, J. and T. Remaley. 2010. *Japanese Wisteria. Fact Sheet.* Plant Conservation Alliance Alien Plant Working Group. http://www.nps.gov/plants/alien/fact/wifl1.htm.

Terrestrial Plants—Vines—
Deciduous—Opposite Leaves

Sweet Autumn Virgin's Bower   *Clematis terniflora*

NAME AND FAMILY
Sweet autumn virgin's bower, sweet autumn clematis, Japanese virgin's bower, leatherleaf clematis, yam-leaved clematis (*Clematis terniflora* DC.); buttercup family (Ranunculaceae). Easily confused with the native virgin's bower (*C. virginiana* L.) and satin curls (*C. catesbyana*).

IDENTIFYING CHARACTERISTICS
Evergreen to semievergreen perennial vine covered with clouds of fragrant four-petaled white flowers (1 in. [2.5 cm] diameter) in late summer to fall. Flowers grow in branched clusters from the leaf axils along the vine's new growth. Leaves are arranged oppositely along the stems and are made up of three to five leaflets each 2 to 3 in. (5 to 7.6 cm) long. Sweet autumn virgin's bower leaflets are more leathery and rounded as opposed to the jagged teeth on most of the leaf edges of Virgin's bower and satin curls. Seed heads are also ornamental, looking like

Feathery hairs aid the seeds of sweet autumn virgin's bower to disperse.

Leathery, rounded leaflets distinguish sweet autumn virgin's bower from native virgin's bower.

silvery-gray rounded puffs of feathery hairs. Sweet autumn virgin's bower climbs to 15 ft. (4.6 m) by twining.

HABITAT AND RANGE

Found along stream banks and in thickets and moist woods from New England south to Florida, and west to Nebraska and Texas.

WHAT IT DOES IN THE ECOSYSTEM

Appreciated by gardeners because its ground-hugging mat shades out weeds and drapes over arbors; sweet autumn virgin's bower also clambers over the ground and over shrubs and small trees, blocking light to the plants below it.

HOW IT CAME TO NORTH AMERICA

Native to Japan and China, sweet autumn virgin's bower was introduced

Vines scramble over surrounding plants.

Four-petaled white flowers bloom in late summer.

as an ornamental plant for its fragrant flowers. Botanists started finding it in natural areas in the early 1950s.

MANAGEMENT

Vines can be pulled or dug up or cut repeatedly. The herbicides glyphosate and triclopyr, among others, are used to treat sweet autumn virgin's bower as a foliar spray or cut-stem treatment.

FOR MORE INFORMATION

Langeland, K. and M. Meisenberg. 2009. "Herbicide evaluation to control *Clematis terniflora* invading natural areas in Gainesville, Florida." *Invasive Plant Science and Management* 2(1):70–73.

Meisenberg, M., K. Langeland, and K. Vollmer. 2008. *Japanese Clematis,* Clematis terniflora *(D.C.) Ranunculaceae.* SS AGR 309. University of Florida IFAS. http://edis.ifas.ufl.edu/ag315.

Old Man's Beard   *Clematis vitalba*

NAME AND FAMILY

Old man's beard, traveler's joy (*Clematis vitalba* L.); buttercup family (Ranunculaceae). Looks similar to the native western clematis (*C. ligusticifolia* Nutt.).

IDENTIFYING CHARACTERISTICS

White clusters of flowers grace this deciduous perennial vine most of the summer. Flowers are borne on branched clusters growing from the leaf axils along upper ends of the vines. Each flower has many stamens and styles, four white to greenish-white sepals (like petals) and are only 0.75 in. (1.9 cm) wide. The many styles persist attached to the seed, forming feathery white balls and giving the plant its name of old man's beard. Seeds gradually fall off the plant from fall into winter. Seeds are dispersed by wind, water, and animals. Oppositely

The puffs of seeds give old man's beard its name.

Small, four-petaled white flowers bloom in summer.

arranged leaves are usually divided into five smooth-edged leaflets, but the end leaflet may have three lobes. The native western clematis grows east of the Cascades and has toothed leaflets. Vines can climb to 90 ft. (27 m).

HABITAT AND RANGE

Found in coastal and lowland areas, forest openings, and waste places west of the Cascades in British Columbia, Washington, and Oregon. Also reported in Ontario and Maine. It can tolerate some shade but prefers sunny spots.

WHAT IT DOES IN THE ECOSYSTEM

Vines climb tall trees and form such a dense canopy that they block light to the vegetation below. The weight of the vines can break branches. On the ground, vines form a blanket several feet thick. In New Zealand, the variety and abundance of understory trees and shrubs were greatly reduced after invasion by old man's beard. Flowers are both insect- and wind-pollinated and plants can vegetatively reproduce from stem or root fragments.

HOW IT CAME TO NORTH AMERICA

Native to Europe, old man's beard was introduced to North America as an ornamental plant. Although documented in natural areas since the early 1960s, it has only recently become a concern to land managers, particularly in the Pacific Northwest.

MANAGEMENT

Vines can be hand pulled or dug out. Vines can also be cut in winter and regrowth sprayed with glyphosate or triclopyr.

FOR MORE INFORMATION

Global invasive species database. http://www.issg.org/database/species/ecology.asp?si=157&fr=1&sts=.

Ogle, C. C., G. D. La Cock, G. Arnold, and N. Mickleson. 2000. "Impact of an exotic vine *Clematis vitalba* (F. Ranunculaceae) and of control measures on plant biodiversity in indigenous forest, Taihape, New Zealand." *Austral Ecology* 25:539–551.

King County Noxious Weed Control Program Best Management Practices. 2010. "Old Man's Beard." http://your.kingcounty.gov/dnrp/library/water-and-land/weeds/BMPs/Old-mans-beard-Clematis-vitalba-control.pdf.

Swallow-Worts   *Cynanchum* spp.

NAME AND FAMILY

Black swallow-wort, strangling vine, dog-strangling vine (*Cynanchum louiseae* Kartesz & Gandhi or *Vincetoxicum nigrum* [L.] Moench) and pale swallow-wort, dog-strangling vine, European swallow-wort (*C. rossicum* [Kleopov] Barbarich or *V. rossicum*); milkweed family (Asclepiadaceae). Black and pale swallow-worts are difficult to distinguish except by flower characteristics.

IDENTIFYING CHARACTERISTICS

These perennial vines distinguish themselves by loose clusters of small, dark-purple (almost black), star-shaped flowers with a central disk carrying pale-yellow organs. The flowers grow from the leaf axils. Black swallow-wort flowers from late spring to late summer, forming green fruit pods longer and thinner (1 to 3 in. [2.5 to 7.6 cm] long and 0.5 in. [1.3 cm] wide) than those of common milkweed. In late summer the pods dry and split, releasing feathered brown seeds like those of milkweed. Vines can reach 6 ft. (1.8 m) in height. Leaves are opposite along the stem, broad at the base and tapering to a fine point, 2.5 to 5 in. (6.2 to 12.7 cm) long and up to 3 in. (7.6 cm) wide. Black swallow-wort can be distinguished from pale swallow-wort by its darker purple petals and hairs

Black swallow-wort flowers have dark-purple petals.

Black swallow-wort's long narrow seed pods.

When pods open, wind disperses swallow-wort seeds.

on the inner surface of the petals. Pale swallow-wort bears pale-purple to reddish-purple petals, sometimes with a tinge of yellow.

HABITAT AND RANGE
Swallow-worts have colonized New England, the Great Lakes states, and the provinces of Ontario and Quebec in Canada. Swallow-worts will form dense stands in moist but not wet open lands, along roadsides, and in wooded shade.

WHAT IT DOES IN THE ECOSYSTEM
Wind disburses the seeds and many seeds are polyembryonic—capable of producing several seedlings. Vines grow from an underground crown with several buds. Shedding up to 2,000 seeds per square yard (2,400 seeds/m2), the plant also spreads by

Vines form a dense carpet.

rhizomes that send up new plants from underground buds. Seeds are often carried inadvertently to new places in hay bales. The denser the stand of black swallow-wort, the fewer the grassland birds in the area. Black swallow-wort climbs on any adjacent plant and also suppresses regeneration of trees. In northern New York, black swallow-wort threatens rare communities of lichens, mosses, and other plants growing in thin soils over bedrock. Black swallow-wort is listed as the most common invasive plant in the Lake Ontario basin and is in the top 20 list of invasives of New York natural areas. Its tangled density leads to the name "dog-strangling vine" and it can indeed tangle up a dog that charges in. Many insects use the flowers for pollen and nectar. As the former Latin name, *Vincetoxicum*, suggests, swallow-wort's leaves and roots are toxic. Monarch butterfly larvae that feed on swallow-worts have higher mortality than if they feed on native milkweeds. Initial studies indicate that toxins that exude from the roots can inhibit establishment of native plants.

HOW IT CAME TO NORTH AMERICA
Black swallow-wort arrived from western Europe and was first recorded in an Ipswich, Massachusetts, botanical garden in 1854, and ten years later pegged as spreading to natural areas. Pale swallow-wort is native to Ukraine and surrounding areas of Europe and Asia and was recorded in New York state in 1897 and in Toronto, Canada, in 1889 but may have entered earlier. Reports say these species were introduced as garden plants.

MANAGEMENT
Small populations can be controlled by digging up the crowns from which the plants sprout. Pulling or mowing more mature plants will not prevent resprouting but can prevent seed dispersal if done early enough. Herbicides that con-

tain triclopyr or glyphosate sprayed after flowering has begun are effective but also kill nearby vegetation. Prevention is important since stands are hard to eradicate. Disturbed soils in areas where strangling vine is present should be quickly reseeded with other vegetation.

FOR MORE INFORMATION

Douglass, C. H., L. A. Weston and A. DiTommaso. 2009. "Black and pale swallow-wort (*Vincetoxicum ingrum* and *V. rossicum*): The biology and ecology of two perennial, exotic and invasive vines." Invading Nature, Springer Series in *Invasion Ecology* 5:261–277.

Gibson, D. M., S. B. Krasnoff, J. Biazzo and L. Milbrath. 2011. "Phytotoxicity of antofine from invasive swallow-worts." *Journal of Chemical Ecology* 37:871–879.

Lawlor, F. 2009. "Black Swallow-wort." Plant Conservation Alliance Alien Plant Working Group. http://www.nps.gov/plants/alien/fact/cylo1.htm.

Ground Ivy   *Glechoma hederacea*

NAME AND FAMILY
Ground ivy, gill-over-the-ground, creeping Charlie (*Glechoma hederacea* L.); mint family (Lamiaceae). Leaves look similar to garlic mustard (*Alliaria petiolata*) first-year plants and to violets (*Viola* spp.), but ground ivy has trailing stems.

IDENTIFYING CHARACTERISTICS
Trailing over the ground, ground ivy's stems are four-sided and the leaves are opposite. The leaves are heart shaped with scalloped edges and have a slight minty smell instead of the garlic smell of garlic mustard leaves. The flowering stalks can

reach up to 1 ft. (0.3 m) tall with two or more flowers clustered in the leaf axils. Lavender flowers, tubular with spreading petals (0.38 in. [0.9 cm] long), bloom in late spring to early summer. Plants are perennials and leaves remain green through the winter.

HABITAT AND RANGE
Ground ivy is common from Newfoundland into Ontario and throughout the United States except in the Southwest. It prefers the moist ground and shade of floodplains, low woods, and disturbed areas.

Small lavender flowers cluster in the leaf axils.

Ground ivy can form dense mats.

WHAT IT DOES IN THE ECOSYSTEM

Ground ivy forms a ground cover that deters the establishment and growth of other species. Mostly a problem in lawns and cultivated areas, but can form dense colonies in natural areas. Stems root at the nodes and stem fragments are the major source of spread. Ecologists have used ground ivy as a model clonal plant to study how clonal plants forage for nutrients and light in patchy environments.

HOW IT CAME TO NORTH AMERICA

Ground ivy is native to Eurasia. Settlers introduced it in North America as early as the 1800s as an ornamental and as a salad plant and pot herb. For centuries, European herbalists recommended it for a variety of ailments. However, as with some other edibles, this one contains some dangerous chemicals—terpenes that can cause intestinal and liver damage.

Ground ivy growing among grasses.

Rounded leaves have scalloped edges.

MANAGEMENT
Small patches can be hand pulled or raked in damp soil. Chemical control of ground ivy has mixed results because some genetic strains seem to be resistant to different herbicides. Fall application of glyphosate, triclopyr, or 2,4-D may have some success.

FOR MORE INFORMATION
Czarapata, E. J. 2005. *Invasive Plants of the Upper Midwest.* Madison: University of Wisconsin Press.

Kohler, E. A., C. S. Throssell, and Z. J. Reicher. 2004. "Cultural and chemical control of ground ivy *(Glechoma hederacea)." Horticultural Science* 39:1148–1152.

Waggy, Melissa A. 2009. "*Glechoma hederacea." Fire Effects Information System.* U.S. Department of Agriculture, Forest Service, Rocky Mountain Research Station, Fire Sciences Laboratory. http://www.fs.fed.us/database/feis/.

Japanese Hop   *Humulus japonicus*

NAME AND FAMILY
Japanese hop (*Humulus japonicus* Sieb. & Zucc.); hemp family (Cannabaceae). Unfortunately, Japanese hop cannot serve as a substitute for the common hops used to make beer (*H. lupulus*) due to its different chemical makeup.

IDENTIFYING CHARACTERISTICS
The generally five-lobed leaves and the downward prickles on the stems help identify this annual vine. Japanese hop is available as an ornamental climber with variegated leaves. The leaves are green, rough, five- to nine-lobed in a

palm shape, 2 to 5 in. (5 to 13 cm) long and with toothed edges. Upper leaves may have only three lobes, but most have five. Leaves are arranged opposite each other, attached to the stem by a long leaf stalk (petiole). Where the leaves attach to the stem there is a triangular, leaflike bract. Stems are covered with rough, downward-pointing hairs and can grow to 35 ft. (10.7 m) long in one growing season. Individual flowers are greenish in color and five-petaled, with male and female flowers on separate plants. Flower clusters emerge from the leaf axils. Male flowers bloom along upright flower stems 6 to 10 in. (15 to 25 cm) long, whereas female flowers hang down in cone-shaped clusters 0.5 in. (1.3 cm) long. Bracts around female flowers are hairy and long-pointed, giving the appear-

Japanese hop vine leaves are divided into three to five leaflets.

ance of the cone. Seeds form within the female flower clusters. Common hops has rounded as opposed to pointed leaf lobes and is perennial.

HABITAT AND RANGE
The Japanese hop tends to become established on disturbed, open ground. It grows along riversides, in old fields, and along roadsides. It ranges from southeastern Canada south to Georgia and west to Iowa and Kansas.

WHAT IT DOES IN THE ECOSYSTEM
The rapid growth of this vine allows it to outcompete native plants as it grows over them, blocking light and occupying space. Seeds are dispersed by wind and water. The pollen causes allergies in many people. The Japanese hop is used medicinally in Asia.

HOW IT CAME TO NORTH AMERICA
The Japanese hop was introduced from eastern Asia as an ornamental, probably in the mid- to late 1800s.

MANAGEMENT
Vines can be hand pulled, especially in spring, but wear gloves, since some people experience minor skin irritation when handling these plants. A pre-

Long male flower clusters point up and short female flower clusters hang down in cone-shaped clusters.

emergent herbicide can be applied to prevent seeds from germinating. Glyphosate or metsulfuron methyl can be sprayed on plants after germination and up to late summer. Spraying just before seed formation can help reduce the new seed crop. Seeds in the soil can remain dormant for up to three years. The vines are not shade tolerant and need bare ground to germinate, so planting a dense cover crop and establishing shrubs and trees that shade out the vines can prevent invasions.

FOR MORE INFORMATION

Czarapata, E. J. 2005. *Invasive Plants of the Upper Midwest.* Madison: University of Wisconsin Press.

DeNoma, J. S. 2000. "*Humulus* genetic resources." U.S. Department of Agriculture. Agricultural Research Service. http://www.ars-grin.gov/cor/humulus/huminfo.html.

Pannill, P. D., A. Cook, A. Hairston-Strang, and J. M. Swearingen. *Japanese Hop.* Plant Conservation Alliance Alien Plant Working Group. http://www.nps.gov/plants/alien/fact/huja1.htm.

Terrestrial Plants—Herbaceous Plants— Rosette or Basal Leaves Only

Lesser Celandine   *Ficaria verna*

NAME AND FAMILY
Lesser celandine, fig buttercup, pilewort (*Ficaria verna* Huds., formerly *Ranunculus ficaria* L.); buttercup family (Ranunculaceae). Lesser celandine was celebrated in England by the poet William Wordsworth, on whose gravestone the flower appears. Looks similar to native marsh marigold (*Caltha palustris*).

IDENTIFYING CHARACTERISTICS
Lesser celandine is a carpet-like ground cover that breaks out in bright yellow stars (flowers up to 3 in. [7.6 cm] wide) in March and April. The flowers are usually five-petaled, rising on stalks above the irregular but generally heart-shaped or kidney-shaped leaves, which vary greatly in size. When the flowering begins to fade, small bulblets (bulbils) are apparent above the ground while it dies back to its underground tubers in late summer. Leaves reappear in late winter. The similar-looking marsh marigold does not have either the aboveground bulbils or belowground tubers that characterize lesser celandine. Lesser celandine is not related to greater celandine (*Chelidonium majus*).

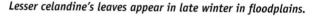

Lesser celandine's leaves appear in late winter in floodplains.

HABITAT AND RANGE

Lesser celandine prefers low forested floodplains but will grow in higher topography. In North America it grows from Newfoundland south to Virginia and Tennessee and west to Ontario, Wisconsin, Indiana, and Missouri, and in the Northwest in British Columbia, Washington, and Oregon.

WHAT IT DOES IN THE ECOSYSTEM

The tubers of lesser celandine send up shoots that form leaves during late winter. A mat of lesser celandine can dominate acres of forest floor, denying light to native plants that would normally use the early spring for their own growth before trees and shrubs leaf out and block direct sunlight. Many native plants that flower and form fruit in early spring cannot compete with lesser celandine. Lesser celandine may also produce chemicals that inhibit the growth of other plants. It spreads mainly by its numerous tubers.

HOW IT CAME TO NORTH AMERICA

Most writers say lesser celandine was introduced as an ornamental but offer no specifics on date, location, or agent. It is native to northern Europe, and cultivars were developed there as early as the late 1500s. Landscapers and horticulturalists still use it and have cultivated many varieties. Like horticulturalists,

Left: *Lesser celandine flowers vary in petal number from five to many petals.*
Right: *Heart-shaped leaves are glossy green.*

Plants form a carpet in very early spring.

the British poet William Wordsworth celebrated its bravery of the cold and damp and its brightening of spring; herbalists touted it as a cure for hemorrhoids, or piles, thus the name pilewort.

MANAGEMENT
Small colonies or scattered plants can be dug out with a trowel, though care should be taken to remove all tubers. Since the foliage dies back in summer, herbicides containing glyphosate approved for use near water should be applied before half the plants are flowering in late winter or in early spring when temperatures are above 50° F (10° C). Since the wet areas that host lesser celandine often provide habitat for amphibians and native spring wildflowers, an early spraying, before the amphibians and native ephemerals emerge would be best.

FOR MORE INFORMATION
Axtell, A. E., A. DiTommaso, and A. R. Post. 2010. "Lesser celandine (*Ranunculus ficaria*): a threat to woodland habitats in the Northern United States and Canada." *Invasive Plant Science and Management* 3(2):190–196.

Swearingen, J. 2010. "Fig buttercup." Plant Conservation Alliance Alien Plant Working Group. http://www.nps.gov/plants/alien/fact/rafi1.htm.

Orange Daylily   *Hemerocallis fulva*

NAME AND FAMILY
Orange daylily, common daylily, ditch lily, outhouse lily, roadside lily (*Hemerocallis fulva* [L.] L.); lily family (Liliaceae). Yellow or lemon daylily (*H. lilioasphodelus* L.) has a more limited distribution and is considered invasive in some Mid-Atlantic and southeastern states.

IDENTIFYING CHARACTERISTICS
The trumpet-shaped orange flowers of daylilies are a sign of summer to most people. Held aloft on 2 to 4 ft. (0.6 to 1.2 m) rounded stalks, each flower has six petals (actually three petals and three sepals), is 3 to 4 in. (7.6 to 10 cm) wide, and opens for just one day. Each cluster of flower usually has five to nine flowers, little cigar-shaped buds that open on different days. Long strap-like leaves (1 to 3 ft. [0.3 to 1 m]), arch and are shorter than the flowering stalks. Thick tuberous roots and rhizomes form underground and are the main means of dispersal. Seeds are seldom fertile. Yellow daylily has yellow flowers instead of orange.

HABITAT AND RANGE
Found on old house sites, along roadsides, streams, in meadows, and in moist open woodlands. Occurs throughout most of southern Canada and the United States except in very dry areas.

Daylily flowers are held in clusters with each flower open for only a day.

Daylilies are often found in moist soils along streams and roadsides.

WHAT IT DOES IN THE ECOSYSTEM

Where established, daylilies form dense clumps that few other plants can penetrate. Daylilies can maintain an optimal position in the soil for survival by contracting their roots or elongating their shoots, a characteristic generally shared by other plants with thickened roots. The long leaves shade out anything that might try to germinate under them. Plants spread by rhizomes and occasionally by seed. Spread is not generally very rapid except when aided by movement of the rhizomes in soils or when gardeners dispose of unwanted plants near natural areas. Flowers and tubers are edible and used in many Asian dishes. Deer also browse the buds and flowers.

HOW IT CAME TO NORTH AMERICA

Introduced from Europe in the 1800s (but native to Asia), the easy-to-grow daylilies became very popular ornamental plants. Horticulturalists developed thousands of cultivars, with some fans saying more than 40,000 are registered. Most daylilies for sale in the nursery trade currently are hybrids that do not spread via rhizomes.

MANAGEMENT

In fall dig up or plow daylilies, then rake to remove all plant parts. Or cut plants close to ground and apply glyphosate to cut surfaces. A newly introduced fungus, daylily rust, could cause significant damage to daylilies.

FOR MORE INFORMATION

Czarapata, E. J. 2005. *Invasive Plants of the Upper Midwest*. Madison, WI: University of Wisconsin Press.

Peat, J. P., and T. L. Petit. 2004. *The Daylily: A Guide for Gardeners*. Portland, OR: Timber Press.

Puetz, N. 1998. "Underground plant movement. V. Contractile root tubers and their importance to the mobility of *Hemerocallis fulva* L. (Hemerocallidaceae)." *International Journal of Plant Sciences* 159(1):23–30.

Hawkweeds	*Hieracium* spp.

NAME AND FAMILY

Orange hawkweed, devil's paintbrush (*Hieracium aurantiacum* L.), mouse ear hawkweed, (*H. pilosella* L.), meadow hawkweed, yellow king devil (*H. caespitosum* Dumort.; formerly *H. pratense*); aster family (Asteraceae). Many native and nonnative hawkweeds are similar to these species. Common catsear (*Hypochoeris radicata*) looks similar to mouse ear hawkweed.

IDENTIFYING CHARACTERISTICS

Hawkweeds are perennial herbaceous plants that form a ground-hugging rosette of leaves. Hawkweeds produce a milky sap. The leaves of orange and mouse ear hawkweed are whitish below and green above, with long white hairs. Leaves are 1 to 5 in. (2.5 to 12.7 cm) long and up to 0.75 in. (1.9 cm) wide,

The ends of the petals are squared, with several rounded teeth.

growing almost exclusively at the base of the bristly stem. Hawkweed flowers are dandelion-like, but the petals are squared on the ends, with several rounded teeth. Orange hawkweed grows 1 to 2 ft. (0.3 to 0.6 m) tall and has orange-red flowers with yellowish centers, held in a cluster of 5 to 30 flower heads at the top of a hairy flowering stalk in summer. Meadow hawkweed has a similar flower arrangement, but flowers are yellow. Mouse ear hawkweed's yellow flowers are held on a stalk 2 to 12 in. (5.1 to 30.5 cm) tall with only one to a few flower heads per stalk. The bracts under the flowers are also hairy. Blooming in summer, flowers are 0.5 to 1.5 in. (1.3 to 3.7 cm) across. Catsear has yellow flowers, but leaves are toothed or lobed.

HABITAT AND RANGE

Mouse ear, orange, and meadow hawkweeds grow in grasslands, pastures, rangelands, and on disturbed open ground. Orange hawkweed has the broadest range, growing across Canada and most of the United States south to Florida and California. Mouse ear and meadow hawkweeds occur from eastern Canada south to Georgia and in the west from Alaska to Oregon.

WHAT IT DOES IN THE ECOSYSTEM

These three nonnative species send out aboveground runners (stolons) from which many daughter plants grow. After new plants appear, the original plant dies. They also spread by seed. Most seeds are produced without fertilization (apomictically). Hawkweeds tend to form dense colonies. Around a group of plants, hawkweed roots reduce available moisture and nutrients and lower the pH of the soil, reducing the ability of other plants to establish. Small barbs on the seeds catch in fur, clothing, and equipment, dispersing the seeds.

Flowers cluster at the end of a hairy stalk.

HOW IT CAME TO NORTH AMERICA

Orange hawkweed came to New England as an ornamental before 1818, and others were introduced for pharmaceutical properties from the middle to the end of the nineteenth century.

MANAGEMENT

Small infestations can be dug out. Mowing is ineffective, as it just causes plants to flower lower to the ground and spread further via the stolons. Hawkweeds are not shade tolerant and can be outcompeted by some grasses and clovers. Application of triclopyr or aminopyralid in spring to early summer, accompanied by overseeding competitive plants, can also reduce hawkweeds.

Plants can form dense colonies on disturbed ground.

FOR MORE INFORMATION

Boswell, C. C., and P. R. Espie. 1998. "Uptake of moisture and nutrients by *Hieracium pilosella* and effects on soil in a dry, sub-humid grassland." *New Zealand Journal of Agricultural Research* 41:251–261.

King County Noxious Weed Control Program Best Management Practices. 2010. "Hawkweed Best Management Practices." http://your.kingcounty.gov/dnrp/library/water-and-land/weeds/ BMPs/hawkweed-control.pdf.

Wilson, L. M. 2006. *Key to the Identification of Invasive and Native Hawkweeds* (Hieracium *spp.*) *in the Pacific Northwest*. British Columbia Ministry of Forest and Range, Forest Practices Branch, Invasive Alien Plant Program. http://www.for.gov.bc.ca/hra/Publications/invasive _plants/Hawkweed_key_PNW_2007.pdf.

Yellow Flag Iris   *Iris pseudacorus*

NAME AND FAMILY
Yellow flag iris, pale yellow iris (*Iris pseudacorus* L.); iris family (Iridaceae). The leaves of different iris species look very similar, but the flowers of yellow flag are distinctive. Whether it is named for the River Lys or, as some say, for French King Louis VII, it is the model for the fleur-de-lis, which adorns the flag of the French kings.

IDENTIFYING CHARACTERISTICS
This is the only completely yellow, large wild iris in North America.

Right: *Yellow flag iris blooms in late spring.* **Bottom:** *Flowers are held on stalks among the swordlike leaves.*

Yellow flag iris grows in wet soils.

Bright-yellow flowers bloom in groups of 2 to 10 atop 1 to 3 ft. (0.3 to 1 m) stems in late spring to early summer. The flower is made up of three drooping deep yellow sepals marked with purple-brown and three smaller upright petals and is about 3 in. (7.6 cm) wide. Swordlike, flattened leaves about ¾ in. (1.9 cm) wide and up to 3 ft. (1 m) long arise in a fan from the base. Leaves die back in winter in colder climates. The leaves and flower stalk grow from a stout rhizome. Oblong brown capsules 2 in. (5 cm) long contain the seeds.

HABITAT AND RANGE
Yellow flag iris generally grows in nutrient-rich marshes, ditches, and other wet areas in water up to 9 in. (23 cm) deep. It can tolerate brackish water,

acidic water, periods of drought, and soils with low oxygen levels. It occurs throughout most of the United States and Canada except in the drier states and provinces.

WHAT IT DOES IN THE ECOSYSTEM

Once established, individual plants continue to increase in size, forming dense clumps that exclude other wetland species. Pieces of rhizomes and seeds float to new areas. The dense mats of rhizomes can trap sediments, creating higher, drier land that favors tree and shrub species. The flowers are visited by bumblebees, butterflies, and hummingbirds in its introduced range. The plant is seldom fed on by animals because of high levels of glycosides in the leaves and rhizomes.

HOW IT CAME TO NORTH AMERICA

Native to Europe, western Asia, and northern Africa, yellow flag iris is commonly sold as a plant for water gardens. It was introduced as an ornamental in the mid-1800s and was found as early as 1868 in a wetland near Poughkeepsie, New York. Horticulturalists have bred many cultivars of yellow flag. It is also used to remove heavy metals in wastewater treatment.

MANAGEMENT

The rhizomes can be dug up or foliage can be sprayed or wiped with glyphosate in summer. Cutting foliage and then applying herbicide to the cut surface is also effective. When handling plants, wear gloves to protect skin from chemicals in the leaves and rhizomes that can cause irritation. Large stands are controlled using mechanical harvesters and chopping machines.

FOR MORE INFORMATION

Cody, W. J. 1961. "*Iris pseudacorus* L. escaped from cultivation in Canada." *Canadian Field Naturalist* 75:139–142.

Ramey, V. 2001. *Non-native invasive aquatic plants in the United States.* Center for Aquatic and Invasive Plants, University of Florida and Sea Grant. http://plants.ifas.ufl.edu/node/205.

Stone, Katharine R. 2009. "*Iris pseudacorus.*" *Fire Effects Information System.* U.S. Department of Agriculture, Forest Service, Rocky Mountain Research Station, Fire Sciences Laboratory. http://www.fs.fed.us/database/feis/.

Terrestrial Plants—Herbaceous Plants— Alternate Leaves—Leaves Entire

Wild Taro — *Colocasia esculenta*

NAME AND FAMILY
Wild taro, elephant's ears, coco yam, dasheen (*Colocasia esculenta* [L.] Schott); arum family (Araceae). Elephant ear (*Xanthosoma sagittifolium* (L.) Schott), introduced from northern South America, is very similar in appearance and distribution.

IDENTIFYING CHARACTERISTICS
The large, ornamental leaves of this plant can be dark green to purple, shaped like an elephant's ear or a fat arrowhead, 2 to 3 ft. (0.6 to 1 m) long and 1 to 2 ft. (0.3 to 0.6 m) wide. Leaves are held on 3 ft. (1 m) long stalks (petioles) that attach in the center of the leaf (peltate). Petioles of native arums (*Peltandra* spp.) and exotic elephant ear attach to the edge of the leaf. Outside their native range, wild taros spread vegetatively and seldom produce flowers. Tiny flowers are held on a fleshy stalk covered by a yellow hood (a spathe and spadix). Underground rhizomes form stout tubers shaped like tops (technically corms) that generally weigh 1 to 2 lbs. (0.45 to 0.9 kg). The invasive variety forms aboveground runners called stolons. Elephant ear has similar leaves but grows to be 6 feet tall.

Leaves are shaped like fat arrowheads, with the leaf stalks attached near the center of the leaf blades.

HABITAT AND RANGE
Grows along rivers and lake shores, ditches and canals. Prefers some shade. Grows from North Carolina to Texas, but mainly considered invasive in Florida. Does not tolerate frost.

Leaves vary in appearance, with many ornamental cultivars available.

WHAT IT DOES IN THE ECOSYSTEM
The huge leaves shade out native vegetation and the vegetative growth forms dense stands. Particularly a problem along waterways where it can change the structure of riparian plant communities. During floods, roots can break off and be carried downstream. Roots are edible after cooking.

HOW IT CAME TO NORTH AMERICA
First introduced to the Americas as food for slaves brought from Africa. The U.S. Department of Agriculture promoted taro in 1910 as an alternative crop to potatoes in southern states. Native to southeastern Asia, where it has been cultivated for at least 6,000 years. Also commonly planted as an ornamental plant.

MANAGEMENT
Small plants can be pulled or dug out. The oxalic acid in the leaves can cause skin irritation. Larger plants can be cut and the regrowth treated with glyphosate, but repeat applications will probably be necessary.

FOR MORE INFORMATION

Langeland, K. A., H. M. Cherry, C. M. McCormick, and K. A. Craddock Burks et al. 2008. *Identification and Biology of Nonnative Plants in Florida's Natural Areas,* 2nd ed. SP 257. Gainesville, FL: University of Florida IFAS.

Wang, J. K. 1983. *Taro: A review of* Colocasia esculenta *and its potentials.* Honolulu, HI: University of Hawaii Press.

Weber, E. 2003. *Invasive Plant Species of the World.* Cambridge: CABI Publishing.

Wild taro threatens native plants along waterways.

Houndstongue   *Cynoglossum officinale*

NAME AND FAMILY
Houndstongue, gypsyflower, beggar's lice (*Cynoglossum officinale* L.); borage family (Boraginaceae). In the Pacific Northwest C. *grande* is also called hound-stongue, or Pacific or Western houndstongue, but this plant is native to western North America.

IDENTIFYING CHARACTERISTICS
The leaves and stem of houndstongue are softly hairy. Leaves alternate along the stem, becoming smaller and almost clasping the stem toward the top. The leaves at the base of the plant, pointed oblongs with a long tapering base, are 4 to 12 in. (10 to 30 cm) long and 0.8 to 2 in. (2 to 5 cm) wide. Elongated, slightly drooping, clusters of around 10 reddish-purple, funnel-shaped, 0.25 in. (0.6 cm) flowers bloom from the upper leaf axils from early to mid-summer. Short, spiny bristles cover the hard, flat-sided fruits. Plants often have a musty odor. They usually grow as biennials, forming a rosette the first year, and growing to 2 ft. (0.6 m) and flowering the second year. Each plant can have multiple stems and the upper parts of the stems branch. A long, black taproot supports the plant.

Right: *Seeds are covered by a hard, bristly cover.* **Bottom:** *Purple flowers bloom in the leaf axils in summer.*

Left: *Leaves become smaller going up the stem.* **Right:** *Flowers generally form on second-year plants.*

Pacific houndstongue has very large leaves, mostly at the base of the plant, and the seedpods have very small barbs.

HABITAT AND RANGE

Houndstongue prefers open, disturbed ground such as along roadsides, in pastures and meadows, at forest edges, and in clear-cuts. It occurs throughout Canada except in Prince Edward Island and Newfoundland and throughout all but the southernmost states of the United States.

WHAT IT DOES IN THE ECOSYSTEM

Houndstongue can sometimes grow in dense stands that inhibit the establishment of native plants, but often it occurs as scattered individuals. The taproot, reaching more than 40 in. (100 cm), allows the plant to withstand droughts and to store food over the winter. The plants contain alkaloids toxic to horses and cattle, and the seedpods catch in fur and can cause injuries to animals' eyes. Seeds are mainly carried to new areas stuck on animal's fur, but are also gravity dispersed and may be wind dispersed. The flowers are pollinated by bumblebees and also provide nectar and pollen for some other insects.

HOW IT CAME TO NORTH AMERICA

Houndstongue probably arrived as a contaminant in crop seeds from Europe. The first herbarium specimen was collected in Ontario, Canada, in 1859. Like other plants in the borage family, houndstongue is still marketed for various herbal and folk remedies, especially for colds and coughs and for poultices, but it does have alkaloids suspected of being carcinogens.

MANAGEMENT

First-year rosettes are difficult to kill if well developed because of the taproot. Small infestations can be hand pulled or the root crown cut below the ground surface. Clipping second-year plants before they set seed will kill many plants, but sometimes regrowth will need to be cut. The herbicides dicamba, metsulfuron, and 2,4-D can kill houndstongue plants if applied before plants bolt and flower. Houndstongue reinfestation can be significantly reduced by establishing good native plant cover. A root-boring weevil, *Mogulones cruciger*, has been released in Canada as a biological control agent.

FOR MORE INFORMATION

Jacobs, J. 2007. *Ecology and Management of Houndstongue (*Cynoglossum officinale *L.).* USDA NRCS Invasive Species Technical Note No. MT-8. ftp://ftp-fc.sc.egov.usda.gov/MT/www/technical/invasive/Invasive_Species_Tech_Note_MT8.pdf.

Upadhyaya, M. K., H. R. Tilsner, and M. D. Pitt. 1988. "The biology of Canadian weeds. 87. *Cynoglossum officinale* L." *Canadian Journal of Plant Science* 68(3):763–774.

Zouhar, K. 2002. "*Cynoglossum officinale.*" *Fire Effects Information System.* U.S. Department of Agriculture, Forest Service, Rocky Mountain Research Station, Fire Sciences Laboratory. http://www.fs.fed.us/database/feis/.

Leafy Spurge   *Euphorbia esula*

NAME AND FAMILY

Leafy spurge, wolf's milk (*Euphorbia esula* L.); euphorb family (Euphorbiaceae). Two species, *E. esula* and *E. pseudovirgata*, as well as several varieties and hybrids, can be found in North America, but they are usually grouped as *E. esula*, and are virtually indistinguishable except by specialist insects. Looks similar to another related invasive, cypress spurge (*E. cyparissias* L.).

Yellow-green bracts surround the tiny flowers of leafy spurge.

A leafy spurge invasion in a prairie.

IDENTIFYING CHARACTERISTICS
Leafy spurge is a perennial plant distinguished by its alternate, linear, 1 to 4 in. (2.5 to 10 cm) long leaves and milky sap. Bluish-green stems appear very early in spring and form dense clusters growing to 3 ft. (1 m) tall. In late spring, and sometimes again in late summer, unusual clusters of flowers bloom that look like they have two to three yellow-green rounded petals. The "petals" are actually bracts about 0.75 in. (1.9 cm) wide; tiny flowers with three parts cluster in the middle. A three-lobed seed capsule holds three seeds, which are dispersed up to 15 ft. (4.6 m) when the capsule explodes after drying. Seeds also float and can disperse along waterways. The brown roots have pinkish buds and the plants have both extensive surface roots as well as taproots that can reach up to 21 ft. (6.3 m) in length. Cypress spurge is a smaller plant with narrower leaves more closely spaced and smaller flower bracts.

HABITAT AND RANGE
Infesting more than 5 million acres (2 million ha) in 35 states and the plains of Canada, leafy spurge occurs from southern Canada south to northern Texas and Virginia, mostly in prairies, rangelands, and pastures, but also along streams and in open woodlands. The highest concentrations are in the Great Plains region.

WHAT IT DOES IN THE ECOSYSTEM
Because of its dense growth, leafy spurge displaces native plants and reduces community diversity. The dense growth occurs both from new buds forming at the crown of the plant and from buds along the roots. The roots exude chemicals that deter the growth of other plants. The extensive root network allows plants

to store energy and recover after damage. The milky sap irritates animals and can poison cattle, so rangelands infested by spurge lose their productivity and monetary value. The flowers are used by many insects because of the large amounts of nectar and pollen produced. The seeds are eaten and spread by sheep, goats, deer, wild turkey, and other birds, as well as by ants.

HOW IT CAME TO NORTH AMERICA
First found in the United States in Newbury, Massachusetts, in 1827, leafy spurge may have been introduced several times accidentally in shipments of seeds or in ship ballast. It was also reported to have arrived in Minnesota in a shipment of oats. Gardeners may have spread the plant for its ornamental bracts. Native to central and eastern Europe.

Stems with narrow leaves hold branched clusters of flowers.

MANAGEMENT
Management of leafy spurge on a large scale uses a combination of methods: the leafy spurge flea beetle and other biological control agents, along with herbicides, grazing regimes, clipping, tilling, and/or burning. Sheep and goats will graze leafy spurge and help in its control. Picloram will control leafy spurge with little harm to grasses, but can travel through the soils and will kill other perennials as well as trees and shrubs. Persons handling leafy spurge should wear gloves since the sap can be a skin irritant. It is important to target new small populations to prevent further spread of the plants because of the extreme difficulty of controlling large infestations.

FOR MORE INFORMATION
Dunn, P. H., and A. Radcliffe-Smith. 1980. "The variability of leafy spurge (*Euphorbia* spp.) in the United States." *North Central WCC Research Report* 37:48–51.

Gucker, Corey L. 2010. "*Euphorbia esula.*" *Fire Effects Information System.* U.S. Department of Agriculture, Forest Service, Rocky Mountain Research Station, Fire Sciences Laboratory. http://www.fs.fed.us/database/feis.

"Team leafy spurge." http://www.team.ars.usda.gov/v2/team.html.

Japanese Knotweed   *Fallopia japonica*

NAME AND FAMILY
Japanese knotweed, Mexican bamboo, Japanese fleece flower, crimson beauty, or Reynoutria (*Fallopia japonica* var. *japonica* (Houtt.) Ronse Decr. Sieb. & Zucc.); buckwheat family (Polygonaceae). Other commonly used Latin names for this species are *Polygonum cuspidatum* and *Reynoutria japonica*. Other native and introduced species of *Polygonum* may look similar in the seedling stages. Japanese knotweed grows much taller and has stouter rhizomes compared with most others. It does closely resemble and can hybridize with giant knotweed, *F. sachalinensis*, another introduced species.

IDENTIFYING CHARACTERISTICS
Growing 4 to 10 ft. (1 to 3 m) tall, Japanese knotweed has smooth, stout, hollow stems, swollen at the joints where the leaf meets the stem. The young stems and leaves have a purplish color. As is characteristic of the family, a membranous sheath surrounds the stem above each joint. The leaves, typically 6 in. (15 cm) long and 3 to 4 in. (7.6 to 10 cm) wide, are broad ovals with pointed tips. Leaves alternate along the stems. The triangular and shiny seeds are about 0.1 in. (0.25 cm) long. Japanese knotweed is an herbaceous perennial, and its stems grow back each year from a large underground rhizome that might stretch 45 to 60 ft. (14 to 18 m) in length. The greenish-white flowers occur in attractive sprays in summer, held aloft along the arching stems. The small white fruits have wings to help the seeds ride winds to new sites.

The sprays of flowers that bloom along the stems cause some to call this plant fleece flower.

Left: *Japanese knotweed flowers in summer, with the flowers held above arching stems.* Right: *Broad, oval leaves with pointed tips alternate along the stems.*

HABITAT AND RANGE

Japanese knotweed occurs in thirty-six states, from Maine to Wisconsin, south to Louisiana, and in several midwestern and western states. In Canada, it is distributed from Newfoundland to Ontario. It grows most commonly along waterways, in low-lying areas, waste places, utility rights-of-way, and around old homesites. It tolerates high temperatures, high salinity, and drought. It generally prefers full sun, but will tolerate shade.

WHAT IT DOES IN THE ECOSYSTEM

Japanese knotweed spreads quickly, forming dense stands that exclude native vegetation. The thick layer of decomposing stems and leaves also mulches out competitors. In riparian areas it can survive severe floods and rapidly recolonize, usurping the role of native species. It is extremely persistent once established. The change in vegetation it causes alters fish and wildlife habitat. Japanese knotweed can also cause structural damage, sprouting through asphalt and foundations.

HOW IT CAME TO NORTH AMERICA

Native to eastern Asia, this species was introduced from Japan to the United States via the United Kingdom as an ornamental, for fodder, and for erosion control in the late 1800s. By 1900 it was spreading around Philadelphia, Pennsylvania, Schenectady, New York, and Atlantic Highlands, New Jersey. In the United Kingdom, where it arrived as early as 1825, it is illegal to spread Japan-

Stems grow quickly in summer from large underground rhizomes.

ese knotweed, and soil removed from infested areas must be disposed of at licensed landfills. Young stems of knotweed are edible, with a flavor similar to that of rhubarb.

MANAGEMENT
Stalks may be cut or pulled repeatedly (two to three times during the growing season) to exhaust the rhizome and kill the plant. This may take up to 10 years in a well-established stand. Stems and rhizomes must be thoroughly dried or burned prior to disposal because they will regenerate. The transfer of fill dirt to new sites often spreads knotweed. Various herbicides effectively control knotweed, but because of the extensive rhizome system, total eradication may require several consecutive years. Cutting stems just before flowering in early summer, followed by spraying foliage regrowth with glyphosate or triclopyr in late summer or recutting and applying herbicide to cut stems may be most effective.

FOR MORE INFORMATION
Aguilera, A. G., P. Alpert, J. S. Dukes, and R. Harrington. 2010. "Impacts of the invasive plant *Fallopia japonica* (Houtt.) on plant communities and ecosystem processes." *Biological Invasions* 12:1243–1252.

Cornwall knotweed forum. http://www.cornwall.gov.uk/default.aspx?page=13789.

Grimsby, J. L., and R. V. Kesseli. 2010. "Genetic composition of Japanese knotweed s.l. in the United States." *Biological Invasions* 12: 1943–1946.

Remaley, T. 2009. *Japanese knotweed.* Plant Conservation Alliance Alien Plant Working Group. http://www.nps.gov/plants/alien/fact/faja1.htm.

Scott, R., and R. H. Marrs. 1984. "Impact of Japanese knotweed and methods of control." *Aspects of Applied Biology* 291–296.

Dyer's Woad   *Isatis tinctoria*

NAME AND FAMILY
Dyer's woad (*Isatis tinctoria* L.); mustard family (Brassicaceae).

IDENTIFYING CHARACTERISTICS
Dyer's woad is most obvious from spring to late summer, when it is covered with yellow, four-petaled flowers with six stamens, four long and two short. Small clusters of flowers bloom atop 1 to 4 ft. (0.3 to 1.2 m) stems. Dyer's woad usually grows as a biennial, beginning as a rosette of succulent, blue-green, hairy leaves 1.5 to 7 in. (3.7 to 18 cm) long. The second year, the plant produces up to 20 stems. Both rosette and stem leaves have a cream-colored midrib. The leaves along the stem alternate and are lance shaped with bases that clasp the stem. Flattened seed pods, 0.38 in. (0.9 cm) long by 0.25 in. (0.6 cm) wide, hang from short stalks at the ends of the stems. As the pods mature they turn black, remaining on the plant. Plants have a long taproot as well as lateral roots.

HABITAT AND RANGE
Usually found on poor, dry soils in the western United States and Canada. Often common on rangelands, roadsides, and in open forests. Seldom found outside of cultivation in eastern states despite its arrival to Plymouth colony in the 1600s.

WHAT IT DOES IN THE ECOSYSTEM
Stands can cover many acres and outcompete native plants. The fruit of dyer's woad contains chemicals that inhibit germination of seeds until the chemicals

In their second year, plants produce many stems topped by yellow flowers.

The seedpods turn black as they mature.

are washed away by rains. By timing germination for the point when soil moisture is optimal, dyer's woad wins an advantage over many other colonizing plants. It reduces crop and forage yields.

HOW IT CAME TO NORTH AMERICA

Native to central Asia and southern Russian grasslands, dyer's woad was cultivated for centuries by Europeans and North American colonists to produce indigo-colored dye. It is inferior to the indigo of India (*Indigofera tinctoria*) but grows in colder climates. The ancient Picts of Scotland who refused to be conquered by the Romans painted their bodies with dye from woad. Botanists first documented naturalized stands in the West, where it arrived from Europe as a contaminant in alfalfa seed, in the early 1900s.

MANAGEMENT

Plants can be hand pulled after they bolt but before they flower. In fields, cultivation in spring and again in late fall will control plants. A combination of the herbicides metsulfuron and 2,4-D applied before plants bolt is an effective chemical control. A native rust, *Puccinia thlaspeos*, which prevents seed production, is establishing in some populations and preventing their spread.

FOR MORE INFORMATION

Farah, K. O., A. F. Tanaka, and N. E. West. 1988. "Autecology and population biology of dyers woad (*Isatis tinctoria*)." *Weed Science* 36:186–193.

Washington State Noxious Weed Control Board. 1999. "Dyer's woad (*Isatis tinctoria* L.)" http://www.nwcb.wa.gov/siteFiles/Isatis_tinctoria.pdf.

Zouhar, Kris. 2009. "*Isatis tinctoria*." *Fire Effects Information System*. U.S. Department of Agriculture, Forest Service, Rocky Mountain Research Station, Fire Sciences Laboratory. http://www.fs.fed.us/database/feis.

Kochia   *Kochia scoparia*

NAME AND FAMILY
Kochia, summer cypress, Mexican fireweed (*Kochia scoparia* [L.] Schrad.); goosefoot family (Chenopodiaceae). Bassia, five hook bassia, fivehorn smotherweed (*Bassia hyssopifolia* [Pall.] Ktze.), a close relative, is also considered invasive but is less widespread.

IDENTIFYING CHARACTERISTICS
This annual bushy plant can grow to 6 ft. (1.8 m) in height with a pyramid shape, and the stems and leaves turn bright reddish-purple in fall. Once dry, the stems break off, forming a tumbleweed that disperses seeds as it rolls around. A taproot generally grows 6 to 8 ft. (1.8 to 2.4 m) down, while lateral roots can extend more than 20 ft. (6 m). The alternate leaves shorten as their position on the stem rises, but they grow to 2.5 in. (6.2 cm) long and are very narrow, less than ¼ in. (0.6 cm) wide. Usually the underside and edges of the leaf are hairy. Small clusters of inconspicuous greenish flowers bloom in the leaf axils in summer. Bassia has hooked spines on the flowers, and the leaves are usually less than 1 in. (2.5 cm) long.

Left: *Purplish stems hold narrow leaves.* **Right:** *Inconspicuous flowers bloom in the leaf axils.*

A bushy kochia plant growing along a driveway.

HABITAT AND RANGE

Found throughout Canada and the United States except in coastal areas and in the most southeastern states of the United States. Most widespread in the plains provinces and states. Salt and drought tolerant, it is often found on rangelands, along roadsides and riverbanks, and in salt-affected fields. It prefers hot, sunny summers.

WHAT IT DOES IN THE ECOSYSTEM

Kochia effectively competes against other plants for water and space, and its leaf litter releases toxins that inhibit the growth of itself and other plants. In areas with severe infestations, it could result in more frequent fires because of the mass of dry stems. In agricultural fields, it reduces crop production severely, but it used to be planted in arid hay fields because its nutritional value is equivalent to that of alfalfa. It is grazed by antelope and deer, and the seeds are eaten by prairie dogs. Eaten in large quantities though, it can be toxic because of oxalates in the leaves and stems. It is an alternate host for beet and tobacco viruses. It has been planted for mine reclamation efforts.

HOW IT CAME TO NORTH AMERICA

Planted as forage and in hay fields where it was known as "poor man's alfalfa" in the southwestern United States. A subspecies introduced around 1900 is also planted as an ornamental because of its red fall color and may contribute to the invasion. Native to Europe and western Asia.

Left: *Lady's thumb has a purple "thumbprint" on the leaf.* **Right:** *Oriental lady's thumb has upright clusters of flowers.*

IDENTIFYING CHARACTERISTICS

The *Polygonum* spp. are all known for their swollen, jointed stems and ocreae, papery sheaths that surround the stem above the nodes. The alternate leaves of these four species are lance shaped. The leaves of pale smartweed have sticky yellow dots or hairs on the undersides of the upper leaves. Lady's thumb leaves have a purplish triangle in the middle of the leaf like a thumbprint. The flowers of these four species cluster tightly along flower spikes. Each five-petaled flower is less than 0.06 in. (0.15 cm) across. Flattened, round to oval seeds are brown to dark brown, and pointed on both ends. Flower clusters grow from the leaf axils and from the ends of the stems, blooming from early summer into fall. Pale smartweed's flower clusters are greenish white to light pink and 0.5 to 4 in. (1.3 to 10 cm) long, with longer clusters nodding. Kiss me over the garden gate also has nodding clusters of flowers, but flowers are dark pink. Lady's thumb and Oriental lady's thumb have upright spikes of white to pink flowers. The ocreae of both lady's thumbs are bristled, with Oriental lady's thumb's bristles as long as the height of the ocrea and lady's thumb's bristles being shorter. The ocrea of Pennsylvania smartweed is not bristled and leaves usually lack the purple mark. Pale smartweed and kiss me over the garden gate can reach 5 ft. (1.5 m) tall. The lady's thumbs tend to sprawl and seldom reach over 2 ft. (0.6 m) tall. All of these species are annuals.

Pale smartweed has nodding clusters of flowers.

HABITAT AND RANGE
Pale smartweed and lady's thumb occur across the United States and Canada. Oriental lady's thumb and kiss me over the garden gate are distributed primarily in eastern and central North America. Smartweeds tend to grow in organic, rich soils with plenty of moisture. They can be found along streambanks, marsh edges, trail edges, and as a weed in cultivated crops.

WHAT IT DOES IN THE ECOSYSTEM
Smartweeds mainly colonize disturbed sites, but they can carpet the ground, potentially outcompeting native plants. They are considered weeds in cultivated crops and can harbor crop diseases.

HOW IT CAME TO NORTH AMERICA
Long used for medicinal properties, the seeds of some species of smartweeds were also used for food, and other parts of the plant used to make dyes. Several species probably arrived in North America with the first European settlers, brought as medicinal or food plants or as contaminants in crop seeds. Kiss me over the garden gate was introduced as an ornamental. Pale smartweed and lady's thumb are native to Europe. Oriental lady's thumb and kiss me over the garden gate are native to Asia. British plant collector Peter Collinson sent seeds of kiss me over the garden gate to John Bartram in Philadelphia, Pennsylvania, and John Custis of Williamsburg, Virginia, in 1736.

MANAGEMENT
These plants can be pulled up, preferably before seeds form. Plants can also be sprayed with glyphosate in summer.

FOR MORE INFORMATION
Cornett, P. 1996. "Naming the flowers . . . according to Jefferson." *Twinleaf Journal* (January). http://www.twinleaf.org/articles/flowers.html.
Mitich, L. 2007. "Pale smartweed and other Polygonums." *Intriguing World of Weeds.* Weed Science Society of America. http://www.wssa.net/Weeds/ID/WorldOfWeeds.htm#x.
Royer, F., and R. Dickinson. 1999. *Weeds of the Northern U.S. and Canada.* Renton, WA, and Alberta: Lone Pine Publishing and University of Alberta Press.

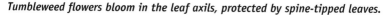

Tumbleweed *Salsola tragus*

NAME AND FAMILY
Tumbleweed, wind witch, Russian thistle (*Salsola tragus* L.); goosefoot family (Chenopodiaceae). Formerly, several species were treated as one, *S. pestifer* A. Nels. The species are adapted to different climates and altitudes and can hybridize. These species include *S. kali* ssp. *tenuifolia, S. australis,* and *S. iberica.* "Salsola" is from the Latin root for salt because the plant has a high salt tolerance. Kochia (*Kochia scoparia,* p. 273) also forms tumbleweeds.

IDENTIFYING CHARACTERISTICS
These are the "tumbling tumbleweeds" that became a symbol of the American West when in the 1940s the Sons of the Pioneers adapted a poem from the University of Arizona literary magazine and made it a hit tune. Anyone who has

Tumbleweed flowers bloom in the leaf axils, protected by spine-tipped leaves.

Bushy plants will eventually dry out and break off to form tumbleweeds.

driven around the semiarid West has seen the skeletal balls, from basketball size to human size, of this annual plant blowing around. These are the mature plants that have auto-detached from the roots to begin rolling around the land, spreading seed. The new seedlings and young plants are bright green and grassy. As they develop a more bushy form, the stems are red or purple striped and grow to 4 ft. (1.2 m) tall. Leaves are alternate, at first dark green and soft and narrow (less than ⅛ in. [0.3 cm] wide). Upper leaves are sharply pointed and 1.25 to 2 in. (3.1 to 5 cm) long. From midsummer to fall small greenish to pink flowers grow from the leaf axils, each protected by a triplet of spine-tipped leaves. Summer cypress tumbleweeds are hairy instead of spiny.

HABITAT AND RANGE

Tumbleweed has taken hold on disturbed soils and in fields and along roadsides throughout the semiarid regions of the United States and Canada. In undisturbed areas it does not compete well against many natives, but it tolerates alkaline and salty soils. It will grow from below sea level in places like Death Valley to almost 9,000 ft. (2,700 m).

WHAT IT DOES IN THE ECOSYSTEM

Tumbleweed takes hold in disturbed environments, and its rolling form of seed dispersal equips it well to spread across the Great Plains and high desert areas of North America. A single plant can shed some 250,000 seeds as it travels. Seeds do not have a protective hard shell or food reserve, but contain miniature plants ready to unfold when temperatures rise above 28° F (−2.2°C). Plants quickly send down long taproots. When dry, they burn easily and can become rolling fireballs. Tumbleweed produces oxalates that increase phosphorous available to other plants, and may be the first part of a restoration cycle. In lands that are left to recover, tumbleweed will die off as other plants take over.

In extreme droughts tumbleweed can provide grazing for wild and domestic animals. It is a preferred food of prairie dogs and the seeds are eaten by some birds, including quail. Canadian farmers used it for silage during a deep drought in the 1930s. Kansas farmers in the same period used over 400,000 tons of tumbleweed for fodder. Tumbleweed harbors several viruses that affect beet crops and can lower wheat yields. It has also been known to cause traffic accidents, and some people are allergic to the pollen.

HOW IT CAME TO NORTH AMERICA
This native of Russia's Siberian steppe and other parts of Eurasia first appeared in Bonne Homme County, South Dakota, in the mid-1870s. It may have been a contaminant in flax or wheat seed carried by Ukrainian and Russian immigrant farmers. Within 30 years it had spread throughout the semiarid western plains and reached the Pacific coast as a contaminant in feeds.

MANAGEMENT
Restoration of natural rangeland conditions will replace tumbleweeds. This can be speeded up by the addition of organic materials to disturbed and depleted soils. The bacteria and fungi in the soils upset tumbleweed colonization. Disking and burning are very short-term controls; and in the long term, the soil disturbance caused by these practices helps spread tumbleweed. Herbicides are generally effective against tumbleweed when plants when plants are young, and preemergent herbicides can be used to keep seeds from germinating.

FOR MORE INFORMATION
Beatley, J. C. 1973. "Russian-thistle *Salsola* species in the western United States." *Journal of Range Management* 26:225–226.
California Department of Food and Agriculture. "Russian thistle." http://www.cdfa.ca.gov/plant/ipc/weedinfo/salsola.htm.
Ontario Ministry of Agriculture and Food. 2003. "Ontario weeds: Russian thistle." Ontario Weeds Gallery. http://www.omafra.gov.on.ca/english/crops/facts/info_russianthistle.htm.
Shinn, C. H. 1895. *The Russian thistle in California.* Agricultural Experiment Station Bulletin No. 107. Berkeley, CA: University of California.
Young, J. A. 1991. "Tumbleweed." *Scientific American* 264(3):82–87.

Green- or White-Flowered Wandering Jew and Boat Lily
Tradescantia spp.

NAME AND FAMILY
Small-leaf spiderwort, green- or white-flowered wandering Jew, (*Tradescantia fluminensis* Vell.); boat lily, Moses in a boat, oyster plant (*T. spathacea* Sw.); dayflower family (Commelinaceae). Most commonly recognized because of their use as houseplants, in the tropical United States these plants are used in landscaping and have become invasive.

IDENTIFYING CHARACTERISTICS

Wandering Jew is an herbaceous plant with trailing stems that root at the nodes. The glossy green leaves with parallel veins are sometimes purplish underneath and they alternate along the stems. The leaves are oblong with pointed tips, 2 in. (5 cm) long and 0.75 in. (1.9 cm) wide, and at the base the leaves clasp the stem. White, three-petaled flowers bloom in clusters at the ends of the stems. Black, three-part capsules hold six seeds. Boat lily is named because the flowers are held in a purple boat-shaped leafy structure called a bract. It also has three-petaled white flowers. Stems are shorter and leaves much larger (6 to 12 in. [15.2 to 30.5 cm] long and 1 to 3 in. [2.5 to 7.6 cm] wide) than in wandering Jew. The leaves are green above and purplish underneath.

Boat lily has long leaves and purplish stems.

HABITAT AND RANGE

Both plants tolerate heavy shade and grow in moist to well-drained soils. Boat lily is more drought tolerant. Wandering Jew is most common in northern and central Florida, but can also be found in Georgia, Alabama, and California. Wandering Jew is very shade tolerant and is found in moist woodlands and hardwood tree islands (hammocks). Boat lily is found in Florida and Louisiana in hardwood hammocks, pinelands, and scrublands. It is not as shade tolerant as wandering Jew but is more drought tolerant.

WHAT IT DOES IN THE ECOSYSTEM

These plants form a dense ground cover on the forest floor that prevents native species from establishing. In New Zealand, wandering Jew changes insect communities and nutrient cycling in forest soils where it naturalizes. Both plants spread rapidly, wandering Jew mainly vegetatively and boat lily vegetatively and by seed. Plants flower year-round and are pollinated by insects.

HOW IT CAME TO NORTH AMERICA

Wandering Jew is native to the tropical forests of Brazil and was introduced as a houseplant. Boat lily was found naturalized in Florida as early as 1933 and was probably introduced as a garden plant from tropical America.

MANAGEMENT

Wandering Jew plants can be pulled, raked, or rolled up. Boat lily can be pulled up, but use gloves because contact with the sap of boat lily plant can cause stinging or a rash in some people. For either species, placing plastic sheeting over plants for two to six weeks during a hot spell will kill the plants, but also everything else under the sheet. Plants can be cut and regrowth treated with triclopyr.

FOR MORE INFORMATION

Langeland, K. A., H. M. Cherry, C. M. McCormick, and K. A. Craddock Burks et al. 2008. *Identification and Biology of Nonnative Plants in Florida's Natural Areas,* 2nd ed. SP 257. Gainesville, FL: University of Florida IFAS.

Standish, R.J., P. A. Williams, A. W. Robertson, N. A. Scott, and D. I. Hedderley. 2004. "Invasion by a perennial herb increases decomposition rate and alters nutrient availability in warm temperate lowland forest remnants." *Biological Invasions* 6:71–81.

Standish, R. J., A. W. Robertson, and P. A. Williams. 2001. "The impact of an invasive weed *Tradescantia fluminensis* on native forest regeneration." *Journal of Applied Ecology* 38:1253–1263.

Yellow Salsify   *Tragopogon dubius*

NAME AND FAMILY

Yellow salsify, yellow goat's beard, western salsify (*Tragopogon dubius* Scop.); aster family (Asteraceae). Two other species of introduced salsify are meadow salsify (*T. pratensis* L.) and salsify or oyster plant (*T. porrifolius* L.).

IDENTIFYING CHARACTERISTICS

This biennial (or sometimes annual or short-lived perennial) plant has green to gray-green, grasslike leaves. Young or first-year plants resemble stalks of grass but have fleshier leaves sometimes covered with long hairs. Leaves can be up to a foot long and the leaf bases clasp the stem. Leaves are arranged alternately along the stem. Older plants send up rounded flowering stalks up to 3 ft. (1 m) tall with an elongated round spearhead bud. When a stalk is cut, it produces milky sap. Large (1.5 to 2.5 in. [3.7 to

Yellow salsify flowers bloom atop tall stalks in summer.

Left: *Seed heads look like giant dandelions.* **Right:** *Below the flowers are long pointed bracts.*

6.2 cm] diameter) lemon-yellow flower heads bloom in summer at the ends of the stalks and have 10 or more green, needlelike bracts as long or longer than the petals underneath the flower head. The receptacle underneath the flower where the seeds form is swollen, distinguishing it from meadow salsify. Meadow salsify generally has only eight to nine bracts as well. Oyster plant has purple flowers. Flowers usually open only on sunny days and close by noon. Fluffy round seed heads are 3 to 4 in. (7.6 to 10 cm) across and look like giant dandelions. The name goat's beard comes from the shape of the fluff attached to the seeds. Plants have long, tapering taproots.

HABITAT AND RANGE
Yellow goat's beard occurs across Canada and the United States except Nunavut and the lower southeastern U.S. It generally grows in pastures, meadows, disturbed ground, and roadsides but tolerates enough shade to be found in open forests and savannahs. It can tolerate extremely dry conditions.

WHAT IT DOES IN THE ECOSYSTEM
Salsify can outcompete native grasses for water and nutrients in meadows, but it does not generally form dense stands. It is not an invasive plant of great importance. The root is a favorite food of pocket gophers, and deer, elk, squirrels, and rabbits eat the aboveground foliage.

HOW IT CAME TO NORTH AMERICA
Native to Europe, salsify was brought by immigrants to North America in the 1900s. Oyster plant is particularly known for its edible root, reputed to taste like oysters or parsnips.

MANAGEMENT
Plants can be dug up or sprayed with a systemic herbicide.

FOR MORE INFORMATION
Gucker, Corey L. 2008. *"Tragopogon dubius."* *Fire Effects Information System.* U.S. Department of Agriculture, Forest Service, Rocky Mountain Research Station, Fire Sciences Laboratory. http://www.fs.fed.us/database/feis.
Ontario Ministry of Agriculture and Food. *Ontario weeds.* Publication 505.
Upadhyaya, M. K., M. Q. Qi, N. H. Furness, and R. S. Cranston. 1993. "Meadow salsify and western salsify: Two rangeland weeds of British Columbia." *Rangelands* 15(4):148–150.

Woolly Mullein   *Verbascum thapsus*

NAME AND FAMILY
Woolly mullein, common mullein, flannel plant, velvet plant, velvet dock (*Verbascum thapsus* L.); figwort family (Scrophulariaceae). Moth mullein (*V. blattaria* L.) is another less invasive species of mullein.

IDENTIFYING CHARACTERISTICS

The soft, green, furry leaves and the 3 to 6 ft. (1 to 1.8 m) high flower stalks of woolly mullein are familiar to almost anyone who has observed old fields and meadows. Woolly mullein usually lives for two years. The first-year rosette has soft, downy, gray-green leaves 4 to 12 in. (10 to 30.5 cm) long and 1 to 5 in. (2.5 to 12.7 cm) wide, which lie close to the ground. In the second year, if the rosette reached a diameter of at least 3.5 in. (8.7 cm), a fuzzy, 4 to 10 ft. (1.2 to 3 m) flowering stalk grows, with leaves alternating along the stem, getting smaller the

Flowers open a few at a time in a spiral up the flowering stalk.

closer they are to the flower spike. Five-petaled yellow flowers about 1 in. (2.5 cm) in diameter open only a few at a time in spirals from the bottom to the top of the dense flower stalk during the summer, making for a long display. If a flower is not pollinated by a bee by the end of a day, it will self-pollinate. Dry capsules contain numerous small brown seeds, most of which fall within a few feet of the parent plant. Roots are taproots. Moth mullein is not hairy, the leaves are toothed along the edges, and the flowers are much showier, arranged on an open spike with white to yellow petals, sometimes pink.

HABITAT AND RANGE
It occurs from southern Canada through most of the United States. Woolly mullein needs sun and tolerates very dry land, with as little as 3 to 6 in. (7.6 to

15.2 cm) of rain a year. It tends to grow in meadows, disturbed field edges, burned areas, roadsides, pastures, and forest openings. Seeds need bare ground to germinate.

WHAT IT DOES IN THE ECOSYSTEM
Mullein is a weak competitor in most natural areas, but its seeds can remain viable for a century or more. (Viable seeds have been reported from 700-year-old archeological digs.) Thus, disturbing a mullein-free area can activate very old seed. In fact, because other plants easily displace mullein, it depends on the longevity of its seeds for survival of the species. In disturbed soil, mullein can dominate until other plants are established.

HOW IT CAME TO NORTH AMERICA
Woolly mullein, native to Europe and Asia, was first introduced in the mid-1700s to Virginia, where it was used as a fish poison because the leaves contain rotenone (also used as an organic insecticide.) Since Europeans had used it for over 2,000 years as an herbal, colonists undoubtedly brought it to North America as a remedy for

Leaves become smaller going up the stalk of the plant.

Leaves of rosettes are soft and fuzzy.

coughs and diarrhea and other ills. It was first described in Michigan in 1839 and on the Pacific coast in 1876. After it arrived, Native American tribes found many uses for woolly mullein, from smoking to use as a drug for internal ailments of several kinds. Despite its medicinal uses, readers should note that it contains not only the insecticide rotenone, but coumarin (often used in rat poison), which has been linked to liver damage and internal bleeding.

MANAGEMENT

Plants can be hand pulled or hoed up in loose soils. The U.S. Department of Agriculture introduced a European seed-feeding weevil before 1937 to the United States to control mullein. Glyphosate or triclopyr can be sprayed on plants in early spring before most other plants emerge.

FOR MORE INFORMATION

Hoshovsky, M. C. 1986. "Element stewardship abstract for *Verbascum thapsus*." The Nature Conservancy. http://wiki.bugwood.org/Verbascum_thapsus.
Remaley, T. "Common mullein." Plant Conservation Alliance, Alien Plant Working Group. http://www.nps.gov/plants/alien/fact/veth1.htm.

Terrestrial Plants—Herbaceous Plants—Alternate Leaves—Leaves Toothed or Lobed

Russian Knapweed   *Acroptilon repens*

NAME AND FAMILY
Russian knapweed, hardheads (*Acroptilon repens* [L.] DC.); aster family (Asteraceae). It looks similar to other knapweeds.

IDENTIFYING CHARACTERISTICS
This perennial, herbaceous plant has stiff, thin stems that emerge in spring and grow to 3 ft. (1 m). When young, the stems are covered with soft, gray hairs. Lower leaves on the stems are oblong to lance shaped and heavily lobed. Leaves become smaller going up the stem and are oblong, with those on the upper part of the stem having toothed edges. The ground-hugging rosette leaves can be very lobed to almost unlobed, 2 to 4 in. (5 to 10 cm) long. The flower heads are urn shaped and hold tubular pink to purple flowers that turn straw colored with age. Flowers bloom summer to fall and occur individually at the tips of the stems. The black to dark-brown roots can grow more than 20 ft. (6 m) below the soil surface. Buds that can develop into new shoots grow alternately along the roots. Russian knapweed can be distinguished from other knapweeds in the *Centaurea* genus (including spotted knapweed) by the scales on the underside of the flower head. In Russian knapweed, these scales (the involucres) have a papery, finely hairy tip that others do not.

HABITAT AND RANGE
Found throughout southwestern Canada and southern Ontario, into the Midwest and western United States. Russian knapweed tends to take hold

Urn-shaped seed head holds many seeds.

Dense stands of Russian knapweed exclude other vegetation.

in arid lands and disturbed areas because it cannot tolerate aggressive competitors. Once established, however, it persists and populates rangelands, roadsides, riverbanks, and the edges of agricultural fields. It is native to central Asia and Asia Minor but now occurs on every continent but Antarctica.

WHAT IT DOES IN THE ECOSYSTEM

Russian knapweed forms dense stands that can persist for more than 75 years. Its deep root system uses water other plants can't reach, and it releases chemicals into the soil that prevent the establishment of other plants. Most grazers do not eat Russian knapweed, and it can be toxic to horses if eaten over a period of time. Bighorn sheep do graze on it, and birds and mice eat the seeds. It spreads by both seed and root sprouts

Purple flowers bloom at the tips of the stems in summer.

or root fragments. Russian knapweed is classified as a noxious weed in Canada and by many western states.

HOW IT CAME TO NORTH AMERICA

Russian knapweed arrived in Canada around 1900 as a contaminant in Turkestan alfalfa seed. It probably arrived in California between 1910 and 1914 the same way. Now it often spreads as a contaminant in hay, straw, and fill dirt.

MANAGEMENT

Pulling is only practical in loose, sandy soils for small infestations. Cut-

Along the stems, oblong leaves alternate.

ting knapweed two to three times annually can control its spread by reducing seed production, but seldom eliminates plants long-term. For large areas, several systemic herbicides, in particular picloram, are effective when applied as plants are setting flower buds and flowering. Two biocontrol insects, a gall-forming nematode and a seed gall mite are available in some states for controlling Russian knapweed.

FOR MORE INFORMATION

Jacobs, J. and K. Denny. 2006. *Ecology and Management of Russian Knapweed* [Acroptilon repens *(L.) DC]*. USDA NRCS Invasive Species Technical Note No. MT-7. ftp://ftp-fc.sc.egov.usda .gov/MT/www/technical/invasive/Invasive_Species_Tech_Note_MT7.pdf.

Watson, A. K. 1980. "The biology of Canadian weeds. 43. *Acroptilon (Centaurea) repens* (L.) DC." *Canadian Journal of Plant Science* 60:993–1004.

Zouhar, Kristin L. 2001, "*Acroptilon repens*." *Fire Effects Information System*. U.S. Department of Agriculture, Forest Service, Rocky Mountain Research Station, Fire Sciences Laboratory. http://www.fs.fed.us/database/feis.

Goutweed   *Aegopodium podagraria*

NAME AND FAMILY

Goutweed, bishop's weed, snow-on-the-mountain, ground elder (*Aegopodium podagraria* L.); carrot family (Apiaceae). Several native species may have similar leaves but a different color or arrangement of flowers (golden Alexander [*Zizia aurea*], Canadian honewort [*Cryptotaenia canadensis*], or anise root [*Osmorhiza claytonii*]).

IDENTIFYING CHARACTERISTICS
This creeping perennial forms a dense ground cover no more than 3 ft. (1m) tall. The lower leaves are usually made up of three groups of three leaflets, each 1 to 3 in. (2.5 to 7.6 cm) long with toothed edges and sometimes with irregular lobes. Leaves alternate along the stems. The upper leaves are smaller and have fewer leaflets. In June, flat, 2 to 4 in. (5 to 10 cm) wide clusters of tiny white flowers somewhat like those of another member of the carrot family, Queen Anne's lace, bloom on stalks held above the leaves. The seeds are small and brown. It spreads via rhizomes (underground stems).

HABITAT AND RANGE
Goutweed grows across Canada from southwestern British Columbia, southern Manitoba to Nova Scotia, and in the United States from Maine to Georgia and into the Midwest. It also occurs in the northwestern United States. It thrives in disturbed areas with moist soils and shade such as disturbed forest understories, roadsides, and forest edges. Known to be an indicator of nitrogen-enriched soils in Europe.

WHAT IT DOES IN THE ECOSYSTEM
It can grow densely enough to prevent other species from establishing. The seedlings need light to establish, but colonies can spread by underground stems in shaded areas. Eaten by the larvae of some butterflies.

HOW IT CAME TO NORTH AMERICA
Goutweed came to North America as an ornamental plant. The variegated form is less aggressive than the green form. Providence, Rhode Island, records the earliest known naturalized occurrence in 1863. It is native to most of Europe and northern Asia, where it grows in deciduous and southern boreal forests. As "pod" (foot) in both parts of its Latin name implies, it was sometimes used to treat gout. It is still used as a salad green and potherb.

Goutweed can have variegated or solid green leaves and it blooms in early summer.

MANAGEMENT

Small patches can be pulled up or dug up, but care must be taken to remove all rhizomes from the site or they will create new plants. Covering plants in spring with black plastic for several months can also kill them. Systemic herbicides like glyphosate are effective sprayed on larger infestations. Frequent mowing may also discourage goutweed along roadways. Cutting followed by heavy mulching is also effective.

FOR MORE INFORMATION

"*Aegopodium podagraria.*" 2003. In *Invasive Plant Atlas of New England,* edited by L. J. Mehrhoff, J. A. Silander Jr., S. A. Leicht, E. S. Mosher and N. M. Tabak. Storrs, CT: Department of Ecology & Evolutionary Biology, University of Connecticut. http://www.ipane.org.

Garske, S., and D. Schimpf. "Goutweed." Plant Conservation Alliance, Alien Plant Working Group. http://www.nps.gov/plants/alien/fact/aepo1.htm.

Garlic Mustard   *Alliaria petiolata*

NAME AND FAMILY

Garlic mustard (*Alliaria petiolata* [M. Beib.] Cavara & Grande); mustard family (Brassicaceae). Young garlic mustard plants look similar to several native species, including violets (*Cardamine* spp.) and, in the West, fringecup (*Tellima grandiflora*) and piggyback plant (*Tolmiea menziesii*). It also looks similar to the nonnative ground ivy (*Glechoma hederacea*, p. 244).

Garlic mustard's white four-petaled flowers bloom in spring.

A dense stand of second-year plants lines a path in a park.

IDENTIFYING CHARACTERISTICS
Garlic mustard gets its name from the pungent odor of the crushed new leaves. The seeds of garlic mustard germinate very early in spring and form a rosette of leaves the first year. Rosette leaves are kidney shaped with scalloped edges. Second-year plants bolt in mid spring sending up a flowering stalk or stalks, 1 to 3 ft. (0.3 to 1 m) tall, with small clusters of white, four-petaled flowers, each flower 0.25 in. (0.6 cm) in diameter. Leaves alternate along the flower stalk and are more triangular and toothed than in the rosette. Seedpods are skinny, 1 to 2.5 in. (2.5 to 6.2 cm) long; turning tan colored by midsummer, they split along the seams to reveal small black seeds. Plants die after flowering. Seeds disperse by gravity and by movement of water and soil. Distinguished from native plants with rosettes of rounded leaves by the garlicky odor and in winter by the taproot, which has an S curve just where it starts below the leaves. The garlic odor also distinguishes it from ground ivy.

HABITAT AND RANGE
Grows in forests and along forest edges, riverbanks, and roadsides. Often establishes in disturbed habitats then spreads into less disturbed habitat. Most widespread from southern Ontario and Quebec south to Virginia and west to Wisconsin, but occurs as far south as Georgia, west to Oregon and Washington, and even in southern Alaska.

Left: *Seedlings germinate in spring and form a rosette that overwinters the first year.* Right: *A large rosette just before bolting.*

WHAT IT DOES IN THE ECOSYSTEM

Garlic mustard can colonize relatively undisturbed forest understories, where it competes for light and space with many spring-blooming wild flowers and tree seedlings. Garlic mustard inhibits the growth of mycorrhizal fungi, important to many native plants that use the fungi to obtain nutrients from the soil. Garlic mustard also threatens two native butterfly species whose larvae feed on native mustard plants. Because garlic mustard becomes so abundant, the butterflies lay their eggs on the plants, but the larvae fail to survive due to the different chemistry of the garlic mustard leaves. The rare butterfly West Virginia white, *Pieris virginiensis*, is considered to be particularly threatened by the spread of garlic mustard.

HOW IT CAME TO NORTH AMERICA

Probably introduced by European settlers in the early 1800s as a food because of its availability in early spring and high vitamin A and C content. Leaves were boiled in soups and eaten in salads. Young roots are sometimes used like horseradish. Seeds can make a hot mustard. It was first recorded by a botanist in 1868 on Long Island, New York.

MANAGEMENT

Plants are easily pulled up when soils are moist. For large areas, second-year plants can be cut at ground level in spring or sprayed in late fall or very early spring with glyphosate. Plants that have begun to flower should not be left on site because seeds may still develop. Plants cut above ground level are likely to send up new flowering shoots.

FOR MORE INFORMATION

Nuzzo, V. 2000. "Element stewardship abstract for *Alliaria petiolata*." The Nature Conservancy. http://wiki.bugwood.org/Alliaria_petiolata.

Rogers, V. L., K. A. Stinson, and A. C. Finzi. 2008. "Ready or not, garlic mustard is moving in: *Alliaria petiolata* as a member of eastern North American forests." *BioScience* 58:426:436.

Common Burdock   *Arctium minus*

NAME AND FAMILY

Common burdock (*Arctium minus* Bernh.); aster family (Asteraceae). This widespread burdock has many other names, including beggar's buttons, cockle button, cuckold dock, stick button, clot bur, and wild rhubarb. One of four non-native burdocks. The others are not as widespread.

IDENTIFYING CHARACTERISTICS

Common burdock often looks like rhubarb, with broad, heart-shaped dark green leaves up to a foot long (0.3 m) and 9 in. (23 cm) wide. Leaves are usually woolly with whitish undersides. These leaves die back the first winter. In its second year, burdock grows 3 to 4 ft. (1 to 1.2 m) tall and forms 0.75 in. (2 cm) diameter flower heads. The tubular flowers are typical of thistles, tightly bunched, usually pink to violet but enclosed in tough bracts that close over the flower head as seeds form, becoming the familiar burs. Seeds are dark gray to brown and about 0.25 in. (0.6 cm) long. Burdock is a biennial, but in poor conditions a plant may live four years before flowering and dying. Great burdock (*A. lappa*) grows almost twice as high, with larger flower heads and grooved leafstalks.

HABITAT AND RANGE

Common burdock, a native in most of Europe and Turkey, has invaded all of the lower 48 states and all but the northernmost Canadian provinces. It is apparently rare or absent from the Florida peninsula. It prefers moist but well-drained soils and tolerates acidic to alkaline conditions. It prefers sunny, open areas.

WHAT IT DOES IN THE ECOSYSTEM

Burdock aggressively colonizes disturbed soils and newly cleared lands, sometimes forming dense colonies.

Burdock flowers grow in clusters at the ends of the branches.

Tightly bunched purple flowers are surrounded by hooked bracts that form the burdock burs.

Plants flower in their second year, usually with large leaves alternating along the flowering stalks.

Burdock is self-fertilizing. The flower heads produce over 10,000 seeds per plant, enclosed in the brown bracts with Velcro-like hooks. The broad long leaves of the first year grow close to the ground and shade out almost all other plants, preparing the soil for more burdock seed. The flowers attract numerous butterflies as well as birds.

HOW IT CAME TO NORTH AMERICA

Since burdock was valued for medicinal purposes, English and French colonists introduced it to North America in the seventeenth century, and, with its ability to hitch rides, it spread rapidly. Among the products that have been made from burdock are coffee (dried roots), potherbs, and paper (inner bark fibers). It became a common New England weed by 1700, then throughout the east by the 1800s. Native Americans near the Pacific coast noted it as a new plant in the 1930s.

MANAGEMENT
Because burdock is a biennial, cultivating or digging usually destroys first-year plants and prevents seed formation. Numerous herbicides approved for broadleaf plant control will kill burdock. Mowing severely reduces seed formation.

FOR MORE INFORMATION
Gross, R. S., P. A. Werner, and W. R. Hawthorn. 1980. "The biology of Canadian weeds. 38. *Arctium minus* (Hill) Bernh. and *A. lappa* L." *Canadian Journal of Plant Science* 60:621–634.

2012. "Common Burdock." *Ohio Perennial and Biennial Weed Guide.* The Ohio State University, Extension. http://www.oardc.ohio-state.edu/weedguide/singlerecord.asp?id=900.

Absinthe   *Artemisia absinthium*

NAME AND FAMILY
Absinthe, old man, common wormwood (*Artemisia absinthium* L.); aster family (Asteraceae). Leaves look similar to those of cultivated chrysanthemums. The natives, plains wormwood (*A. campestris*) and biennial wormwood (*A. biennis*), can be confused with absinthe.

IDENTIFYING CHARACTERISTICS
The whole plant has a silvery-gray color with multiple stems coming from the base growing to 6 ft. (1.8 m) tall. It is a perennial and the root becomes woody with age. Growth begins in late spring and plants flower in midsummer. Gray-green leaves grow alternately along the stem, 1 to 4 in. (2.5 to 10 cm) long and wide. Lower leaves are divided into two to three narrow segments with rounded tips. Upper leaves are lance shaped, not lobed. Crushed leaves have a very strong sagelike scent. Flower heads less than 0.25 in. (0.6 cm) wide, which grow at the bases of the leaves in the axils, match the leaf color on the outside of the head but have yellowish-brown centers. A single plant can produce 1,500 flower heads and 50,000 seeds. The native plains wormwood has flower heads mostly held on a stalk above the leaves and very narrow lobes on the leaves. Biennial wormwood has reddish stems and the leaves have toothed edges.

HABITAT AND RANGE
Absinthe generally establishes in heavily disturbed areas where there is little competition from other plants. It thrives in dry soils of roadsides, pastures, rangelands, and waste areas. It grows throughout most of Canada as far north as Hudson Bay and into the United States as far south as North Carolina and Missouri.

Left: *Clusters of tiny flowers have yellow-brown centers.* **Right:** *The whole plant has a silvery-gray color.*

WHAT IT DOES IN THE ECOSYSTEM

Absinthe is mainly a threat to prairie restoration projects, where it can interfere with the reestablishment of native species. Chemicals emitted from the roots discourage the growth of other plants. Seeds spread by wind or water and can be carried in soil or hay. Root fragments can also resprout. Absinthe colonizes pastures and is not a favored forage of livestock. When it is in hay eaten by cows, it gives milk an odd flavor. The pollen causes allergies.

HOW IT CAME TO NORTH AMERICA

Seeds of absinthe were offered for sale in the United States by 1832 because of its popularity as a garden and medicinal plant. Native to Europe, it was the flavoring in the drink absinthe and in vermouth. Chemicals in the plant, thujone and santonin, can cause liver damage and hallucinations, among other side effects, by acting on brain receptors that are also involved in epilepsy. Wormwood oil, which contains thujone, is still used in some common herbal "reme-dies" for ailments of appetite, liver, and gall bladder. Absinthe was a fashion-able nineteenth-century drink for European artists and intellectuals, and some researchers say it is possibly the source of Van Gogh's hallucinations and of visions painted by other artists, and also a contributing cause of their deaths.

MANAGEMENT
Where possible, repeated mowing can limit absinthe's growth. Herbicides applied in midsummer, including glyphosate, dicamba, and picloram, control absinthe. Planting vigorous grasses may also lead to a decline in absinthe through competition.

FOR MORE INFORMATION
Evans, J. E., and N. Eckardt. 1987. "Element stewardship abstract for *Artemisia absinthium*." The Nature Conservancy. http://wiki.bugwood.org/Artemisia_absinthium.
Royer, F., and R. Dickinson. 1999. *Weeds of the Northern U.S. and Canada*. Renton, WA, and Alberta: Lone Pine Publishing and University of Alberta Press.

Mugwort   *Artemisia vulgaris*

NAME AND FAMILY
Mugwort, chrysanthemum weed (*Artemisia vulgaris* L.); aster family (Asteraceae). Leaves look similar to those of cultivated chrysanthemums.

IDENTIFYING CHARACTERISTICS
This perennial herb can grow to 5 ft. (1.5 m) in height. The heavily lobed, pointed leaves are arranged alternately along the stem. Leaves are 2 to 4 in. (5 to 10 cm) long and 1 to 3 in. (2.5 to 7.6 cm) wide, covered with white hairs underneath but smooth to some-what hairy above. Crushed leaves smell distinctly of sage. In summer, greenish, inconspicuous clusters of wind-pollinated flowers bloom at the ends of the stems on spikes. Seeds are seldom viable in the North but are an important form of dispersal in the

Leaves are heavily lobed and dark green on the upper side.

Left: *Plants grow densely from an extensive underground root system.* **Right:** *The undersides of the leaves are whitish, with fine hairs.*

southern United States. Long white and tan roots spread out to form colonies of mugwort. Chrysanthemum leaves are not as white and hairy underneath as those of mugwort.

HABITAT AND RANGE
Most common in eastern Canada and the United States, mugwort grows in meadows, along roadsides, and in agricultural fields in open sunny to partly shady areas. It is spreading to the West Coast through the nursery trade.

WHAT IT DOES IN THE ECOSYSTEM
Mugwort forms dense colonies in open areas that restrict the growth of native species, causing a decline in species diversity. The pollen causes allergies in some people.

HOW IT CAME TO NORTH AMERICA
Native to Eurasia, mugwort was thought to have many medicinal and culinary uses and its name may have been derived from its use as an insect repellent against midges. ("Mug" is similar to Old English for midge, but also to Old Norse for marsh, while "wort" is derived from old European words for root.) European settlers probably brought mugwort to North America for its many uses by the mid-1800s. More recently, cultivars with variegated leaves were developed to be sold in nurseries, but these plants can also be aggressive. The essential oils are used in perfume and aromatherapy.

MANAGEMENT
Small infestations can be hand pulled or dug up, but root fragments will resprout. Repeated monthly mowing for several years will control mugwort's spread. Herbicides like clopyralid and glyphosate applied several times during the growing season will control mugwort.

FOR MORE INFORMATION
Barney, J. N., and A. DiTommaso. 2003. "The biology of Canadian weeds. 118. *Artemisia vulgaris* L." *Canadian Journal of Plant Science* 83:205–215.

Barney, J. N., A. DiTommaso, and L. A. Weston. 2005. "Differences in invasibility of two contrasting habitats and invasiveness of two mugwort (*Artemisia vulgaris*) populations." *Ecology* 42(3):567–576.

Asian mustard   *Brassica tournefortii*

NAME AND FAMILY
Asian mustard, desert mustard, Moroccan mustard, African mustard, Mediterranean mustard, Mediterranean turnip, prickly turnip, Sahara mustard, southwestern mustard, Tournefort's mustard, turnip weed, wild turnip, (*Brassica tournefortii* Gouan); mustard family (Brassicaceae).

IDENTIFYING CHARACTERISTICS
This annual mustard begins growing in winter as a ground-hugging basal rosette with 3 to 14 alternate pairs of deeply lobed leaves 3 to 22 inches (8 to 56 cm) long. The leaves themselves have serrated edges with stiff white hairs. Leaves become smaller as they rise on the stem until, near the flower (which is sometimes 40 inches high), the leaves are tiny bracts. These plants grow rapidly and flower early, bearing 6 to 20 small, dull yellow flowers, each with four long petals. Flowers are 0.6 in (1.5 cm) wide. Long, narrow seed pods about 2 inches long with up 75 brown to purplish seeds mature in spring and the plants die after setting seed. A long taproot anchors the plant and accesses deep moisture.

HABITAT AND RANGE
Across the southwest from Texas to California and also Nevada, Asian mustard has rapidly invaded dry desert areas from sandy dunes to alkaline flats and gravelly soils. It is most abundant below 3000 feet (about 1,000 m).

WHAT IT DOES IN THE ECOSYSTEM
Because Asian mustard plants grow closely together in a continuous field, it fills in the spaces between sparsely scattered desert plants and carries fire across what otherwise would have been protective open space. Because it begins to grow early in the season, it uses moisture needed by native plants, then spreads

Many small flowers bloom on slender branches held above large lobed leaves at the base of the plant.

and crowds them out. Seeds are spread by animal droppings and by rodents caching seeds. In the manner of tumbleweed, the dead plant breaks at the ground and wind blows the seeds across the land. Rain-dampened seeds cling to animals, clothes, and vehicles. As with many desert plants, its abundance in any given year depends on rainfall.

HOW IT CAME TO NORTH AMERICA
Asian mustard is thought to have arrived as a weed seed in shipments of date palms to California in the early 1900s. It was first recorded in California in 1927. In its native range of North Africa, the Middle East, and the Mediterranean, it is often eaten. The seeds contain edible oil.

MANAGEMENT
Small populations can be hand pulled but the seed banks must also be controlled. Although controlled burns can seriously set back mustard populations, some seeds survive and populations usually reestablish quickly. Since Asian mustard often appears before most other plants, emergence is a good time to spray. Herbicides include 2,4-D, glyphosate, and others that target broadleaf plants. Although it establishes early, Asian mustard is a weak competitor and might be crowded out by other fast and strong growers.

FOR MORE INFORMATION

Barrows, C. W., E. B. Allen, M. L. Brooks, and M. F. Allen. 2009. "Effects of an invasive plant on a desert sand dune landscape." *Biological Invasions* 11(3):673–686.

Halvorson, William l. and Patricia Guertin, 2003. *Fact sheet for* Brassica tournefortii *Gouan*. USGS Weeds in the West project. http://sdrsnet.srnr.arizona.edu/data/sdrs/ww/docs/brastour.pdf.

USDA NRCS. "Asian mustard." http://developer.usanpn.org/saved-content-2009-02-28-2215UTC/ Brassica_tournefortii_238_253.html.

White Top   *Cardaria draba*

NAME AND FAMILY

White top, heart-podded hoary cress, perennial peppergrass (*Cardaria draba* [L.] Desv.); mustard family (Brassicaceae). Two other Cardaria spp. overlap in range with white top, but are not as often found in natural areas: lens-podded hoary cress (*C. chalapensis* [L.] Handel-Mazetti) and globe-podded hoary cress, Siberian mustard (*C. pubescens* [CA Mey.] Rollins).

IDENTIFYING CHARACTERISTICS

A flat head of white four-petaled flowers, each less than 0.5 in. (1.3 cm) across, blooms in early spring to early summer atop a single main stem seldom more than 3 ft. (1 m) tall. Flower heads are made up of many branches topped with dense flowers. The hoary cress species are best distinguished by the shapes of their seedpods. White top has heart-shaped seedpods, lens-podded hoary cress has a lens-shaped pod, and globe-podded hoary cress has a globe-shaped pod. Broadly oval leaves alternate along the stem and the arrow- or heart-shaped base of the leaf clasps the stem. Leaves are 1.5 to 3 in. (3.7 to 7.6 cm) long and gray green in color, often with toothed edges. Leaves at the base tend to be hairy and more lance shaped than stem leaves and form a rosette. Seeds germinate in fall and plants overwinter as rosettes.

HABITAT AND RANGE

Grows in sunny prairies, shrublands, rangelands, and disturbed, saline soils

White top flowers bloom atop a single main stem.

The four-petaled flowers bunch tightly together.

and along roadsides. Can grow in any soil but most aggressive in moist, disturbed soils. Occurs throughout most of Canada and in all but the southeastern part of the United States, but is most problematic in the western part of the continent.

WHAT IT DOES IN THE ECOSYSTEM

Through a system of underground rhizomes, white top forms dense mats that exclude other plants. White top is a strong competitor for water and nutrients because roots can extend 6 ft. (1.8 m) down into the soil and usually make up most of the biomass of the plant. When large infestations flower, it looks like it has snowed. Flowers are insect pollinated. Considered a noxious weed in many western states, and all three species are considered noxious weeds in Canada. Stands can live for many years.

HOW IT CAME TO NORTH AMERICA

This native of central Europe and western Asia was first reported near Yreka, California, in 1876 and in Ontario, Canada, in 1878. Seen on ballast heaps in New York in 1898. Introduced into the southwestern United States in 1910 in a shipment of alfalfa seed from Turkestan. Seeds and fragments of roots are spread in hay, on farm equipment, and in flowing water.

MANAGEMENT
Repeated cutting, mowing, or tilling can be effective over time. No leaves should be allowed to form. Herbicides such as glyphosate and 2,4-D have shown limited success due to the extensive root system, but can be effective, especially if combined with cutting. Flooding for two months will also kill plants.

FOR MORE INFORMATION
Bossard, C., and D. Chipping. "*Cardaria draba*." California Invasive Plant Council. http://www.cal-ipc.org/ip/management/ipcw/pages/detailreport.cfm@usernumber=23&surveynumber=182.php.

Jacobs, J. 2007. *Ecology and management of whitetop [*Cardaria draba *(L.) Desv.]*. USDA NRCS Invasive Species Technical Note No. MT-12. ftp://ftp-fc.sc.egov.usda.gov/MT/www/technical/invasive/Invasive_Species_Tech_Note_MT12.pdf.

Royer, F., and R. Dickinson. 2004. *Weeds of the Northern U.S. and Canada*. Renton, WA, and Alberta: Lone Pine Publishing and University of Alberta Press.

Musk Thistle   *Carduus nutans*

NAME AND FAMILY
Musk thistle, nodding thistle, nodding plumeless thistle (*Carduus nutans* L.); aster family (Asteraceae). This very large thistle looks similar to another introduced thistle, plumeless thistle (*Carduus acanthoides* L.), and they occur in similar habitats. Can also be confused with native thistles, some of which are endangered.

Left: *Spiny leaves give musk thistle a formidable appearance.*
Right: *The flower head has rows of purplish spiny bracts at the base.*

IDENTIFYING CHARACTERISTICS

Called nodding thistle because the flower heads droop, or nod. This thistle is usually biennial, sometimes annual, producing a rosette of six to eight leaves up to 1 ft. (3 m) long and 6 in. (15 cm) wide the first year. The leaves are covered with woolly hairs. The second year a flowering stalk grows to 8 ft. (2.4 m) tall. Leaves alternate along the flowering stalk and are deeply lobed, each lobe having three to five spiny points and a larger white or yellowish spine at the tip of the leaf. Dark-green leaves have a distinctive white midvein. The stem looks winged and spiny because the bases of the leaves extend down the stem. About the only part of the plant without spines is the stalk of the flower head. The flower head, 0.75 to 3 in. (1.9 to 7.6 cm) in diameter, is made up of many red to purple florets with several rows of purplish spiny bracts under the flowers, and blooms throughout much of the summer. Each stem has one to three flower heads. Plumeless thistle flower stalks are spiny, the flower heads are held upright, and each flower head is usually less than 1 in. (2.5 cm) across.

HABITAT AND RANGE

Found throughout the southern Canadian provinces and the United States in sage brush, open woods, fields, rangelands, roadsides, and abandoned lands.

WHAT IT DOES IN THE ECOSYSTEM

Musk thistle forms dense stands, up to 60,000 plants per acre (150,000 plants/ha), reducing yield in pastures and rangelands drastically by crowding out forage plants. Because it is so spiny, livestock and probably wildlife avoid grazing near it. It appears to release chemicals into the soil that inhibit the growth of other species, but its own seeds' germination is enhanced by decaying musk thistle tissue. Seeds are eaten by birds and small animals and are dispersed both by animals (including ants) and by wind and water. A single plant can produce 11,000 seeds. Provides nectar for bees, bumblebees, and butterflies. Considered a noxious weed in 4 Canadian provinces and 22 states.

HOW IT CAME TO NORTH AMERICA

First noticed in central Pennsylvania in 1852, it probably arrived in North America in ballast water from its native range of Europe, western Asia, and North Africa and was further spread by movement of seeds, plants, and animals for agriculture.

MANAGEMENT

Prevent introduction by limiting disturbance and overgrazing and promoting a healthy growth of plants that will outcompete seedlings for light. Cut plants just before flowering, 1 to 2 inches below ground, using a sharp shovel. Repeated mowing once plants have bolted can reduce spread of thistle, but seldom kills the plants. Goats love to eat the flowers and can significantly reduce seed production. Seedlings or rosettes can be sprayed with glyphosate, dicamba, clopyralid, or 2,4-D in spring. Three biological control insects and one rust fungus are also used to control musk thistle. The introduced seed-feeding weevils have begun damaging rare native thistles in the western U.S., however.

FOR MORE INFORMATION
Royer, F., and R. Dickinson. 2004. *Weeds of the Northern U.S. and Canada.* Renton, WA, and Alberta: Lone Pine Publishing and University of Alberta Press.

Zouhar, K. 2002. *"Carduus nutans." Fire Effects Information System.* U.S. Department of Agriculture, Forest Service, Rocky Mountain Research Station, Fire Sciences Laboratory. http://www.fs.fed.us/database/feis.

Red Star Thistle   *Centaurea calcitrapa*

NAME AND FAMILY
Red or purple star thistle (*Centaurea calcitrapa* L.); aster family (Asteraceae). Named after an ancient weapon, the caltrop, which had four sharp points and was used to defend against warriors on horseback. Difficult to distinguish from Iberian star thistle (*C. iberica* Spreng.).

IDENTIFYING CHARACTERISTICS
Like yellow star thistle (*C. solstilialis*, p. 312), this one distinguishes itself first by its sharp 1 in. (2.5 cm) spines that tip each bract below the flowers and the shorter spines surrounding the base of the flower head. These spines often last into winter. Reddish-purple to lavender-pink flower heads bloom atop 1 to 4 ft. (0.3 to 1.2 m) stems in midsummer to fall. Plants can have many stiff branches. The pale-green stems are spineless, with alternating leaves. Lower leaves and rosette leaves are deeply lobed or even divided, sometimes toothed, but uppermost stem leaves are not lobed and sometimes clasp the stem. Young leaves and stems are covered with cobwebby hairs. Mature leaves can be hairless. Rosette leaves are larger than stem leaves, up to 10 in. (25.4 cm) long, and there are spines in the middle of the rosette. Plants have thick taproots up to 1.5 in. (3.7 cm) in diameter. Usually grows as an annual or biennial, but sometimes a short-

The seed heads of red star thistle are defended by sharp spines.

lived perennial. The dry fruits are less than ⅛ in. (0.3 cm) long and do not have bristles (pappi), unlike many other star thistles. Iberian star thistle tends to have paler flowers and shorter spines, and its fruits have short bristles.

HABITAT AND RANGE
Red star thistle often grows in floodplains, dry forests, and grasslands. It is often found on heavier bottomland soils. It is most common in western states from Washington south to Arizona, but also occurs in some midwestern and Mid-Atlantic states.

WHAT IT DOES IN THE ECOSYSTEM
Forms dense stands, usually establishing first in disturbed areas. Out-competes native plants for light, space, water, and nutrients. Reduces the quality of forage on rangelands since animals avoid eating it due to the spines and bitter taste.

HOW IT CAME TO NORTH AMERICA
Native to the region between the Black and Caspian seas, red star thistle was probably accidentally introduced as a contaminant in seeds or hay in 1886 near Vacaville, California. In Washington State it was first noticed in 1929.

Reddish-purple flowers bloom at the ends of the stems.

Each plant has many stiff stems.

MANAGEMENT

Small patches can be hand pulled or cut below the soil level. Mowing simply encourages more rosettes to form. Herbicides containing 2,4-D or dicamba sprayed on seedlings and rosettes are effective chemical treatments. Planting perennial grasses may help reduce reinvasions.

FOR MORE INFORMATION

Pitcairn, M. J., J. A. Young, C. D. Clements, and J. Balciunas. 2002. "Purple starthistle (*Centaurea calcitrapa*) seed germination." *Weed Technology* 16:452–456.

Randall, J. M. "*Centaurea calcitrapa*." California Invasive Plant Council. http://www.cal-ipc.org/ip/management/ipcw/pages/detailreport.cfm@usernumber=26&surveynumber=182.php.

Roche, C. T., and B. F. Roche, Jr. 1990. "Purple starthistle (*Centaurea calcitrapa* L.) and Iberian starthistle (*Centaurea iberica* Trev. ex Sprengel)." PNW 350. Pacific Northwest Extension: Washington, Oregon, Idaho.

Yellow Star Thistle   *Centaurea solstitialis*

NAME AND FAMILY
Yellow star thistle (*Centaurea solstitialis* L.); aster family (Asteraceae). Other invasive Centaurea spp. with yellow flowers include: tocalote; maltese or napa star thistle (*C. melitensis*), which is a larger plant with shorter spines; and Sicilian star thistle (*C. sulphurea*), which has much larger flowers. None of these has invaded North America nearly as successfully as yellow star thistle.

IDENTIFYING CHARACTERISTICS
Yellow star thistle is a winter annual or sometimes a biennial plant growing 6 in. (15 cm) to 5 ft. (1.5 m) in height depending on soil and moisture conditions. It is called a winter annual because the dusky-green shoots appear during the winter months. In warmer climates the bright-yellow flowers surrounded by sharp spines at their base (the star) appear as early as May and continue until December. Each flower head can have 30 to 100 flowers. Flower heads of many small flowers are usually solitary at the ends of each stem, surrounded by stiff spikes. Particularly vigorous plants may also have flower stalks where the stems branch. Each flower head has two types of seeds. Along the edge of the flower head are darker seeds with no hairs (pappi) that often are not dispersed

Left: *Yellow star thistle plants have a dusky-green color.* Right: *The yellow flowers are defended by bract with long spines.*

Dense stands of yellow star thistle are particularly common in western North America.

until late winter. The central seeds are lighter in color and have short, stiff white hairs. They are dispersed as the flower's parts dry and fall off. Seeds are only dispersed by falling and secondarily by movement of soil, animals, floods, or equipment. The leaves and stems are blue-gray to gray-green and covered with wooly hairs. Leaves at the base of the plant have deep lobes and grow 2 to 3 in. (2.5 to 7.6 cm) long. The edges of the leaves can be smooth, toothed, or wavy. Leaves along the flowering stalk are arranged alternately and are short (0.5 to 1 inch [1.3 to 2.5 cm]) and narrow with few or no lobes. The bases of the leaves extend down the stem, giving the stem a winged appearance. The plant has a taproot that can grow 3 ft. (1 m) into the soil to collect moisture.

HABITAT AND RANGE
Yellow star thistle grows in annual grasslands, on rangelands, along roadsides, in scoured floodplains, and in orchards and vineyards. It prefers sites with full sun, deep well-drained to dry soils, and rainfall of 10 to 60 in. (8 to 150 cm) a year. Its efficient taproot allows it to thrive in the hot, dry summers of the West. It is most common in the western United States and Canada, but occurs in 41 U.S. states and several Canadian provinces. In California alone it has invaded 15 million acres (6 million ha). It does poorly in shade.

WHAT IT DOES IN THE ECOSYSTEM
In annual and perennial grasslands, yellow star thistle consumes significant amounts of soil moisture, outcompeting native plants. It can deplete soil moisture to a depth of 6 feet (1.8 m). It also grows so densely that it excludes the

Rosette leaves are long and narrow with lobes.

growth of other plants. When eaten by horses it can cause "chewing disease," which can eventually destroy a horse's ability to chew or swallow. Fields are often so thick with thistle that western beekeepers have begun to market the distinctive honey. Beekeepers also value the plant as one of the last flowers of the fall season.

HOW IT CAME TO NORTH AMERICA

Although native to Eurasia (Balkan-Asia Minor, the Middle East, and south-central Europe) and the Mediterranean region of southern Europe and northern Africa, yellow star thistle probably arrived in North America around 1849 as a contaminant in Chilean alfalfa seed (which would have originally come from Spain). Later, as demand for alfalfa boomed in the United States, farmers imported alfalfa seed from other parts of Europe and central Asia, giving rise to new introductions. By 1915 it had rendered some large wheat fields worthless. In the 1930s and 1940s it became very visible in pastures and rangelands and along roadsides. It has since spread within North America through the transport of grain and the movement of farm equipment, road maintenance equipment, and other vehicles.

MANAGEMENT

Small patches can be hand pulled or repeatedly mowed. Intense grazing during May and June by cattle, sheep, or goats also reduces survival and seed production in plants. Protein content in yellow star thistle is relatively high during this time of year, making good forage. In rangelands, other competitive species may be planted to reduce the abundance of yellow star thistle; however most restoration to date has used other nonnative grasses as competitors. Applying mulch also helps control star thistle. Six biological control insects that feed on various parts of the plant are also established in the western United States. Many herbicides are effective against young plants of yellow star thistle, including glyphosate, 2,4-D, clopyralid, and picloram, but some plants have become resistant to herbicides.

FOR MORE INFORMATION

Bossard, C. C., J. Randall, and M. C. Hashovsky. 2000. *Invasive Plants of California's Wildlands.* Berkeley: University of California Press.

Dremann, C. 1996. "Grasses and mulch control: Yellow-star thistle." *Restoration and Management Notes* 14(1):79.

Jacobs, J., J. Mangold, H. Parkinson, and M. Graves. 2011. *Ecology and management of yellow starthistle* (Centaurea solstitialis *L.*). USDA NRCS Invasive Species Technical Note No. MT-32. ftp://ftp-fc.sc.egov.usda.gov/MT/www/technical/invasive/Invasive_Species_Tech_Note_MT32.pdf.

Zouhar, Kris. 2002. "*Centaurea solstitialis.*" *Fire Effects Information System.* U.S. Department of Agriculture, Forest Service, Rocky Mountain Research Station, Fire Sciences Laboratory. http://www.fs.fed.us/database/feis.

Spotted Knapweed   *Centaurea stoebe*

NAME AND FAMILY
Spotted knapweed (*Centaurea stoebe* ssp. *micranthos* (Gugler) Hayek); aster family (Asteraceae). Former Latin names of this species were *C. biebersteinii* and *C. maculosa*. There are several other similar introduced species of knapweeds, including *C. diffusa*, *C. nigra*, *C. jacea*, *C. nigrescens*, and *C. trichocephala*. All of these species are considered to be introduced, weedy plants and can be treated similarly to spotted knapweed.

IDENTIFYING CHARACTERISTICS
Often first noticed because of its profusion of purple bachelor's button–like flowers blooming in summer. Spotted knapweed is a biennial or short-lived perennial with a stout taproot that can penetrate deep into hard soils for sustaining moisture and nutrients. In its first year, it has a rosette of deeply lobed leaves (about 8 in. [20 cm] long) borne on short stalks. The flowering stalks have alternate leaves, with the leaves deeply lobed toward the base and smaller and with fewer lobes toward the top. Plants can grow up to 3 ft. (1 m) tall. The closed flower heads are egg shaped and are surrounded by green bracts with brown to black triangular tips

Left: *Plants thrive in sun or partial shade.* Right: *The purple flowers bloom in summer.*

with fringed ends. The black tips with fringed edges distinguish this species from other similar knapweeds. When the flowers open, the petals are purple to pink and sometimes white with 25 to 35 flowers in a head. Flower heads are single or borne in clusters of two or three at the branch ends. Seeds simply fall from the flower heads, then spread as water or animals move soil; seeds may also be dispersed when the ripe plants are harvested with hay or other crops. A single plant produces as many as 1,000 seeds, giving itself a big stake in the dispersal lottery.

HABITAT AND RANGE

Spotted knapweed is expanding its range in the United States and Canada. It thrives on over 4 million acres of pasture and rangeland in Washington, Montana, Idaho, Oregon, and California and occurs in all but five U.S. states. In Canada it occurs from Nova Scotia to British Columbia but is most common in southwestern Canada. It establishes readily on disturbed ground, but can also invade open forests and prairies. Its taproot allows it to tolerate drought, accessing water deep in the soil.

WHAT IT DOES IN THE ECOSYSTEM

Spotted knapweed outcompetes native plant species, particularly in disturbed habitats. It reduces native plant and animal diversity with chemicals (allelo-

chemicals called catechins) that poison would-be competitors. Catechin induces the production of harmful reactive oxygen in neighboring plant roots, eventually killing their cells. The chemical remains in the soil, making restoration difficult. Spotted knapweed also decreases forage production for livestock and wildlife that do not feed on it. Because it is taprooted rather than having more shallow fibrous roots like many native grasses and herbaceous plants, it can degrade soil and water resources by increasing erosion, surface runoff, and stream sedimentation. Spreading and shallow fibrous roots hold soil in place better than taproots.

Leaves along the flowering stalks are silvery gray.

The black-tipped bracts with fringed edges help distinguish spotted knapweed from other knapweeds.

HOW IT CAME TO NORTH AMERICA

Spotted knapweed probably arrived in the United States in the ballast of ships or as a contaminant in alfalfa seeds in the 1890s. It was first recorded in British Columbia in 1893. It was spread in alfalfa and hay before the seriousness of its spread was noticed. It is now designated as a noxious weed in most western states.

MANAGEMENT

Mowing soon after flowers open and before seeds mature can control populations, but seeds remain viable for five to eight years in the soil. Eliminating the first small infestations prevents expensive treatments later. Small infestations can be hand pulled or clipped repeatedly. Wear gloves, because plants cause skin irritation in some people. Burning plants with a propane torch is effective. The herbicide 2,4-D will kill existing plants, but will not affect the seeds in the soil. Picloram will persist in soil for up to four years but is very expensive to apply. Clopyralid and dicamba are also used to treat spotted knapweed. Four insect species have been introduced in the western United States as biological control agents of knapweed. Long-term grazing by sheep and goats can also control spotted knapweed.

FOR MORE INFORMATION

Bais, H. P., R. Vepachedu, S. Gilroy, R. M. Callaway, and J. M. Vivanco. 2003. "Allelopathy and exotic plant invasion: From molecules and genes to species interactions." *Science* 301:1377–1380.

Minteer, C. 2008. "*Centaurea stoebe* ssp. *micranthos.*" http://wiki.bugwood.org/Centaurea_stoebe_ssp._micranthos.

Zouhar, Kris. 2001. "*Centaurea maculosa.*" *Fire Effects Information System.* U.S. Department of Agriculture, Forest Service, Rocky Mountain Research Station, Fire Sciences Laboratory. http://www.fs.fed.us/database/feis.

Celandine   *Chelidonium majus*

NAME AND FAMILY
Celandine, greater celandine, rock poppy, swallowort (*Chelidonium majus* L.);
poppy family (Papaveraceae). Early in its growth it may be difficult to distinguish from the native celandine poppy or yellow hornpoppy (*Stylophorum
diphyllum*).

IDENTIFYING CHARACTERISTICS
This biennial herb is distinguished by its orange to yellow sap and brittle stem.
Celandine matures at a height of 1 to 2 ft. (0.3 to 0.6 m). In early summer to
midsummer it bears yellow flowers 0.5 to 0.75 in. (1.3 to 1.9 cm) across with
four petals spread out flat beneath the stamens. Plants can continue flowering
into fall. Clusters of long (0.75 to 1 in. [1.9 to 2.5 cm]), smooth seedpods follow the flowers. The alternate leaves are deeply lobed and each major lobe has
many lesser lobes. The celandine poppy has flowers more than 1 in. (2.5 cm)
across; seedpods are oval and hairy and the juice darker, like saffron.

HABITAT AND RANGE
Celandine grows across southern Canada and across the northern half of the
United States south to Georgia and west to Utah. It prefers moist soils and sun-
light and often grows in floodplain forests, wet meadows, and as a city weed.

Left: *Oblong, smooth seedpods distinguish celandine from the native celandine
poppy, which has oval, hairy pods.* **Right:** *Four-petaled yellow flowers of celandine
bloom mainly in early spring.*

The leaves are lobed and flowers grow at the tips of the stems.

WHAT IT DOES IN THE ECOSYSTEM
Celandine can outcompete some natives, but the literature shows no major effects in natural areas yet. Leaves appear early in spring and do not die back until late fall. Ants disperse the seeds, and leaves broken off the plant can root. The roots contain a strong, sometimes fatal poison, isoquinoline alkaloid. Some writers say all parts of the plant are toxic if eaten. As with many toxics, its active qualities also give it a long history in herbal medicine.

HOW IT CAME TO NORTH AMERICA
New England herb gardens were growing celandine in the late 1600s and almost certainly colonists brought it to North America as an herbal remedy. The medicinal uses of celandine go as far back as Aristotle's claim that mother swallows used the sap to treat their offspring's eyes. Centuries ago the yellow juice, resembling bile, gave the plant a place in treating liver diseases but recently it has since been fingered as a possible cause of liver damage. Many herbal pharmacies sell it to address a variety of complaints. It is native to Europe and eastern Asia.

MANAGEMENT
Plants can be hand pulled, but try to minimize soil disturbance and replant quickly. Wear gloves, as the sap may cause skin irritation. Glyphosate can be used to treat celandine as well.

FOR MORE INFORMATION
Sanders, J. 2003. *The Secrets of Wildflowers.* Guilford, CT: Lyons Press.
Smith, L. "Invasive Exotic Plant Management Tutorial for Natural Lands Managers." Mid-Atlantic Invasive Plant Council. http://www.dcnr.state.pa.us/forestry/invasivetutorial/greater_celandine.htm.

Canada Thistle   *Cirsium arvense*

NAME AND FAMILY
Canada thistle (*Cirsium arvense* [L.] Scop.); aster family (Asteraceae). Outside the United States, Canada thistle is also known as creeping thistle or California thistle. Other purple-flowered thistles include bull thistle (*C. vulgare*) and musk thistles (*Carduus* spp.). There are many native thistles, some rare, that can be confused with Canada thistle, so proper identification is important.

IDENTIFYING CHARACTERISTICS
This perennial thistle grows 1.5 to 5 ft. (0.5 to 1.5 m) tall and is distinguished from other thistles by its extensive horizontal roots, by the dense growth of clones from those roots, and by having male and female flower heads on separate plants. The flower heads are small (less than ⅛ in. [0.25 cm]) and lavender to rose-purple in color, sometimes white. Flowering is triggered by day length, but plants in more northern latitudes bloom for longer periods of time. A single plant produces an average of 1,500 seeds. The leaves vary in appearance and size but are generally irregularly lobed with many spines along the edges and are arranged alternately along the stem. They can be 2 to 6 in. (5 to 15 cm) in length. The stem is often slightly hairy and ridged.

HABITAT AND RANGE
Canada thistle grows between 37 and 58 to 59° N around the northern hemisphere and is particularly problematic in the northern United States and southern Canada. It grows in a wide range of wet to moist habitats, including grasslands, river floodplains, prairie marshes, roadsides, pastures, and drainage ditches.

WHAT IT DOES IN THE ECOSYSTEM
Canada thistle competes with and displaces native vegetation, lowering plant and animal species diversity and changing species composition. Because of its economic threat to farmers and ranchers, most states consider it a noxious weed. It reduces crop yields and

Clusters of small purple flower heads bloom atop Canada thistle plant.

The leaves are deeply lobed.

2 ft. (0.6 m) in diameter in late spring to midsummer. The hollow stem is 2 to 4 in. (5 to 10 cm) in diameter. The stem is covered with purple blistery spots and bristles, while native cow parsnip is not as purple or hairy. Giant hogweed leaves are deeply incised and up to 5 ft. (1.5 m) wide, alternating along the stem. Leaves of both cow parsnip and poison hemlock are smaller. Fruits are about 0.38 in. (0.9 cm) flattened ovals with ridged margins. By late summer plants die back to the roots.

HABITAT AND RANGE
Giant hogweed is a northern plant most commonly found in the east from Newfoundland south to Maryland and in the west from British Columbia south into Oregon. It grows in the rich, moist soils of floodplains and stream-banks and in disturbed areas.

WHAT IT DOES IN THE ECOSYSTEM
Giant hogweed is so large it can outcompete many natives, but it is mainly a problem as a threat to human health in North America. If sap from giant hog-weed gets on exposed skin, exposure to sunlight causes burning and blistering caused by chemicals called furocoumarins. Blisters often give way to purple or black scars. Giant hogweed can cause increased erosion along stream banks because its roots do not hold soil well.

HOW IT CAME TO NORTH AMERICA
Giant hogweed is native to the Caucasus Mountains where Asia and Europe meet between the Caspian and the Black seas. Early in the twentieth century, North American ornamental gardeners planted it as a curiosity. Gardeners in

the United States cultivated it as early as 1917. Iranian cooks use the seeds, "golpar," in cooking, and many seeds have come into the United States with Iranian visitors and immigrants.

MANAGEMENT
The first mandate of any management plan should be to avoid skin contact. In many states, the state Department of Agriculture will help you control giant hogweed. Individual plants can be dug up, but since roots are thick and deep, be sure to cut roots at least 6 in. (15.2 cm) or deeper below the surface. Herbicides will kill young plants, but the area should be inspected every year for at least three years. Cattle and pigs graze on plants without harm.

The towering plants like the moist, rich, soils of streambanks.

FOR MORE INFORMATION
Michigan Department of Agriculture. *Giant Hogweed.* http://mipn.org/MDA_Hogweed_Brochure .pdf.

Page, N. A., R. E. Wall, S. J. Darbyshire, and G. A. Mulligan. 2006. "The biology of invasive alien plants in Canada. 4. *Heracleum mantegazzianum* Sommier & Levier." *Canadian Journal of Plant Science* 86:569–589.

Westbrooks, R. G. 1991. "Federal noxious weed inspection guide." Whiteville, NC: Whiteville Plant Methods Center. http://www.invasivespecies.org/FNWDetail.CFM?RecordID=82.

Dame's Rocket   *Hesperis matronalis*

NAME AND FAMILY
Dame's rocket, dame's violet, mother-of-the-evening (*Hesperis matronalis* L.); mustard family (Brassicaceae). Sometimes confused with phlox.

IDENTIFYING CHARACTERISTICS
In late spring, the fragrant violet flowers of dame's rocket stand out along roadsides and in forests. The flowers are four-petaled, 0.5 in. (1.3 cm) across, and held in oblong clusters on 3 ft. (1 m) high flowering stalks. Occasionally flowers are white or pink. Plants are short-lived perennials that reproduce by seed. Seeds are held in cylindrical pods 1 to 5 in. (2.5 to 13 cm) long. Leaves alternate along

the stems and are lance shaped with teeth along the edge, 2 to 6 in. (5 to 15 cm) long, and often hairy on both sides. Plants also have a rosette of leaves at the base that overwinters. Phlox have five petals that are fused into a tube and leaves are opposite.

HABITAT AND RANGE

Occurs in damp, shaded places such as along wooded streams and floodplains and shady roadsides. Occurs throughout Canada and the United States except in the southernmost states.

Right: *The four-petaled flowers of Dame's rocket range in color from white to dark pink.* **Bottom:** *Dame's rocket grows among other plants in damp, shaded places.*

WHAT IT DOES IN THE ECOSYSTEM
Dame's rocket can form locally dense stands that may displace native species. It is primarily of concern because its widespread use in commercially sold meadow flower mixes and its high seed production are introducing it to many new areas. It is also an alternate host for several crop mosaic viruses.

HOW IT CAME TO NORTH AMERICA
Colonists introduced it from Europe in the 1600s as an ornamental plant or perhaps as a medicinal plant since it had a long history in European and Asian herbal medicine. Seeds are often included in meadow and wildflower seed mixes, even in those labeled as native seed mixes.

MANAGEMENT
Plants can be hand pulled, although if there is a seed bank, several years of pulling may be required. Plants that have flowered should be bagged or burned to prevent seed formation. Plants can also be sprayed with glyphosate in late fall when other plants are dormant.

FOR MORE INFORMATION
Royer, F., and R. Dickinson. 2004. *Weeds of the Northern U.S. and Canada*. Renton, WA, and Alberta: Lone Pine Publishing and University of Alberta Press.

Black Henbane   *Hyoscyamus niger*

NAME AND FAMILY
Black henbane, stinking nightshade, hog's bean (*Hyoscyamus niger* L.); nightshade family (Solanaceae).

IDENTIFYING CHARACTERISTICS
The greasy hairs that cover the thick, upright stems and most of the plant distinguish this annual or biennial that grows up to 3 ft. (1 m) tall. The putrid smelling leaves are up to 8 in. (20 cm) long, 6 in. (15 cm) wide with shallow lobes between pointed tongues. After the first-year or early-season rosette has formed, leaves grow alternately on the stems. Five-

Plants can live in semiarid areas.

Left: *Cream-colored flowers with purple throats bloom all summer.* **Right:** *The leaf bases clasp the stem and the stems are covered with greasy hairs.*

lobed flowers are cream to creamy green, 2 in. (5 cm) in diameter, with purplish throats. They are borne on spikes rising from the leaf axils and bloom from spring till early fall. The tiny black seeds develop in the tubular, urn-shaped calyx of the flower.

HABITAT AND RANGE

Black henbane is found throughout the United States and southern Canada, especially in pastures, along fencerows, on road shoulders, and in abandoned fields, cutover areas, and other disturbed sites. It can live in semiarid areas and prefers sandy, gravelly soils or loams and clays that are well drained.

WHAT IT DOES IN THE ECOSYSTEM

Black henbane is poisonous to most mammals and grazers usually avoid it. A single plant can produce as many as half a million seeds. Naturalizes in a variety of environments but is generally not noted as a plant that takes over large areas. Considered a serious problem on Bureau of Land Management (BLM) government lands in Wyoming.

HOW IT CAME TO NORTH AMERICA

This Mediterranean native was a well-known powerful herbal in Europe and one of the first alien plant introductions in colonial America. All parts of the plant contain poisonous alkaloids that can cause seizures, trembling, and insanity, symptoms similar to those caused by its much more toxic relative, belladonna. In the medieval era, before the widespread use of hops, German brewers flavored beer with black henbane.

MANAGEMENT
Tough stems and roots make hand pulling a hard job. Pulling and mowing should be done before seeds mature. Herbicides containing glyphosate can provide excellent control.

FOR MORE INFORMATION
Belliston, N., R. Whitesides, S. Dewey, J. Merritt and S. Burningham. 2010. *Noxious Weed Field Guide to Utah.* Uintah County Weed Department and Utah State University Extension Service. http://www.utahweed.org/PDF/FieldGuide_Ed4.pdf.
Whitson, T. D., L. C. Burrill, S. A. Dewey, D. W. Cudney, B. E. Nelson, R. D. Lee, and R. Parker. 1992. *Weeds of the West.* Jackson, WY: Western Society of Weed Science.

Himalayan Balsam — *Impatiens glandulifera*

NAME AND FAMILY
Himalayan balsam, policeman's helmet, bobby tops, copper tops, ornamental jewel weed, gnome's hatstand, (*Impatiens glandulifera* Royle); touch-me-not family (Balsaminaceae).

IDENTIFYING CHARACTERISTICS
This annual herb grows 3.3 to 6.5 ft (1 to 2 m) tall, with simple oblong to elliptical leaves growing opposite each other or in a whorl of three. Leaves are 2 to 9 inches long (5 to 23 cm) and 0.75 to 2.75 in. (2 to 7 cm) wide, with sharply toothed edges. The green to reddish stems branch and are hollow and hairless. The helmet-shaped flowers range from white through pink and purple and have five petals, some of which are fused, forming a flower that ressembles an English policeman's helmet. The flowers grow in small clusters from among the leaves and bloom from midsummer into fall. The plants flower in summer and form green seed pods 0.75 to 1.5 inches (2 to 3 cm) long and 0.25 (8 mm) broad. Each pod contains 4 to 16 seeds. These pods burst open if disturbed and can fling seeds up to 24 ft. (about 7 m). Native jewelweed, *Impatiens capensis*, has yellow to orange flowers.

HABITAT AND RANGE
Himalayan balsam prefers moist areas near bodies of water but grows in a wide range of soils and sites up to elevations of 9,000 ft. It has established populations from Alaska down the West Coast to California, in the east from New-

foundland south to New York, and in the Midwest from Manitoba to Michigan and Indiana.

WHAT IT DOES IN THE ECOSYSTEM

In riparian areas it can alter water flow or stop flow altogether, while its shallow, weak root system and fall dieback can leave these areas more prone to erosion. It competes aggressively by producing numerous seeds, adapting to harsh conditions, and growing rapidly. It produces nectar at several times the flow of many native plants and may attract insects away from pollinating native species. Scientists debate its impact on natives, with some studies saying it displaces mainly other exotics, and others indicating a 25 percent overall reduction in diversity of species. Attracts many butterflies.

HOW IT CAME TO NORTH AMERICA

This native of Pakistan, India, and Nepal was brought to Europe and North America as an ornamental, an easily grown and inexpensive showy and richly colored flower. Seeds were sent to England in 1839 and the plant quickly spread in the British Isles. It was first recorded in Canada in Ottawa in 1901.

Dark pink flowers hang from this tall plant.

MANAGEMENT

Himalayan balsam is best destroyed in spring and fall, but its destruction often lays the ground bare for invasions of Japanese knotweed or other invasives. Himalayan balsam thrives in nutrient-rich wetlands, and since the nutrients are often from pollution, reduction in these nutrient loads gives natives a better chance of recovering. Small infestations can be hand-pulled or mowed before flowering. Sheep and cattle eat the entire plant. Herbicides containing glyphosate applied before flowering will kill balsam. In grassy areas, herbicides with triclopyr, 2,4-D, or metsulfuron will kill balsam but leave most grasses growing. Germany is using an economic approach to control—trying to develop food products made from the nectar-rich flowers. Seed pods, seeds, and young shoots are also edible.

FOR MORE INFORMATION

Clements, D. R., K. R. Feenstra, K. Jones, and R. Staniforth. 2008. "The biology of invasive alien plants in Canada. 9. *Impatiens glandulifera* Royle." *Canadian Journal of Plant Sciences* 88(2):403–417.

Global Invasive Species Database. 2009. "*Impatiens glandulifera*." http://www.issg.org/database/species/ecology.asp?fr=1&si=942.

Tansy   *Jacobaea vulgaris* and *Tanacetum vulgare*

NAME AND FAMILY

Tansy ragwort, tansy butterweed, stinking Nanny, mare's fart (*Jacobaea vulgaris* Gaertn., formerly *Senecio jacobaea* L.), and common tansy, golden buttons (*Tanacetum vulgare* L.); aster family (Asteraceae). May be confused with native tansy species in some areas.

IDENTIFYING CHARACTERISTICS

Tansy ragwort starts from a fibrous taproot as a small rosette of medium- to dark-green ruffled, broad, lobed leaves at ground level. From the rosette grow one to several stems that reach up to 6 ft. (1.8 m) high. The coarse and sometimes purplish stems branch toward the top. Leaves are finely lobed, with each lobe bearing small teeth, giving the leaves a lacy or ruffled appearance. They alternate along the stems, decreasing in size going up. Leaf undersides are hairy. Flower

Left: *A large cluster of bright tansy ragwort flowers.* **Right:** *Common tansy flowers lack ray flowers.*

clusters form in summer to fall at the tops of the stalks, blossoming into arrays of disk and ray yellow flowers 0.5 in (1.3 cm) long. Seeds have hairs that stick out of the top of the flower (pappi). Plants are usually biennial but sometimes grow as annuals or short-lived perennials. Common tansy is a perennial that has clusters of yellow buttonlike flowers 0.5 (1.2 cm) wide (and no ray flowers), held on stems up to 7 ft. (2 m) high. Its leaves are similar to those of tansy ragwort. The leaves of both species have a pungent odor when crushed. Native tansy species are generally smaller in stature.

Lacy leaves alternate along the stem.

HABITAT AND RANGE

Tansy ragwort prefers dry, open areas with some sun exposure and well-drained soils. It will not grow where the water table is high. Often found in pastures and on rangeland, along roads, in forest clearcuts, and in disturbed lands. It grows across southern Canada and into the Pacific Northwest, in California, Illinois, and Michigan, and from Pennsylvania north to Maine. Common tansy grows throughout most of the U.S. and in all Canadian provinces except Nunavut. It is considered most problematic in northern regions where it grows on rangelands, along rivers, and in open forests.

WHAT IT DOES IN THE ECOSYSTEM

Tansy ragwort in pastures and on rangelands can reduce forage by up to 50 percent, and it can invade cut-over forest land, where it outcompetes native species. To retard competition it releases chemicals that are toxic to other plants and it also reduces soil microbe diversity. Alkaloids in the plant, especially the flowers, damage the livers of cattle and horses, sometimes fatally (usually when they eat it in contaminated hay). In the 1970s in Oregon it caused thousands of livestock deaths. A single plant can produce 150,000 seeds. Common tansy is eaten by grazing animals, but may be toxic to cattle. It can grow in dense stands, presumably displacing native species. It has also been reported as restricting water flow in streams in Alaska.

HOW IT CAME TO NORTH AMERICA

Tansy ragwort, a native of Europe, Asia, and Siberia, may have been introduced to Canada in the 1850s in ship's ballast, and it may also have been brought in as a medicinal herb, although it can cause liver damage in humans as well as in livestock. Common tansy was established in the United States by the early 1600s, considered an indispensable medicinal plant in colonial gardens. It is native to Eurasia.

MANAGEMENT

Since common tansy and tansy ragwort can spread rapidly, open areas should be closely watched and new plants hand pulled, carefully removing the root since pieces of it will sprout. Bag and burn flowering plants. Mowing and scything only cause the roots to send up more shoots. New Zealanders have used sheep to control tansy ragwort since the toxins in it do not affect sheep. Sheep are also very effective at controlling common tansy. Herbicides containing glyphosate or 2,4 D applied to rosettes kill ragwort, but also kill other broadleaf plants. The "tansy tiger" caterpillar (of the cinnabar moth) introduced in the 1960s often consumes tansy ragwort leaves but does not eliminate host plant populations. However, these caterpillars, working with tansy ragwort flea beetles and seed head flies, which attack other parts of the plant, almost eliminated Oregon's tansy ragworts in the 1980s. The plant made a comeback when wet falls allowed the plant to recover after the predators stopped work. Maintaining healthy stands of grass and avoiding overgrazing will prevent tansy ragwort from establishing.

FOR MORE INFORMATION

Coombs, Eric, 2011. *Tansy Ragwort*. Oregon Department of Agriculture. http://extension.oregon state.edu/douglas/sites/default/files/documents/tragwortupdate2011.pdf.
Gucker, Corey L. 2009. *"Tanacetum vulgare."* *Fire Effects Information System*. U.S. Department of Agriculture, Forest Service, Rocky Mountain Research Station, Fire Sciences Laboratory. http://www.fs.fed.us/database/feis.
Jacobs, J. and S. Sing. 2009. *Ecology and management of tansy ragwort* (Senecio jacobeae *L.*). USDA NRCS Invasive Species Technical Note MT-24. http://citeseerx.ist.psu.edu/viewdoc/download?doi=10.1.1.180.6162&rep=rep1&type=pdf.
"Ragwort: Myths and Facts." 2012. http://www.ragwort.jakobskruiskruid.com.

Prickly Lettuce   *Lactuca serriola*

NAME AND FAMILY
Prickly lettuce, horse thistle, compass plant, wild opium (*Lactuca serriola* L.); aster family (Asteraceae). Can be confused with sow thistles (*Sonchus* spp.).

IDENTIFYING CHARACTERISTICS
This common annual or biennial weed is easily identified by its prickly stems that yield a bitter white sap like the sap in an overripe lettuce stalk. The plant begins with a rosette of rounded leaves close to the ground, then sends up a prickly stalk that bears alternate leaves that are deeply lobed, often shaped like a fleur-de-lis, with prickly edges and pointed lobes clasping the stem. Leaves have a prominent white midvein and the underside of the vein is heavily prickled. Leaves often extend in north-south directions to collect maximum sun, thus earning the plant the name "compass plant." A domed crown made up of 0.5 in. (1.3 cm) wide, creamy yellow flower heads blooms from midsummer to late summer, maturing to finely tufted, gray-brown, dandelion-like seeds. Each tightly packed flower head is made up of many single-petaled flowers (ray flowers). The plant can grow to 6 ft. (1.8 m) tall and has a long taproot.

Sow thistles look similar but do not have prickles on the undersides of the leaves along the midveins.

HABITAT AND RANGE
This European native has naturalized across Canada and the United States. It is abundant on roadsides, in no-till farm fields, in abandoned fields, and even on sand dunes. Prefers to do most of its growing in the wet season, especially in the relative mild and very wet winters of western states.

WHAT IT DOES IN THE ECOSYSTEM
Prickly lettuce is mainly a nuisance in no-till crop areas, but it is an aggressive colonizer and will take over spaces

Prickly lettuce plants tower over surrounding plants.

Left: *The leaves have a prominent white midvein and prickly edges.* **Right:** *Plants have a milky sap.*

that might become habitat for slower-growing native plants. It is a drought-tolerant and prolific seed producer whose small, tightly packed flowers produce an average of 20 seeds each, and in total some 45,000 seeds per large plant. Seeds germinate soon after release and seldom last more than three years in the soil.

HOW IT CAME TO NORTH AMERICA
Prickly lettuce is native to the Mediterranean region and was introduced to Canada and the United States in the late 1890s, probably as a contaminant in seed. It quickly spread across the continent. Extracts of the sap are used in herbal remedies for insomnia and restlessness because of its opiate-like compounds.

MANAGEMENT
Plants can be pulled by hand before they set seed, and cultivation is effective in crop areas. Mowing is usually ineffective because the basal rosette is very close to the ground, and when mowing cuts a stem, a new stem appears. While prickly lettuce is acquiring resistance to some herbicides, most practitioners find it is killed by isoxaben and glyphosate herbicides. To prevent herbicide resistance from developing, the same herbicide should not be used every year.

FOR MORE INFORMATION
Mallory-Smith, C., D. Thill, and M. Dial. 1990. "Identification of sulfonylurea herbicide-resistant prickly lettuce (*Lactuca serriola*)." *Weed Technology* 4:163–168.
Uva, R. H., J. C. Neal, and J. M. DiTomaso. 1997. *Weeds of the Northeast.* Ithaca: Cornell University.

Perennial Pepperweed — *Lepidium latifolium*

NAME AND FAMILY
Perennial pepperweed, tall whitetop (*Lepidium latifolium* L.); mustard family (Brassicaceae). Several other problem weeds, including the perennials hoary cress (*L. draba*), lens-podded whitetop (*L. chalepensis*), and hairy whitetop (*L. appelianum*) and the annuals/biennials field pepperweed (*L. campestre*) and clasping pepperweed (*L. perfoliatum*), resemble perennial pepperweed but do not grow more than 2 ft. (61 cm) tall and have leaves that clasp the stems.

IDENTIFYING CHARACTERISTICS
This perennial plant forms dense stands with stems reaching 1 to 5 ft. (0.3 to 1.5 m) tall. It begins to grow in late fall to early spring, forming a rosette of leaves for several weeks before flowering and setting fruit from late spring into summer. By midsummer to late summer plants die back, leaving stems standing with dry seedpods. Rosette leaves can be 4 to 12 in. (10 to 30 cm) long and 1 to 2 in. (2.5 to 5 cm) wide with toothed edges. The rosette leaves have long (1.5 to 6 in.) stalks. The alternate leaves on the stems are much smaller (1 to 3 in. [2.5 to 7.6 cm]), oblong, and have no or few teeth along the margins. Leaves are green to gray green. Flat, dense clusters of flowers are made up of thousands of white, four-petaled flowers. Each flower produces a flattened, rounded to oval pod (0.06 in. [0.15 cm] diameter) covered with long hairs, held on a stalk and bearing two seeds. Seeds appear to be short-lived and have low germination, but 1 acre of dense infestation can produce six billion seeds. Taproots grow long enough to reach the water table and become very thick.

A dense stand of perennial pepperweed in moist soil.

Oblong leaves alternate along the flowering stalk.

HABITAT AND RANGE

Pepperweed grows in British Columbia, Alberta, Quebec, coastal New England, some midwestern states, and all the states west of the Rocky Mountains. It colonizes riverbanks, floodplains, coastal wetlands, and marshes, and will grow in hay meadows, on rangelands, and along roadsides.

WHAT IT DOES IN THE ECOSYSTEM

The thick layer of leaf litter laid down by dense colonies of pepperweed inhibits establishment of native plants like willows and cottonwoods. It reduces habitat available to endangered species like the salt marsh harvest mouse. Pepperweed also acts as a "salt pump," bringing up salts from deep underground and depositing them near the surface, favoring salt-tolerant species in a community. In tidal marshes, pepperweed changes soil characteristics such as pH and organic matter concentrations. Despite dense growth, each plant puts down only one central root. Pepperweed replaces willows, whose mats of roots do a better job of resisting erosion.

HOW IT CAME TO NORTH AMERICA

Native to southeastern Europe and southwestern Asia, perennial pepperweed arrived in the United States as a contaminant in seeds for agriculture, appearing in California in 1936. It was also grown in the United States as a cut flower. It is spread by agricultural equipment, in hay, by waterfowl, and through discarded dried flower arrangements.

MANAGEMENT

Because of the deep taproots, pepperweed is difficult to pull up. It resprouts after cutting or mowing. Herbicides containing glyphosate, chlorsulfuron, metsulfuron-methyl, and imazapyr sprayed at low concentrations on plants as flower buds are forming can control pepperweed, but not all of these can be used near water. Cutting plants combined with herbicide application tends to be more effective. Long-term flooding may reduce infestations. For native vegetation to recover, leaf litter and salinity may need to be reduced.

FOR MORE INFORMATION

California Invasive Plant Council. "*Lepidium latifolium*." http://www.cal-ipc.org/ip/management/plant_profiles/Lepidium_latifolium.php.

Francis, A. and S. I. Warwick. 2007. "The biology of invasive exotic plants in Canada. 8. *Lepidium latifolium* L." *Canadian Journal of Plant Science* 87:639–658.

Renz, M. J. 2000. "Element stewardship abstract for *Lepidium latifolium* L." The Nature Conservancy. http://wiki.bugwood.org/Lepidium_latifolium.

Zouhar, K. 2004. "*Lepidium latifolium*." *Fire Effects Information System*. U.S. Department of Agriculture, Forest Service, Rocky Mountain Research Station, Fire Sciences Laboratory. http://www.fs.fed.us/database/feis.

Ox-Eye Daisy   *Leucanthemum vulgare*

NAME AND FAMILY

Ox-eye daisy, moon daisy, marguerite, dog daisy (*Leucanthemum vulgare* Lam.); aster family (Asteraceae). Cultivated Shasta daisies (*Chrysanthemum maximum* L.) look almost identical to the ox-eye daisy.

IDENTIFYING CHARACTERISTICS

The white-petaled flower heads with yellow centers, 1 to 2 in. (2.5 to 5 cm) across, make ox-eye daisy very recognizable from spring into late summer. One

Left: *Flowers are held on long stalks that bloom all summer.*
Right: *Flowering stalks arise from a rosette at the base of the plant.*

The familiar ox-eye daisy is smaller than cultivated Shasta daisy flowers.

to 40 stems rise up 1 to 3 ft. (0.3 to 1 m) from a single plant's shallow, creeping roots. The dark-green leaves are toothed to lobed, 1 to 4 in. (2.5 to 10 cm) long, and decrease in size going up the stem. Leaves at the base of the stem are more spoon-shaped. Up to 200 seeds are produced per flower head. Shasta daisy flowers are larger, more than 2 in. (5 cm) across.

HABITAT AND RANGE
Commonly found along roadsides and in pastures and fields, it also invades prairies, meadows, and open woodlands. Does particularly well in heavy, moist soils, but will grow in many soil types and in sun to part shade. Found throughout Canada and the United States.

WHAT IT DOES IN THE ECOSYSTEM
Ox-eye daisy can form dense stands that displace native vegetation. One study found that larger numbers of root-feeding nematodes were associated with plots containing ox-eye daisy, and could be contributing to the lower overall plant diversity in those plots. Ox-eye daisy is not eaten by most animals, and it forms a long-lasting seed bank (20 years or more). Plants harbor several crop diseases. Seeds are spread in movement of soils by vehicles and shoes. Seeds are also dispersed by birds and by livestock on hooves and caught in fur.

HOW IT CAME TO NORTH AMERICA
Ox-eye daisy was introduced from Europe as an ornamental plant in the 1800s. It is native from Scandinavia through Russia and into Central Asia. It is commonly sold as a wildflower and in meadow mixes. Its seeds also have contaminated hay and grain seed shipments.

MANAGEMENT
Plants can be pulled or dug out. Mulching heavily will also cause plants to rot over winter. Repeated grazing by sheep and goats can help control populations. Herbicides containing glyphosate, 2,4-D, picloram, or imazapyr applied just as plants begin to flower can also be effective, but some of these herbicides remain in the soil.

FOR MORE INFORMATION

Alvarez, M. *"Leucanthemum vulgare."* California Invasive Plant Council. http://www.cal-ipc.org/
    ip/management/ipcw/pages/detailreport.cfm@usernumber=59&surveynumber=182.php.
Jacobs, J. 2008. *Ecology and management of ox-eye daisy* (Leucanthemum vulgare *Lam.*).
    USDA NRCS Invasive Species Technical Note No. MT-19. http://www.msuextension.org/
    ruralliving/Dream/PDF/oxeye.pdf.
Royer, F., and R. Dickinson. 2004. *Weeds of the Northern U.S. and Canada.* Renton, WA, and
    Alberta: Lone Pine Publishing and University of Alberta Press.
Ruijven, J., G. B. De Deyn, and F. Berendse. 2003. "Diversity reduces invisibility in experimental
    plant communities: The role of plant species." *Ecology Letters* 6:910–918.

Scotch Thistle   *Onopordum acanthium*

NAME AND FAMILY
Scotch thistle, cotton thistle (*Onopordum acanthium* L.); aster family (Aster-aceae). The Scots reportedly adopted this thistle as a national emblem after invading Vikings lost the advantage of surprise when they stumbled into a stand of thistles. Considered a noxious weed in many states and provinces.

IDENTIFYING CHARACTERISTICS
Scotch thistle usually grows as a biennial and has groups of two dark-pink to lavender flower heads atop 2 to 8 ft. (0.6 to 2.4 m) stems in midsummer. Under the 2 in. (5.1 cm) wide, flat-topped flower heads, stiff, spiny bracts circle around a globe-shaped base. The stems are thick (to 4 in. [10 cm] wide at the base), woody, and branched, with spiny wings that extend up to the base of the flower heads. Woolly leaves grow alternately along the stems and are armed with big yellowish spines. Leaves can reach 2 ft. (0.6 m) long and a foot wide near the base, getting smaller toward the upper part of the stems. First-year rosettes put down a taproot as long as 1 ft. (0.3 m).

HABITAT AND RANGE
Scotch thistle grows in sagebrush com-munities, along riversides and road edges, and in rangelands and prairies.

*Rosettes have spiny leaves
covered in hairs.*

Occupying most of the United States and Canada, Scotch thistle is considered particularly problematic on dry rangelands in the western United States and in scattered locations elsewhere.

WHAT IT DOES IN THE ECOSYSTEM

Dense, spiny stands of thistles on lands used for cattle diminish forage and form barriers along waterways. Seeds are both wind dispersed and dispersed by animals.

HOW IT CAME TO NORTH AMERICA

Scotch thistle was probably introduced to North America as an orna-

Left: *Scotch thistle plants growing on rangelands can diminish forage and block access to waterways.* Bottom: *Spiny bracts encircle the globe-shaped bases of the large flower heads.*

mental plant in the late 1800s. It is native to Europe and western Asia. Immature flower heads used to be eaten like artichokes.

MANAGEMENT
Small patches of thistle can be eliminated by cutting plants below the soil surface. Goats will eat thistle. Second-year growth depletes nutrients in the root and the best time for mowing is before the second-year growth forms flower heads and seeds. Herbicides, such as picloram, dicamba, 2,4-D, and metsulfuron, can be applied in early spring or to rosettes in fall.

FOR MORE INFORMATION
Beck, K. G. 1999. "Biennial thistles." In *Biology and Management of Noxious Rangeland Weeds*, edited by R. L. Sheley and J. K. Petroff. Corvallis, OR: Oregon State University Press.

Cavers, P. B., M.M. Qaderi, P. F. Threadgill and M. G. Steel. 2011. "The biology of Canadian Weeds. 147. *Onopordum acanthium* L." *Canadian Journal of Plant Sciences* 91:739–758.

Global Invasive Species Database. 2012. "*Onopordium acanthium*." http://www.issg.org/database/species/ecology.asp?si=295&fr=1&sts=.

Creeping Buttercup   *Ranunculus repens*

NAME AND FAMILY
Creeping buttercup (*Ranunculus repens* L.); buttercup family (Ranunculaceae). Tall buttercup (*R. acris* L.) is another introduced buttercup that is considered invasive in some parts of North America and occurs in habitats similar to those occupied by creeping buttercup.

IDENTIFYING CHARACTERISTICS
Creeping buttercup has the typical five- to seven-petaled, glossy yellow, 1 in. (2.5 cm) flowers of buttercups, flowering from late spring into early summer. This perennial buttercup's stems creep along the ground, rooting at the nodes, but flowering stems rise up to 2 ft. (1.5 m) tall. Seeds are held in a rounded head. Leaves are divided into three-toothed or lobed leaflets attached to the stem by long leafstalks. The whole leaf is 0.5 to 3.5 in. (1.3 to 8.7

Flowers have five to seven glossy yellow petals.

Creeping buttercup leaves are divided into three leaflets and are attached to the stem by a long stalk.

cm) long and 0.75 to 3.75 in. (1.9 to 9 cm) wide. Leaves on the upper stems are smaller. In areas with mild winters, plants overwinter as rosettes with short, swollen stems. Stems connecting the rooted nodes die in winter, resulting in many individual plants the following year. Tall buttercup does not have creeping stems.

HABITAT AND RANGE
Distributed across much of Canada and the United States in damp soils along riverbanks and freshwater wetlands, and in fields, pastures, and forest edges. Common in wet meadows and pastures of the Pacific Northwest. Tolerates long periods of saturated soils and all kinds of temperate-zone weather except long, dry desert summers.

WHAT IT DOES IN THE ECOSYSTEM
Mats of creeping buttercup can prevent native plants from establishing on disturbed sites. Plants mostly spread by creeping stems, but also produce seeds that are eaten and dispersed by birds and rodents or which catch on fur by means of a small hook. A single plant can form a mat of over 4 square yards in one year. It changes soil chemistry by depleting potassium. Seeds also float and can be dispersed by water. Seeds can persist in the soil for decades. Creeping buttercup is considered a crop weed and is mildly toxic to livestock.

HOW IT CAME TO NORTH AMERICA

Native to Europe but now distributed globally, creeping buttercup may have been introduced deliberately as an ornamental plant or accidentally with agricultural goods as early as the 1700s. It is currently sold as a ground cover by some nurseries.

MANAGEMENT

Plants can be hand pulled if roots can be removed. Repeated mowing will also limit the plant's spread. The herbicides 2,4-D and MCPA are effective for controlling creeping buttercup.

FOR MORE INFORMATION

King County, Washington Noxious Weeds. 2011. "Creeping buttercup." http://www.kingcounty
.gov/environment/animalsAndPlants/noxious-weeds/weed-identification/creeping-buttercup
.aspx.
Lovett-Doust, L. 1981. "Population dynamics and local specialization in a clonal perennial,
Ranunculus repens L. I.: The dynamics of ramets in contrasting habitats." *Journal of Ecology*
69:743–755.

Tall Tumble Mustard   *Sisymbrium altissimum*

NAME AND FAMILY

Tall tumble mustard, Jim Hill mustard (*Sisymbrium altissimum* L.); mustard family (Brassicaceae). From the same genus, London rocket (*Sisymbrium irio*) and Indian hedge mustard (*S. orientale*) are considered invasive in California.

IDENTIFYING CHARACTERISTICS

This is the tallest of the mustards, able to grow nearly 5 ft. (1.5 m) tall. Tall tumble mustard grows as a winter annual or biennial. Leaves alternate along the flowering stem, from 0.5 to 5 in. (1.3 to 12.7 cm) long, getting smaller farther up the stem. Lower leaves have alternating pointed lobes along a midrib. Lobes become finer

*A cluster of four-petaled flowers tops
a tall tumble mustard plant.*

on higher leaves. Four-petaled white to yellow flowers are held on a stalk, flowering from late spring through summer. Flowers form 2 to 4 in. (5 to 10 cm) long narrow pods containing 120 or more seeds. Plants grow a thick taproot up to 2 ft. (0.6 m) long. Indian hedge mustard has leaves with more oval lobes, and London rocket is a shorter plant with many branches from the base of the plant.

HABITAT AND RANGE
Tall tumble mustard occurs throughout most of North America in fields, disturbed open areas, and open forests. It is most invasive in the West where it can grow in ponderosa pine forests and in dry shrublands.

WHAT IT DOES IN THE ECOSYSTEM
Tumble mustard can outcompete native vegetation and reduces crop yields. A single plant can produce 1.5 million seeds. It is called tumble mustard because the dry stems break off and are blown across the landscape, slowly releasing seeds over several months. The seeds have a mucilaginous coat that sticks to fur and feathers. Seeds survive in the seedbank for at least seven years. Animals such as deer and jackrabbits will eat some tumble mustard. The young leaves can be eaten in salads or cooked as greens. Native Americans ate the seeds.

HOW IT CAME TO NORTH AMERICA
Native to Europe and Asia, tumble mustard probably arrived as a contaminant in seeds or ballast. It was first seen in 1878 on grounds in Philadelphia where ships discarded ballast. It is known as Jim Hill mustard because it spread along railway tracks; Jim Hill was a U.S. railroad owner known as the "Empire Builder."

MANAGEMENT
Small patches can be controlled by hand pulling the rosettes in fall or early spring. It can also be controlled with broadleaf systemic herbicides but native plants should be planted following control to prevent reinfestation.

FOR MORE INFORMATION
Howard, Janet L. 2003. "*Sisymbrium altissimum.*" *Fire Effects Information System.* U.S. Department of Agriculture, Forest Service, Rocky Mountain Research Station, Fire Sciences Laboratory. http://www.fs.fed.us/database/feis.

Mitich, L. W. "Tumble mustard, tansy mustard." Weed Science Society of America. http://www.wssa.net/photo&info/weedstoday_info/mustards.htm.

Sow Thistle   *Sonchus arvensis*

NAME AND FAMILY
Perennial sow thistle, field sow thistle (*Sonchus arvensis* L.); aster family (Asteraceae). The species is divided into two subspecies, *S. arvensis* ssp. *arvensis* and *S. arvensis* ssp. *uliginosus* (Bieb.) Nyman. Listed as a noxious weed in many states and provinces. There are also two annual sow thistles not native to the United States that are considered agricultural weeds. Can be confused with prickly lettuce (*Lactuca serriola* L., p. 335), another introduced weedy species.

IDENTIFYING CHARACTERISTICS
Sow thistle is a perennial plant that looks like a huge dandelion with 1 or more stalks up to 6 ft. (1.8 m) tall. Stems may branch near the top. The often hollow stems and the leaves contain a milky sap. Leaves are larger and more crowded at the base of the stems and sparser toward the upper part of the stem, and are arranged alternately. Leaves are generally oblong to lance shaped, often lobed; they have prickly edges, clasp the stem, and do not have a leaf stalk. Plants produce both vertical roots that can reach down 5 to 10 ft. (1.5 to 3 m) and horizontal roots that spread to 3 to 6 ft. (1 to 1.8 m) in a single growing season. Plants flower in midsummer, forming clusters of yellow flower heads, each 1.5 to 2 in. (3.7 to 5 cm) across, at the ends of the stems. Under

Sowthistle flowers look like dandelions but occur in clusters atop tall stems.

the flowers are green bracts with sticky hairs. Seeds are wind dispersed and sometimes animal dispersed with the aid of pappi, hooked bristles attached to the seeds. Prickly lettuce looks very similar to sow thistle, but has prickles on the undersides of the leaves along the midveins.

HABITAT AND RANGE
Occurs throughout Canada and the United States except in most southeastern states. It is often associated with moist soils and can tolerate saline conditions. Found along riverbanks and lake edges, roadsides, disturbed ground, woods, and meadows. It grows in desert areas but mainly in depressions where moisture sometimes gathers.

WHAT IT DOES IN THE ECOSYSTEM
Perennial sow thistle displaces native species by its rapid spread via roots and seeds. Few researchers have reported on sow thistle in natural areas, but it does threaten desert tortoise habitat by displacing native plants the turtles rely on for food. Plants are occasionally eaten by rabbits, pronghorn antelope,

Left: *Leaf bases clasp the stem.* **Right:** *Sowthistle leaves have prickly edges.*

sheep, and cattle, and birds will feed on seeds. In agricultural fields plants reduce yields of some crops and harbor some crop pests.

HOW IT CAME TO NORTH AMERICA
Sow thistle probably arrived as a contaminant in crop seeds from Europe or western Asia. It was first noted in Pennsylvania in 1814. The milky latex is mostly made of oils that are being investigated for oil production.

MANAGEMENT
Control of sow thistle is difficult because plants have extensive root systems and are relatively resistant to herbicides. Tilling helps to control sow thistle in fields. 2,4-D, and other herbicides that act on the plant hormone auxin show some control of sow thistle, especially if applied while plants are growing vigorously. In Canada, three biocontrol insects were introduced, but show limited success at controlling sow thistle.

Flowers will form a round puff of seeds that are dispersed by wind.

FOR MORE INFORMATION
Brooks, M. L., and T. C. Esque. 2002. "Alien plants and fire in desert tortoise (*Gopherus agassizii*) habitat of the Mojave and Colorado deserts." *Chelonian Conservation Biology* 4(2):330–340.

Lemna, W. K., and C. G. Messersmith, 1990. "The biology of Canadian weeds. 94. *Sonchus arvensis* L." *Canadian Journal of Plant Science* 70:509–532.

McWilliams, J. 2004. "*Sonchus arvensis*." *Fire Effects Information System*. U.S. Department of Agriculture, Forest Service, Rocky Mountain Research Station, Fire Sciences Laboratory. http://www.fs.fed.us/database/feis.

"Table distinguishing perennial and annual sow thistles." Dow AgroSciences. http://www.dowagro.com/PublishedLiterature/dh_0046/0901b80380046853.pdf?filepath=ca/pdfs/noreg/010-20370.pdf&fromPage=GetDoc (accessed January 15, 2012).

Terrestrial Plants—Herbaceous Plants—Alternate Leaves—Leaves Divided

Crown Vetch   *Coronilla varia*

NAME AND FAMILY
Crown vetch (*Coronilla varia* L.); pea family (Fabaceae). Many native and nonnative vetches look similar in the vegetative state, but the flower clusters of crown vetch are distinctive, and most true vetches (*Vicia* spp.) have tendrils for climbing.

IDENTIFYING CHARACTERISTICS
A crown of white and pink to lavender cloverlike flowers gives this plant its name. Flowers are arranged in 1 in. (2.5 cm) wide clusters on long stalks that grow from the leaf axils, flowering through the summer. Divided leaves have 15 to 25 pairs of oval leaflets (0.75 in. [1.9 cm] long) with one leaflet on the end as well. Leaves, 2 to 6 in. (5 to 15 cm) long, grow alternately along the stem. Slender, 2 in. (5 cm) long seedpods mature in late summer, each containing 3 to 12 brown seeds. Stems trail along the ground, growing to 6 ft. (1.8 m) long. Plants tend to form mounds 6 to 12 in. (15 to 30 cm) high. Plants are perennials whose aboveground stems die back in winter.

Crown vetch flowers bloom in rounded clusters through the summer.

Green pods blend into the background of compound leaves.

HABITAT AND RANGE

Often spreads from erosion control or decorative plantings along roadsides into prairies, pastures, woodland edges, and along streambanks. Prefers full sun and will grow in most soil types. Occurs throughout most of southern Canada and the United States.

WHAT IT DOES IN THE ECOSYSTEM

The stems grow over native vegetation shading it out. Crown vetch also fixes nitrogen, thereby increasing nitrogen levels in the soil. In communities adapted to nutrient-poor soils, this increase in nitrogen may lead to a change in the composition of the community. Crown vetch threatens communities occupied by rare and endangered plant species in the southeastern United States. Spreads by underground roots and by seed. Seeds can remain viable in the soil for many

Thick mats of crown vetch prevent other plants from growing.

years. Crown vetch is toxic to horses because it contains nitroglycosides, but other livestock and wildlife eat it with impunity. It is used as a host plant by several native butterfly species and as cover by ground-nesting birds and some rodents. The method of seed dispersal of crown vetch is uncertain, but crown vetch seed has germinated in deer droppings and may be dispersed by other animals as well.

HOW IT CAME TO NORTH AMERICA
Crown vetch was widely planted as a ground cover and for erosion control beginning in the 1950s. It is also used as a green fertilizer crop on fields. It is native to Europe, Asia, and northern Africa.

MANAGEMENT
Small patches can be controlled with repeated hand pulling or by covering them with black plastic during the growing season. Mowing in late spring and again in late summer for several years will also control plants. In fire-adapted areas, controlled burns in late spring can reduce infestations. Clopyralid, metsulfuron, 2,4-D, triclopyr, or glyphosate applied to the leaves in late spring are also effective.

FOR MORE INFORMATION
Czarapata, E. J. 2005. *Invasive Plants of the Upper Midwest*. Madison, WI: University of Wisconsin Press.
Southeast Exotic Pest Plant Council. "Crown Vetch." *Southeast Exotic Pest Plant Council Invasive Plant Manual*. http://www.se-eppc.org/manual/COVA.html.

Bicolor Lespedeza   *Lespedeza bicolor*

NAME AND FAMILY
Bicolor lespedeza, shrubby lespedeza, shrubby bushclover (*Lespedeza bicolor* Turcz.); pea family (Fabaceae). Other invasive *Lespedeza* species include Chinese lespedeza (*L. cuneata*) and Thunberg's bushclover (*L. thunbergii*). There are native *Lespedeza* species as well.

IDENTIFYING CHARACTERISTICS
Growing as a somewhat woody perennial or shrub, bicolor lespedeza has arching branches that grow from 3 to 10 ft. (1 to 4 m) tall. In summer it is covered with upright clusters of 5 to 15 pealike, dark-pink flowers, each flower less than ½ in. (1.3 cm) long. Flower clusters grow from the upper leaf axils and branch tips. Leaves made up of three leaflets alternate along the stems. Each leaflet is oval with a hairlike tip. The lower surface is a lighter green than the upper surface. Fruit is a small, flat pod that holds a single seed. Stems often remain upright

Bicolor lespedeza can grow as a shrub with arching branches.

Three leaflets compose a single leaf.

through winter. Chinese lespedeza grows as an herbaceous plant with flowers in the leaf axils instead of in an elongated cluster. Thunberg's bush-clover looks very similar to bicolor lespedeza, but the flower clusters tend to droop rather than remaining upright.

HABITAT AND RANGE
Bicolor lespedeza occurs in eastern North America from Massachusetts and Michigan south to Florida and Texas. Often found in forests, old fields, and along roadsides. Bicolor lespedeza is killed back by frost but will resprout from the rootstock.

WHAT IT DOES IN THE ECOSYSTEM
Many game birds such as quail like to eat seeds of legumes, so many seed suppliers offer bicolor lespedeza for planting in wildlife food plots. Rabbits, deer, and gophers browse on the plant. It can spread to form dense stands, especially in disturbed areas, that reduce the growth and survival of tree seedlings and that increase nitrogen levels in the soil. Flowers are insect pollinated or self

Small seed pods mature at the ends of the branches.

pollinated and seeds are dispersed by birds and other animals. Seeds can remain viable in the soil for decades.

HOW IT CAME TO NORTH AMERICA

Introduced as an ornamental plant from Japan in 1856, bicolor lespedeza is also used for game food plots, for soil stabilization and enrichment, and for mine reclamation. The Civilian Conservation Corps and Soil Conservation Service grew millions of seedlings for restoration projects from the 1930s through the 1950s. Beekeepers like it because the plant provides late-season pollen for honeybees.

MANAGEMENT

Fires often increase the density of bicolor lespedeza. Small plants can be pulled out, but larger plants are difficult to pull because of the spreading root system. Plants can be sprayed in summer with glyphosate, triclopyr, or metsulfuron. Mowing one to three months before herbicide application can increase the effectiveness of the herbicides.

FOR MORE INFORMATION

Gucker, Corey L. 2010. *"Lespedeza bicolor." Fire Effects Information System.* U.S. Department of Agriculture, Forest Service, Rocky Mountain Research Station, Fire Sciences Laboratory. http://www.fs.fed.us/database/feis.

Miller, J. H., E. B. Chambliss, N. J. Loewenstein. 2010. *A field guide for the identification of invasive plants in southern forests.* General Technical Report SRS-119. Asheville, NC: U.S. Department of Agriculture, Forest Service, Southern Research Station. http://www.srs.fs .fed.us/pubs/gtr/gtr_srs119.pdf.

Chinese Lespedeza   *Lespedeza cuneata*

NAME AND FAMILY

Chinese lespedeza, hairy lespedeza, Chinese or Himalayan bush clover, sericea lespedeza (*Lespedeza cuneata* [Dum.-Cours.] G. Don); pea family (Fabaceae). Other invasive *Lespedeza* species include bicolor lespedeza (*L. bicolor*) and Thunberg's bush-clover (*L. thunbergii*). There are native *Lespedeza* species as well.

IDENTIFYING CHARACTERISTICS

This perennial legume forms a bushy clump from 1 to 5.5 ft. (0.3 to 1.6 m) tall with as many as 20 stems. Each stem is covered by leaves divided into three leaflets. Leaves are arranged alternately along the stem. Chinese lespedeza can

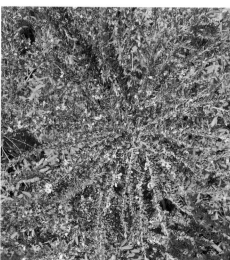

Left: *Purple and white lespedeza flowers are held close to the leaf axils.* **Right:** *Sprays of branches grow from the base of the plant.*

be distinguished from all other lespedezas by the wedge-shaped bases of the leaflets. Leaflets are only 0.5 to 1 in. (1.3 to 2.5 cm) long and are covered by dense hairs, giving them a gray-green hue. Flowers from midsummer to fall, Chinese lespedeza has two flower types. The showier flowers are small, creamy white flowers with purple throats nestled in clusters of two to four among the leaves along upper parts of the branches. These flowers are insect pollinated. The other flowers do not have petals and are self pollinated. They are mixed in with the insect-pollinated flowers. Bicolor lespedeza and Thunberg's bush-clover are woodier and larger than Chinese lespedeza and the flowers are arranged in elongated clusters.

HABITAT AND RANGE
Chinese lespedeza occurs throughout the United States and into Canada. It grows in meadows and prairies, along roadsides, and in pastures. It tolerates poor soil conditions and prefers sunny sites. Stands can be damaged by late-spring freezes.

WHAT IT DOES IN THE ECOSYSTEM
Plants develop into large stands through spreading root systems. A single plant can persist for more than 20 years. Because it spreads vigorously, it displaces native plants and hinders their colonization. It is competitive during droughts because a deep taproot allows it to persist. The tannins and other chemicals inhibit the growth of other plants and make older lespedeza plants unpalatable

Left: *Lespedeza forms dense stands, especially in meadows and open areas.* **Right:** *Small, oblong leaflets make up a lespedeza leaf.*

to grazers. It has been frequently planted to stabilize roadsides. In meadows it provides cover for ground-nesting birds, and bobwhite quail eat the seeds.

HOW IT CAME TO NORTH AMERICA

Native to Asia and Australia, Chinese lespedeza was brought first to Arlington, Virginia, in 1899. It spread in the southeastern United States, planted as forage for livestock and for erosion control. Wildlife managers often recommended it for meadow plantings to encourage quail.

MANAGEMENT

Because of the extensive root system and because seeds remain viable in the soil for many years, eradicating Chinese lespedeza takes persistence. Hand pulling is ineffective because of the extensive root system, but mowing for several successive years just before plants begin to flower can reduce the vigor of plants and keep them from spreading. Goats preferentially graze on Lespedeza over prairie grasses and can significantly reduce seed production. Herbicides, including metsulfuron methyl, triclopyr, clopyralid, and glyphosate, can be sprayed on plants in early summer to midsummer just before flowering, but often this is when surrounding grasses and wildflowers would also be susceptible to herbicides. Spot treatment can limit unintended side effects.

FOR MORE INFORMATION

Gucker, Corey. 2010. (Revised from Munger, Gergory T., 2004). *"Lespedeza cuneata."* Fire Effects Information System. U.S. Department of Agriculture, Forest Service, Rocky Mountain Research Station, Fire Sciences Laboratory. http://www.fs.fed.us/database/feis.

Miller, J. H., E. B. Chambliss, N. J. Loewenstein. 2010. *A field guide for the identification of invasive plants in southern forests.* General Technical Report SRS-119. Asheville, NC: U.S. Department of Agriculture, Forest Service, Southern Research Station. http://www.srs.fs .fed.us/pubs/gtr/gtr_srs119.pdf.

Remaley, T. 1998. "Chinese Lespedeza." Plant Conservation Alliance, Alien Plant Working Group. http://www.nps.gov/plants/alien/fact/lecu1.htm.

Birds-Foot Trefoil   *Lotus corniculatus*

NAME AND FAMILY
Birds-foot trefoil, birdsfoot deervetch (*Lotus corniculatus* L.); pea family (Fabaceae).

IDENTIFYING CHARACTERISTICS
Bright, yellow to orange flowers cover these low-growing perennial plants from spring until frost. Stems grow to nearly 2 ft. (0.6 m), usually sprawling on the ground. Compound leaves are made up of three leaflets, about 0.5 in. (1.3 cm) long and less than 0.13 in. (0.3 cm) wide, with two leaflike stipules at the base of the leafstalk. Leaves alternate along the stem. The flowers grow in rounded clusters of two to eight flowers. Each flower is sweet pea–shaped and up to 0.67 in. (1.7 cm) long. Petals sometimes have red streaks. Seeds are held in cylindrical brown to black pods, 0.25 to 1.75 in. (0.6 to 4 cm) long, forming from midsummer into fall. The name birds-foot trefoil comes from the seedpods, which fan out like a bird's toes. Plants have taproots that can reach 3 ft. (1 m) long, branched underground roots, and aboveground runners.

These low-growing plants produce copious seeds.

Bird's foot trefoil's yellow flowers grow in round clusters.

HABITAT AND RANGE
Found in moist, open areas such as meadows, pastures, roadsides, and riverbanks. Can withstand drought but not flooding. Established throughout Canada and the United States except in the southernmost states.

WHAT IT DOES IN THE ECOSYSTEM
Birds-foot trefoil forms dense mats that shade out other plants. It is mainly of concern where prescribed burns used to maintain prairies also enhance the germination of birds-foot trefoil seeds, which in turn compete with native grasses and forbs. As a legume, it fixes nitrogen, and it is a good forage plant for deer, elk, and geese.

HOW IT CAME TO NORTH AMERICA
A highly nutritious and delicious (to cattle) forage plant, bird's-foot trefoil was introduced by chance from Europe, but is now widely cultivated in pastures and for hay and used in erosion control along highways. Different cultivars have been selected to grow in various climates.

MANAGEMENT
Plants can be dug up, taking care to remove all root fragments. Extensive colonies can be kept mowed to 2 in. or less for several years, but this also sets back native plants. Herbicides containing clopyralid, which targets legumes and plants in the aster family, can be sprayed on plants. Some plants are resistant to glyphosate herbicides.

FOR MORE INFORMATION
Czarapata, E. J. 2005. *Invasive Plants of the Upper Midwest*. Madison, WI: University of Wisconsin Press.

Frame, J. *"Lotus corniculatus."* Food and Agriculture Organization. http://www.fao.org/ag/AGP/AGPC/doc/GBASE/DATA/PF000344.htm.

Terrestrial Plants—Herbaceous Plants— Opposite or Whorled Leaves— Leaves Entire

Ice Plant   *Carpobrotus* spp.

NAME AND FAMILY
Highway ice plant, Hottentot fig (*Carpobrotus edulis* (L.) N. E. Br.), and sea fig, ice plant (*C. chilensis* (Molina) N.E. Br.); fig-marigold family (Aizoaceae). These two species often hybridize and the hybrids are also considered invasive. A third related species, crystalline or common ice plant (*Mesembryanthemum crystallinum* L.) is also invasive in California.

IDENTIFYING CHARACTERISTICS
Ice plants are succulent plants with crowded three-sided (triangular in cross-section) opposite leaves, 1.5–3 in. (3.7–7.6 cm) long, that form dense mats. Leaves may be tinged with red. At every node, fibrous roots form that can take root, allowing the plant to creep along the ground. Stems generally grow less than 2 ft. (0.6 m) tall. Flowers year-round but peak flowering is in late spring.

Left: *The yellow flower of a highway ice plant.* Right: *Succulent leaves form a dense mat.*

Reddish carpets of ice plants grow over dunes in California.

Highway ice plant has yellow to pink flowers, 2.5 to 6 in. (6.2 to 15 cm) diameter with many narrow petals, whereas sea fig has purple flowers 1.5 to 2.5 in. (3.7 to 6.2 cm) in diameter. Hybrids tend to have pink flowers that are intermediate in size. Rounded fruits ripen from green to purple red and can contain up to 1,500 small black seeds. Crystalline ice plant has spoon-shaped, flat, fleshy leaves and smaller white- to pink-petaled flowers.

HABITAT AND RANGE
Common in coastal habitats from southern Oregon south to Rosarita, California, on dunes, coastal bluffs, coastal prairie, and maritime chaparral. Establishes along roads, trails, and on gopher mounds. Its range is limited by its intolerance to frost. Also introduced to Florida.

WHAT IT DOES IN THE ECOSYSTEM
Once established, ice plants form dense colonies, outcompeting other plants for space, water, and light. They directly compete with several rare and endangered plant species in California. They slow the growth of established shrubs and suppress growth of seedlings. They can also lower the pH of sandy loam soils and cause a buildup of organic matter that can lead to faster invasion by other species that would otherwise not grow on such soils. Successful establishment depends on disturbance and on grazing pressure, particularly from rabbits. The conical fruit is yellowish when ripe and is eaten by deer, rabbits, rodents, and humans. The seeds survive digestion and spread in droppings.

HOW IT CAME TO NORTH AMERICA
Highway ice plant was brought from coastal South Africa to stabilize railroad embankments in the early 1900s. By the early 1970s, thousands of acres of ice plant had been planted along railroads and roadsides. Highway ice plant and sea fig are also popular garden groundcovers and rock garden plants.

MANAGEMENT
Plants are easily pulled by hand, and mats can be rolled up like a carpet, but must be removed from the site or mulched on-site so they will not reroot. Glyphosate sprayed on plants can be effective, but follow-up treatments will probably be necessary. Replant or mulch to prevent reestablishment of ice plant seedlings.

FOR MORE INFORMATION
Albert, M. *"Carpobrotus edulis."* California Invasive Plant Council. http://www.cal-ipc.org/ip/management/ipcw/pages/detailreport.cfm@usernumber=25&surveynumber=182.php.
Conser, C. and E. F. Connor. 2009. *"Assessing the residual effects of Carpobrotus edulis invasion: implications for restoration."* *Biological Invasions* 11(2):349–358.
D'Antonio C.M. 1993. *"Mechanisms controlling invasion of coastal plant communities by the alien succulent Carpobrotus edulis."* *Ecology* 74(1): 83–95.

Common St. John's Wort   *Hypericum perforatum*

NAME AND FAMILY
Common St. John's wort, Klamath weed (*Hypericum perforatum* L.); mangosteen family (Clusiaceae). Listed as a noxious weed by several western states. There are native species of Hypericum, often called St. John's wort, that could be confused with common St. John's wort.

IDENTIFYING CHARACTERISTICS
Sunny yellow flowers bloom on this 3 to 4 ft. (1 to 1.2 m) perennial plant all summer. The 1 in. (2.5 cm) flowers have five distinct petals, each with tiny black dots along the margins, and a puff of yellow stamens in the center. They occur in clusters of 25 to 100 flowers at the ends of the branches. Small, 1 to 2 in. (2.5 to 5 cm) leaves grow opposite along the reddish branches. Branches can be woody near the base. The name "perforatum" comes from the translucent dots that can be seen if the leaf is held up to the light. The lower branches will often retain leaves through the winter. Three-part pods hold many seeds. Plants have a long taproot and rhizomes. The native species that look similar to common St. John's wort lack the black dotted petals and the stems have four ridges instead of the two ridges on stems of common St. John's wort.

HABITAT AND RANGE

Common St. John's wort grows in rangelands, fields, waste areas, and along forest edges. It likes sun and well-drained soils. Grows throughout the United States and Canada, although it is uncommon in the Canadian prairie regions. Native to Europe, northern Africa, and Asia.

WHAT IT DOES IN THE ECOSYSTEM

Plants spread by above and below ground creeping stems and by seed, forming dense colonies that exclude native species. Grazing animals tend to avoid eating St. John's wort, but if it is eaten, chemicals in the plant can cause photosensitivity in light-colored animals. By the 1950s, more than 2 million acres of rangelands in the western United States were unusable because of colonization by common St. John's wort. Biological control beetles introduced in 1944 ate their way through many of the plants. A single plant can produce 15,000 to 30,000 seeds, which are dispersed by wind, water, and movement of soil and on animal fur.

HOW IT CAME TO NORTH AMERICA

German immigrants to Philadelphia introduced common St. John's wort in 1696. They used the plant in mystic rites, attributing to it the ability to exorcise the devil and ward off demons of melancholy among other things. Its medicinal

Left: *Yellow flowers bloom on common St. John's wort most of the summer.* Right: *Plants form dense stands in sunny areas.*

Left: *Seedpods open when dry to release seeds.* **Right:** *Flowers have five petals and a puff of stamens in the center.*

uses date back to Ancient Greece. One story claims that St. John's wort was used by the Pied Piper to lure rats out of Hamelin. It spread as European settlers moved across the continent, probably reaching California around 1900. Common St. John's wort is still commonly used as a remedy for mild depression.

MANAGEMENT
Small plants can be pulled or dug up. Gloves should be used when handling the plants. Herbicides glyphosate, picloram, or 2,4-D can be sprayed on plants in summer. Repeated cultivation in fields can control common St. John's wort. At least five insect biological control agents have been released in the western United States and Canada to control the plant. These include two leaf-feeding beetles that were the first biological control insects ever deliberately released in North America.

FOR MORE INFORMATION
Mitich, L. 1994. "Common St. Johnswort." In *Intriguing world of weeds,* no 46. *Weed Technology* 8(3):658–661.

Schooley, J. "St. Johnswort and *Chrysolina* spp. beetles." Ontario Ministry of Agriculture, Food and Rural Affairs. http://www.omafra.gov.on.ca/english/crops/facts/info_sjwbeetles.htm.

Zouhar, Kris. 2004. "*Hypericum perforatum.*" *Fire Effects Information System.* U.S. Department of Agriculture, Forest Service, Rocky Mountain Research Station, Fire Sciences Laboratory. http://www.fs.fed.us/database/feis.

Purple Loosestrife   *Lythrum salicaria*

NAME AND FAMILY
Purple loosestrife (*Lythrum salicaria* L.); loosestrife family (Lythraceae). Six other alien loosestrifes grow in North America but are not nearly as invasive or evident. However, *L. virgatum* hybridizes with *L. salicaria* and is usually grouped with it in states and provinces with noxious weed laws against purple loosestrife. There are native species of loosestrife as well.

IDENTIFYING CHARACTERISTICS
Perennial loosestrife is abundant in damp and marshy areas, where a single rootstock sends up 30 to 50 erect stems that form a plant mass up to 8 ft. (2.4 m) high and 5 ft. (1.5 m) wide. Flowers with five to seven petals form in summer and are deep pink to purplish (occasionally pink or white) and arranged on long spikes. The plants continue to flower into fall. Leaves are narrow with smooth edges and arranged oppositely or in whorls along an angular stem. When leaves dry in autumn they turn bright red. No other wetland plant forms uniform stands with this color of flower.

HABITAT AND RANGE
Loosestrife has invaded most Canadian provinces and all the lower 48 states except Florida. It is most heavily concentrated in northeastern North America.

Plants form dense stands along waterways.

Left: *Purple loosestrife flowers bloom in summer and have five to seven petals.*
Right: *Flowers occur in whorls on long flowering stalks.*

WHAT IT DOES IN THE ECOSYSTEM

Loosestrife's prolific seeding, its tolerance of a variety of water regimes and soils, its ability to produce as many as 2 million seeds a season, and its reproduction from broken pieces have allowed it to spread across the continent and outcompete many natives. When trampled by animals or humans, damaged areas produce new shoots and root buds. Leaf size will change to maximize light availability. In some places it has replaced 50 percent of the native species. Once established, exclusive loosestrife stands maintain themselves for over 20 years. Loosestrife now occupies some half a million acres. Among the plants and animals whose populations have been seriously reduced by loosestrife invasions are flowering rush (*Butomus umbellatus*), the threatened bulrush (*Scirpus longii*), the rare bog turtle (*Clemmys muhlenbergii*), and the black tern (*Chlidonias niger*). Muskrats may aid the spread of loosestrife by eating the roots of its competitors such as cattails. Loosestrife benefits few foraging animals, but it has a long history as a medicinal herb and even a potherb. Loosestrife can be a major source of nectar for honeybees in some areas but has been shown to reduce pollination of nearby native species.

HOW IT CAME TO NORTH AMERICA

Genetic analyses indicate many sources contributed to the invasion of loosestrife. Studies of coastal and riverine sources of ballast for European sailing ships strongly suggest that purple loosestrife, native to Europe and Asia, arrived in colonial North America as soon as European colonization began. One of the

Flowers open from the bottom to the top of the spike.

first reports of it was along the Northeast coast in 1814. Bedding and feed for imported livestock and even sheep wool might also have carried loosestrife seed to North America. Horticulturalists also imported seed for gardens. By the early 1800s it was so common some botanists considered it native. It is still sold in the nursery trade except in a few states where its sale is banned. Although cultivars labeled as sterile are available from nurseries, scientists showed that these cultivars do produce fertile seeds when they are pollinated by a different cultivar. Because plants are insect pollinated, cross-pollination is likely.

MANAGEMENT
To date, no effective means exists to eliminate large, established stands of purple loosestrife. Killing stands requires preventing seed germination, denying nutrients to existing roots, and preventing dispersal by plant fragments. Early detection and elimination of loosestrife while protecting natives is the most effective control tactic. Small infestations can be pulled up. Cut off flowers to reduce seed production. Sheep grazing can reduce infestations. Herbicides with glyphosate will kill loosestrife and are best used in midsummer to late summer, but these are broad-spectrum chemicals that also affect other plants in the wetlands. Four insects were released as biological control agents. The two most obvious are species of *Galerucella* beetles that feed on leaves, stems, and buds. They can cause a 90 percent reduction in the biomass of the plants in a growing season and appear to be reducing the rate of spread of purple loosestrife.

FOR MORE INFORMATION

Blossey, B. 2002. "Purple Loosestrife." In *Biological Control of Invasive Plants in the Eastern United States,* edited by R. Van Driesche et al. U.S. Department of Agriculture Forest Service Publication FHTET-2002-04. http://wiki.bugwood.org/Archive:BCIPEUS/Lythrum _salicaria.

Bunch, Q. 1977. "Purple loosestrife: A honey plant." *American Bee Journal* 117:398.

Hovick, S. M., D. E. Bunker, C. J. Peterson and W. P. Carson. 2011. "Purple loosestrife suppresses plant species colonization far more than broad-leaved cattail: experimental evidence with plant community implications." *Journal of Ecology* 99:225–234.

Mexican Petunia   *Ruellia tweediana*

NAME AND FAMILY

Mexican petunia, Britton's wild petunia, Mexican bluebell (*Ruellia tweediana* Grisebach); Acanthus family (Acanthaceae). One of some 150 species of Ruellia named in honor of the fifteenth-century French herbalist Jean de la Ruelle, with this species honoring James Tweedie, a nineteenth-century Edinburgh gardener. May be confused with the native wild petunia (*R. caroliniensis*) or softseed wild petunia (*R. malacosperma*), which is native to Texas but a garden escapee in Florida. *Ruellia* spp. are not related to the annual garden petunias.

IDENTIFYING CHARACTERISTICS

Often planted for its tubular, five-petaled, purple-blue flowers. Cultivars have flower colors ranging from blue to pink to white. Flowers bloom from spring until frost, occurring singly or in small clusters near the ends of the branches. Seeds mature in small cylindrical capsules. The stems of this perennial reach 3

Mexican petunia is planted for its five-petaled purple flowers.

ft. (1 m), often arching over and rooting where the tips touch the ground. Stems are dark purple and leaves have purple veins. The leaves are lance shaped with smooth or wavy edges, ¾ in. (1.9 cm) wide and up to 12 in. (30.5 cm) long. Wild petunia and softseed wild petunia have shorter (5 to 6 in. [12.7 to 15.2 cm]) leaves.

HABITAT AND RANGE
Grows from South Carolina west to Texas. It dies back to the ground if temperatures fall below 20°F (–6.7°C) but can resprout. Prefers sun and moist soils, but is drought tolerant and can grow in most kinds of soil. Principally found on disturbed sites such as along drainage ditches, but also found along lakeshores and in wooded areas.

WHAT IT DOES IN THE ECOSYSTEM
Mexican petunia can form dense mats that shade out other plants. It spreads through root and stem fragments and by seed. Although it is promoted to gardeners as a butterfly-attracting plant, no evidence indicates that butterflies use the plant for nectar or as a host plant in natural areas in the United States. Certain butterflies do use native species of ruellia as host plants. Horticulturists are currently developing ornamental cultivars that will be sterile hybrids.

HOW IT CAME TO NORTH AMERICA
Introduced as an ornamental plant from Mexico. Found naturalizing in Florida as early as 1958.

MANAGEMENT
Plants can be hand pulled or dug out, but it is difficult to remove all roots, and seeds may persist in the soil. Larger plants can be sprayed with herbicides containing glyphosate, triclpyr or 2,4-D.

FOR MORE INFORMATION
Hammer, R. L. 2002. "Mexican bluebell (*Ruellia tweediana* Griseb.): A pretty invasive weed." Florida Exotic Pest Plant Council. *Wildland Weeds* 5(spring). http://www.se-eppc.org/pubs/ww/bluebellSpring2002.pdf.

Hupp, K. V. S., A. M. Fox, S. B. Wilson, E. L. Barnett, and R. K. Stocker. 2009. "Natural Area Weeds: Mexican petunia *(Ruellia tweediana)*." University of Florida IFAS Extension. http://edis.ifas.ufl.edu/ep415.

Wilson, S. B., C. P. Wilson, and J. A. Albano. 2004. "Growth and development of the native *Ruellia caroliniensis* and invasive *Ruellia tweediana*." *Horticultural Science* 39:1015–1019.

Terrestrial Plants—Herbaceous Plants—Opposite Leaves—Leaves Toothed

Beefsteak plant   *Perilla frutescens*

NAME AND FAMILY
Beefsteak plant, perilla, shiso (*Perilla frutescens* [L.] Britton); mint family (Lamiaceae).

IDENTIFYING CHARACTERISTICS
This annual herb has square, purple stems that reach 3 ft (1 m) in height. The toothed, broad leaves come to a tapering point. They are arranged oppositely along the stems and often have purple undersides. Crushed leaves have a musky, minty odor (or some describe it as the odor of raw beef). In midsummer, beefsteak plant sends up hairy, purple flowering stalks, 6 in. (15 cm) long or more, with closely spaced, small, pink to white and lavender flowers each opening above a broad green bract.

HABITAT AND RANGE
Grows from Ontario, Canada, south to Florida and west to Texas and Minnesota. Usually found in areas where there has been soil disturbance, in open areas, in woodlands, along streams, and in pastures. Able to tolerate wet or dry soils.

WHAT IT DOES IN THE ECOSYSTEM
Once considered a roadside weed, beefsteak plant has been spreading rapidly in natural areas. It may release chemicals toxic to other plants and it

Spikes of small flowers are held above ruffled leaves.

is toxic to cattle if cut and dried in hay in late summer. Seeds are spread short distances by wind and longer distances in movement of soil.

HOW IT CAME TO NORTH AMERICA
Native to Asia, beefsteak plant is used as a culinary herb and medicinal plant. Oil extracted from the seeds is used in cooking, dietary supplements, and as a finish for furniture similar to linseed oil.

MANAGEMENT
Can be hand pulled. Herbicides containing glyphosate can be applied in late summer.

FOR MORE INFORMATION
Imlay, M. 2011. "You don't want a stake in this! Maryland Invasive Species Council Invader of the Month." http://www.mdinvasivesp.org/archived_invaders/archived_invaders_2011_06.html.

Swearingen, J., B. Slattery, K. Reshetiloff, and S. Zwiker. 2010. *Plant invaders of Mid-Atlantic natural areas,* 4th ed. National Park Service and U.S. Fish and Wildlife Service. Washington, DC. http://www.nps.gov/plants/alien/pubs/midatlantic/pefr.htm.

Terrestrial Plants—Herbaceous Plants— Opposite Leaves—Leaves Divided

Poison Hemlock   *Conium maculatum*

NAME AND FAMILY
Poison hemlock (*Conium maculatum* L.); carrot family (Apiaceae). Similar in appearance to native cow parsnip (*Heracleum maximum* Bartr.) and water hemlock (*Cicuta maculata* L.) and to the nonnative giant hogweed (*H. mantegazzianum* Sommier & Levier, p. 324). In ancient Greece in 399 B.C., Socrates chose an extract of poison hemlock as his means of execution.

IDENTIFYING CHARACTERISTICS
Growing 3 to 10 ft. (1 to 3 m) tall, poison hemlock has a ribbed, hollow stem mottled with purple spots. In spring, stems are topped with many umbrella-shaped heads, 2 to 2.5 in. (5 to 6.2 cm) in diameter, made up of small five-petaled white flowers. Leaves are opposite and triangular in outline, 8 to 16 in. (20 to 40 cm) long, and have a ferny look because they are finely divided,

The umbrella-shaped flower heads are made up of many small, five-petaled flowers.

Left: *Purple spots mottle the stems.* **Right:** *Leaves are finely divided, giving the foliage a ferny appearance.*

with many small, toothed leaflets. The petiole, the leaf's stem, often forms a sheath around the stem. Leaves have a foul odor when crushed. The leaves can be used to distinguish poison hemlock from water hemlock and from giant hogweed. In poison hemlock, the leaf veins run to the tips of the teeth, whereas in water hemlock, the leaf veins run to the notches between the teeth. Cow parsnip and giant hogweed have much larger leaves that are not finely divided. Poison hemlock has a long, white, fleshy root. It usually grows as a biennial, with a rosette of leaves forming the first year, flowering during the second year.

HABITAT AND RANGE

Poison hemlock grows throughout southern Canada and the United States, usually along streams, pond edges, and wet meadows and along roadsides. Prefers partial shade and moist soils. Considered a noxious weed in several western states.

WHAT IT DOES IN THE ECOSYSTEM

Poison hemlock forms dense stands in moist soils and because it grows to such a large size it outcompetes other plants for space and light. When eaten, it is toxic to most animals and can cause death due to alkaloids in all parts of the plant. A single plant can produce more than 30,000 seeds and seeds remain viable in the soil for several years. Seeds are dispersed from late summer into winter by water and wind, and when caught in fur, clothing, shoes, and equipment.

HOW IT CAME TO NORTH AMERICA
Poison hemlock was introduced to North America in the 1800s as a garden plant because of its ferny foliage.

MANAGEMENT
Plants can be hand pulled before seed set. Wear gloves when handling the plant, as it can cause dermatitis in some people, and avoid breathing in particles of the plant. Mowing in spring kills second-year plants, and mowing again in late summer will kill seedlings and any regrowth. Herbicides such as 2,4-D and glyphosate can be applied in late spring.

Poison hemlock grows tall in moist soils.

FOR MORE INFORMATION
Drewitz, J. "*Conium maculatum*." California Invasive Plant Council. http://www.cal-ipc.org/ip/management/ipcw/pages/detailreport.cfm@usernumber=32&surveynumber=182.php.
Weber, E. 2003. *Invasive Plant Species of the World*. Cambridge: CABI Publishing.

Puncture Vine   *Tribulus terrestris*

NAME AND FAMILY
Puncture vine, caltrop, Mexican sandbur, horny goat weed (*Tribulus terrestris* L.); caltrop family (Zygophylaceae). A second weedy species, called puncture vine, caltrop, or burrweed (*T. cistoides* L.) grows as a perennial in the southeast.

IDENTIFYING CHARACTERISTICS
Called puncture vine for its formidable spined burrs, which are capable of puncturing bicycle tires. They are shaped something like a caltrop, an ancient weapon, with each burr consisting of five parts, each tipped with two stout spines. Each 0.38 to 0.75 in. (1 to 2 cm) diameter burr contains two to five seeds. Puncture vine is a low-spreading annual (sometimes perennial) that has trailing stems arising from a taproot. The 1 to 3 in. (2.5 to 7.6 cm) hairy leaves

Five-petaled yellow flowers bloom at the leaf axils.

The spiny burs can injure feet and puncture bicycle tires.

occur opposite along the stems and are divided into 8 to 18, 0.25 in. (0.6 cm) long oblong leaflets. Usually one of the pairs of leaves is noticeably larger than the other. Seeds germinate in spring and flowers can bloom within three weeks of seed germination. Five-petaled yellow flowers bloom from the leaf axils. *Tribulus cistoides* has showy flowers with petals up to 1 in. (2.5 cm) long held on 1.1 in. (3 cm) long stalks.

HABITAT AND RANGE

Puncture vine occurs throughout much of the United States (except New England) into British Columbia, Canada, but needs warm temperatures for seed germination. It is considered problematic mainly in the West on dry, sandy soils, but will grow in many soil types. It colonizes disturbed lands along roadsides, on scoured floodplains, and in pastures and fields. *Tribulus cistoides* grows from Florida to Georgia and west to Texas on sandy sites and in pinelands.

WHAT IT DOES IN THE ECOSYSTEM

The plants form a dense ground cover that can prevent other plants from establishing, but the main concern is the potential injury burrs can cause to livestock and lightly soled feet. The plants are also toxic to grazers, particularly sheep. Seeds can remain viable in the soil for up to five years and a single plant can produce up to 200 to 5,000 seeds in a single growing season. Seeds are dispersed when the pods stick to animals or tires. Flowers are insect pollinated.

HOW IT CAME TO NORTH AMERICA

Native to the Mediterranean, puncture vine was accidentally introduced to the United States, probably with livestock imports. Herbal medicines made from the plant claim to enhance testosterone levels and serve as a general tonic. The plant does contain steroidal saponins that scientists are testing for various effects.

Plants can form a dense ground cover.

Compound leaves grow opposite along the vining stems.

MANAGEMENT

Plants can be hand pulled or hoed before seeds begin to form. Picloram or chlorsulfuron can be applied in late winter to prevent seed germination. Glyphosate, imazapyr, or 2,4-D can be sprayed on growing plants. Two biological control insects, both weevils, that feed on the seeds and stems of both species; they have been introduced in the western United States and occasionally appear in the southeast.

FOR MORE INFORMATION

California Department of Food and Agriculture. "Puncturevine." http://www.cdfa.ca.gov/phpps/ipc/weedinfo/tribulus-terrestris.htm.

Langeland, K. A., H. M. Cherry, C. M. McCormick, and K. A. Craddock Burks et al. 2008. *Identification and Biology of Nonnative Plants in Florida's Natural Areas,* 2nd ed. SP 257. Gainesville, FL: University of Florida IFAS.

Washington State Noxious Weed Control Board. "Puncturevine." http://www.nwcb.wa.gov/detail.asp?weed=137.

Terrestrial Plants—Grasses and Sedges— Leaves Angular in Cross Section (Sedges)

European Lake Sedge   *Carex acutiformis*

NAME AND FAMILY
European lake sedge, lesser pond sedge (*Carex acutiformis* Ehrh.); sedge family (Cyperaceae). This sedge is difficult to distinguish from several other common wetland sedge species and identification should be confirmed by an expert. It most closely resembles tussock sedge (*C. stricta*), aquatic sedge (*C. aquatilis*), and lake bank sedge (*C. lacustris*).

IDENTIFYING CHARACTERISTICS
One to several stout stems arise from underground rhizomes and these stems grow 1.5 to 4 ft. (0.5 to 1.2 m) tall. The stems will feel triangular at the base. Leaves are about 0.13 in. (0.3 cm) wide and if cut crosswise, will be M shaped. Male and female flowers are held on separate spikes, with two to four female spikes and two to five male spikes. Each 1 to 3 in. (2.5 to 7.6 cm) long, cylindrical spike is densely packed with flowers. *Carex* fruits are enclosed by a special bract called the perigynium. The perigy-

Fruits grow in a tight cylindrical cluster.

nium in this species is shaped like a broad triangle and the surface is smooth with 12 to 18 raised ridges or nerves. The perigynium is 0.06 to 0.19 in. (0.15 to 0.45 cm) long and the beak at the end is less than one-fourth the length of the perigynium. The leaves of European lake sedge stay green much later into fall than those of the native sedges.

HABITAT AND RANGE

Although presently occurring at a few sites in Canada and the northeastern and midwestern United States, European lake sedge is increasingly recognized in new sites. It can grow in water more than 1.5 ft. (0.46 m) deep and on dry sites. Found in wet meadows and marshes.

WHAT IT DOES IN THE ECOSYSTEM

This sedge spreads rapidly by underground stems, creating dense stands. It competes against native vegetation for space and nutrients, often becoming the dominant plant, and the dying leaves smother other growing plants. Because it is difficult to distinguish from native sedges, it can spread undetected.

HOW IT CAME TO NORTH AMERICA

European lake sedge probably arrived accidentally in hay shipped from Europe. Native to Eurasia and northern Africa.

MANAGEMENT

For small infestations, pull or dig up plants. Herbicides containing glyphosate can be sprayed on plants in fall after most other plants are dormant.

FOR MORE INFORMATION

Catling, P. M., and B. Kostiuk. 2003. "*Carex acutiformis* dominance of a cryptic invasive sedge at Ottawa." *Botanical Electronic News* 315:1–6.

Ling Cao. 2012. "*Carex acutiformis.*" *USGS Nonindigenous Aquatic Species Database.* http://nas.er .usgs.gov/queries/factsheet.aspx?SpeciesID=2704.

Asiatic Sand Sedge   *Carex kobomugi*

NAME AND FAMILY

Asiatic sand sedge, Japanese sedge (*Carex kobomugi* Ohwi.); sedge family (Cyperaceae). Can be confused with other sedge species.

IDENTIFYING CHARACTERISTICS

Triangular stalks with distinct edges distinguish nearly all sedges from grasses. Asiatic sand sedge is a coarse-looking sedge, about a foot high (0.3 m), with light green leaves. There are very fine teeth along the edges of the leaves that you can feel with your fingers. Plants send out rhizomes (underground stems) from which new shoots grow. Sand burial seems to stimulate the growth of rhizomes. Plants flower and set seeds from April to June. Male and female flowers are on different plants. The flowers are greenish, many held in a stubby spike on a triangular stalk, below the tallest leaves. Fruits are small, triangular, and

Asiatic sand sedge traps sand, but may invade habitat of native plants.

nutlike, enclosed in a paperlike sack. Other sedge species growing in the same habitat flower in late summer to fall and do not have teeth on the margins of the leaves.

HABITAT AND RANGE
Occurs from Massachusetts south to North Carolina along the coast, on dunes and in the area between the high tide line and the foot of the dunes.

WHAT IT DOES IN THE ECOSYSTEM
Asiatic sand sedge invades habitat of the federally endangered sea-beach amaranth (*Amaranthus pumilus*). It also out-competes taller American beach grass, coastal spurge, and sea oats. Dunes invaded by sand sedge tend to be lower and wider than dunes colonized by beach grass, and they are more vulnerable to erosion. It spreads mainly from root fragments but also from seeds if male and female plants are present.

HOW IT CAME TO NORTH AMERICA
Native to northeastern Asia, it was first discovered in 1929 at Island Beach in New Jersey. Seeds were probably carried in the ballast of cargo ships. Because it resists disease and trampling, it was planted intentionally for erosion control, starting in the 1970s.

MANAGEMENT
Small infestations can be dug out, with care not to leave behind any fragments. New shoots can be very sharp, so wear gloves. Glyphosate sprayed on larger colonies during the summer and/or fall will also eradicate Asiatic sand sedge. Areas should be monitored for resprouts. Dunes should be replanted with native species to protect against erosion.

FOR MORE INFORMATION

Lea, C., and G. McLaughlin. "Asiatic Sand Sedge." Plant Conservation Alliance, Alien Plant Working Group. http://www.nps.gov/plants/alien/fact/cako1.htm.

Small, J. A. 1954. "*Carex kobomugi* at Island Beach, New Jersey." *Ecology* 35:289–291.

Wootton, L. 2009. "Why *Carex kobomugi* should be removed from New Jersey's coastal dunes." http://gcuonline.georgian.edu/wootton_l/rationale.htm.

Wootton, L., S. Halsey, K. Bevaart, A. McGough, J. Ondreika, and P. Patel. 2005. "When invasive species have benefits as well as costs: managing *Carex kobomugi* (Asiatic sand sedge) in New Jersey's coastal dunes." *Biological Invasions:* 7:1017–1027.

Deep Rooted Sedge   *Cyperus entrerianus*

NAME AND FAMILY

Deep rooted sedge (*Cyperus entrerianus* Boeckeler); sedge family (Cyperaceae). Often confused with a native *Cyperus* species, woodrush flatsedge (*C. luzulae* auct. non [L.] Rottb. ex Retz.). Until recently, many botanists thought it to be the same species or a subspecies of woodrush flatsedge.

IDENTIFYING CHARACTERISTICS

Deep rooted sedge is a very robust, perennial sedge with triangular stems 16 to 30 in. (40 to 75 cm) tall. Glossy leaves about 1 ft. (0.3 m) long have dark purple, almost black, leaf bases. All leaves emerge from the base of the plant. The flowers are held in a tight, rounded cluster about 1 in. (2.5 cm) diameter at the end of a stem, appearing in midsummer into fall. Each head is made up of 30 to 50 spikelets that look like a miniature oat. They start out pale green and mature to a light tan color. Each spikelet holds 16 to 20 flowers. Below the

The angle of the bracts under the inflorescence helps distinguish deep rooted sedge from native woodrush flat sedge.

inflorescence are six to eight leaflike bracts, 6 to 16 in. (15 to 30 cm) long that stick out at a 45- to 60-degree angle from the stem. Known as deep rooted sedge because of the thick, almost black, underground stems (rhizomes) that tend to be set deeply in the soil. Woodrush flatsedge grows from 8 to 16 in. (20 to 40 cm) tall, the floral bracts are held horizontally to the stem, and several other fine characteristics related to the flowers differ.

HABITAT AND RANGE
Found principally in the southeastern coastal plain from Florida and Georgia west to Texas. It has been found in Missouri and there is concern that it will spread as far north as Virginia. It prefers wet areas such as floodplains, lowland forests, roadside ditches and marshes and has a clear pattern of following highway rights-of-way.

WHAT IT DOES IN THE ECOSYSTEM
Because of its resemblance to native sedges, this species was overlooked until recently. It appears to be spreading rapidly, and in a given year a large plant can produce more than a million seeds that are dispersed by water, mowing, construction equipment, and movement of soil. Although it generally establishes in disturbed sites, it does displace native vegetation in undisturbed sites. It is naturalizing rapidly in habitat used by the endangered Attwater's prairie chicken in coastal Texas.

HOW IT CAME TO NORTH AMERICA
First found in Pensacola County, Florida, in 1941. Its introduction was probably accidental from Mexico or South America, possibly as a contaminant in rice seeds or brought by migratory birds.

MANAGEMENT
In fields, repeated tilling can control deep rooted sedge. Glyphosate applied to actively growing plants will kill them. Hexazinone is also an effective herbicide.

FOR MORE INFORMATION
Rosen, D. J., R. Carter, R., and C. T. Bryson. 2006. "The recent spread of *Cyperus entrerianus* (Cyperaceae) in the southeastern United States and its invasive potential in bottomland hardwood forests." *Southeastern Naturalist* 5:333–344. http://www.valdosta.edu/~rcarter/Rosen.Carter.Bryson.2006.pdf.

"*Cyperus entrerianus* Boeckeler." *Flora of North America* 23:155. www.efloras.org/florataxon.aspx?flora_id=1&taxon_id=242357653.

Rosen, D. J., C. T. Bryson, R. Carter, and C. Jacono. *Control/suppression of deeprooted sedge.* http://www.invasive.org/eastern/other/contol-deeprootedsedge.pdf.

Terrestrial Plants—Grasses and Sedges— Leaves Not Angular (Grasses)— Plants Form Distinct Clumps, Usually over 2 Ft. (0.6 M) Tall

European Beach Grass — *Ammophila arenaria*

NAME AND FAMILY
European beach grass (*Ammophila arenaria* [L.] Link); grass family (Poaceae). Other common beach grasses include American beachgrass (*A. breviligulata*), introduced on the West Coast but native to the Atlantic and Great Lakes, and American dunegrass (*Leymus mollis*), native to both coasts.

IDENTIFYING CHARACTERISTICS
Growing on sandy, coastal dunes, European beach grass is a perennial grass that spreads via stiff rhizomes (underground stems). Thick, waxy leaves grow from stiff, upright clumps of stems. Leaves can be 1 to 3.5 ft. (0.3 to 1.1 m) long and 0.1 to 0.25 in. (0.2 to 0.6 cm) wide, and end in sharp tips. The upper side of the leaf is light green and smooth and the underside is ridged and has a whitish coating. The edges of the leaves are often rolled down. The flowers form a dense, spikelike cluster held on a stiff stem. American dunegrass

European beach grass replaces native dune and beach grasses.

The grasses trap sand but cause a steep, wave-vulnerable foredune to form.

has darker-green, wider, and less stiff leaves compared with European beach grass. American beachgrass looks very similar but can be distinguished by its shorter ligule (a membranous appendage where the leaf meets the stem) (0.06 to 0.12 in. [0.15 to 0.3 cm] in American, 0.5 to 1.5 in. [1.3 to 3.7 cm] in European).

HABITAT AND RANGE
Most common on the West Coast from British Columbia to San Diego County in California, but also found on the east coast in Maryland and Pennsylvania. Most common on unstable dunes but also occurs on stabilized dunes.

WHAT IT DOES IN THE ECOSYSTEM
European beach grass forms dense stands that lower the native plant diversity on beaches. Insect diversity and the diversity of rare insects also decline. Several species of rare native

Seeds are held in a closely packed spike.

European beach grass forms large clumps.

dune plants are excluded by European beach grass. By reducing the West Coast's open sand area, European beach grass has also decreased available nesting space for the threatened western snowy plover. Its density and hold on the dunes also result in a steeper foredune, stabilizing dunes but reflecting more wave energy onto the beach. It can withstand being buried under more sand (up to 3.5 ft. [1.1 m]) than native beach grasses, although it has lower salinity tolerances.

HOW IT CAME TO NORTH AMERICA
Native to the coasts of Europe and North Africa, European beach grass was planted near San Francisco's Golden Gate Park in 1869 to stabilize dunes.

MANAGEMENT
Repeated digging and sifting sand with rakes to remove root fragments from spring to fall can eliminate European beach grass. Glyphosate mixed with a surfactant can be sprayed or wiped onto the leaves. Bulldozing has been used in large-scale projects.

FOR MORE INFORMATION
Apteker, R. *"Ammophila arenaria."* 2000. In *Invasive Plants in California's Wildlands,* edited by C. C. Bossard, J.M. Randall, and M. C. Hoshovsky. Berkley, CA: University of California Press. http://www.cal-ipc.org/ip/management/ipcw/online.php.

Buell, A. C., A. J. Pickart, and J. D. Stuart, 1995. "Introduction history and invasion patterns of *Ammophila arenaria* on the North Coast of California." *Conservation Biology* 9:1587–1593.

Pickart, A. J. 1997. "Control of European beachgrass (*Ammophila arenaria*) on the west coast of the United States." *California Exotic Pest Plant Council Symposium Proceedings.* http://www.cal-ipc.org/symposia/archive/pdf/1997_symposium_proceedings1934.pdf.

Wiedemann, A. M., and A. Pickart. 1996. "The *Ammophila* problem on the Northwest coast of North America." *Landscape and Urban Planning* 34:287–299.

Zarnetske, P. L., E. W. Seabloom, and S. D. Hacker. 2010. "Non-target effects of invasive species management: beachgrass, birds and bulldozers in coastal dunes." *Ecosphere* 1(5).

Giant Reed   *Arundo donax*

NAME AND FAMILY
Giant reed, elephant grass, bamboo reed, arundo grass, giant cane, river cane (*Arundo donax* L.); grass family (Poaceae).

IDENTIFYING CHARACTERISTICS
Giant reed is a perennial grass that grows over 20 ft. (6 m) high with bamboo- or cornstalk-like "culms" (stems) topped in late summer by a feathery plume of flowers up to 3 ft. (1 m) long. The creeping root is knotty and 0.75 to 1.5 in. (1.9 to 3.7 cm) thick. Culms are hollow, with separations at the nodes, and can reach a diameter of 1.5 in. (3.7 cm). The flat, smooth leaf blades reach up to 1.5 ft. (0.5 m) long. Flower plumes appear in late summer to early fall.

HABITAT AND RANGE
Giant reed has spread across warmer parts of North America from California to the the Gulf states and north across the Southeast to Virginia. It grows along riverbanks, streams, and ditches where it finds moist or wet soils.

WHAT IT DOES IN THE ECOSYSTEM
Giant reed is a common crop in some countries because under optimum conditions it can produce over 7 tons of biomass per acre (17 tons/ha), which can be used for biofuel, paper, fishing poles, mats and weaving, and the reeds of woodwind instruments. It does this by developing a massive root system that will annually consume some 500 gals./yd.^{2} (2,300 l/m^2) of water. Where it is not wanted, its aggressive growth has destroyed stands of native vegetation, increased fire dangers, and decreased native wildlife habitat. The firm hold of its root masses on riverbanks tends to fix channels of rivers that used to wander naturally and water a broad floodplain. Large stands of giant reed change a territory from flood-dependent to fire-dependent habitat. Giant reed roots survive fire much better than those of other riparian species. Because the

Feathery plumes top giant reed stalks.

Left: *Leaves clasp the stalk like culms.* **Right:** *Long leaves droop downwards from the tall stems.*

plant contains many chemicals toxic to insects and vertebrates (silica, cardiac glycosides, and nerve toxins, for instance), not many herbivores browse it. Water-short California considers the tens of thousands of acres of giant reed on its waterways the greatest threat to water-dependent natural systems.

HOW IT CAME TO NORTH AMERICA

A native of eastern Asia and cultivated in Asia, southern Europe, the Middle East, Latin America, and Africa, giant reed was brought from the Mediterranean to Los Angeles, California, in the 1800s as an ornamental plant. By 1820 it was so abundant along the Los Angeles River that it was used for thatching. People in the Southwest planted it along drainage ditches and canals for erosion control.

MANAGEMENT

Because giant reed spreads so easily from pieces distributed by floods and currents, real elimination has to be planned from the highest invasion point in a river basin to the lowest.

The most effective methods will depend on whether native plants are present and the quantity of giant reed, but no method works for long unless the root mass is killed. Because it is a wetlands-related plant, safe herbicide use is very limited. Broad-spectrum glyphosate formulations for use near water can be used. Late summer to early fall applications work best because plants are moving nutrients into their roots during this period. Also effective is cutting the reed, then waiting three to six weeks to spray regrowth, or applying concentrated herbicide immediately to cut stems. Cutting by hand or machine provides only temporary setbacks to an established stand. Scientists are studying the potential of several biocontrol insects.

FOR MORE INFORMATION

Bell, G. 1997. "Ecology and management of *Arundo donax*: And approaches to riparian habitat restoration in Southern California." In *Plant Invasions: Studies from North America and Europe,* edited by J. H. Brock, M. Wade, P. Pysek, and D. Green, 103–113. Leiden, the Netherlands: Blackhuys Publishers. http://ceres.ca.gov/tadn/ecology_impacts/arundo_ecology.html.

Dudley, T. "*Arundo donax.*" 2000. In *Invasive Plants in California's Wildlands,* edited by C. C. Bossard, J.M. Randall, and M. C. Hoshovsky. Berkley, CA: University of California Press. http://www.cal-ipc.org/ip/management/ipcw/online.php.

McWilliams, Jack. 2004. "*Arundo donax.*" *Fire Effects Information System.* U.S. Department of Agriculture, Forest Service, Rocky Mountain Research Station, Fire Sciences Laboratory. Available: http://www.fs.fed.us/database/feis.

Pampas Grass   *Cortaderia* spp.

NAME AND FAMILY
Andean pampas grass, purple pampas grass, jubata grass (*Cortaderia jubata* [Lem.] Stapf.); grass family (Poaceae). True pampas grass (*C. selloana* [J. A. & J. H. Schultes] Aschers. & Graebn.) is not as widespread yet, but its range is expanding. Landscapers have developed many varieties of true pampas grass.

IDENTIFYING CHARACTERISTICS
The very thick clumps, almost hummocks, with a fountain of tall upright leaves and plumes distinguish this grass in landscaping and natural areas. Young plumes of Andean pampas grass are pink to purplish and turn cream colored. Flowering starts in midsummer and the flower heads stand into winter. Andean pampas grass flowers more quickly than true pampas grass and its flowers rise twice as high as its leaf clump. The 0.5 to 0.75 in. (1.3 to 1.9 cm) wide green leaves can be up to 10 ft. (3 m) long. Any given clump can be 12 ft. (3.6 m) tall and up to 6 ft. (1.8 m) in diameter. Leaves are sharp and can cut the hands or any skin making casual contact. True pampas grass has male and female flowers on separate plants. Its flower plumes are a light violet to silvery white

Left: *Andean pampas grass plumes rise to twice the height of the leaves.* **Right:** *The narrow leaves of true pampas grass form a fountain.*

and female flowers rise to about the same level as the leaves while male flowers rise to twice the height of the leaves. Leaves are blue-green and the tips are bristly and curled.

HABITAT AND RANGE
Andean pampas grass prefers disturbed lands like logging sites, landslides, roadsides, and coastal bluffs and dunes, and grows in California and Oregon. It does not compete well with established grasslands. While common in sunny areas of Redwood National Park and other redwood areas, it will not grow in the dense shade of a redwood canopy. True pampas grass has naturalized in scattered locations across the southern United States from California to New Jersey

WHAT IT DOES IN THE ECOSYSTEM
Where Andean pampas grass has established itself in Oregon and California coastal habitats, it has noticeably diminished native plant numbers and diver-

True pampas grass plumes are silvery.

sity, sometimes transforming brushland into grassland habitat. Changes in populations of vertebrates and invertebrates also followed. Rabbit populations increase in pampas grass areas, but rodent populations decrease. Once established, pampas grass lives up to 15 years. While it does not take hold readily in meadows and shaded forests, its very light, small, and prolifically produced seeds find shelter in many places and germinate within two weeks, without a long dormant period. The massive root system makes it a formidable competitor for water. Because leaves stay green even in dry western summers, it sometimes provides cattle a substitute food for hay.

HOW IT CAME TO NORTH AMERICA

Andean pampas grass is a native of the Andes mountains that lives in semiarid areas, while true pampas grass is from the riverbanks of Argentina, Uruguay, and Brazil. Europeans exploring South America took pampas grasses to Europe for landscaping, and both Andean and true pampas grass took hold in North America as landscape plants in the southwest and California by the mid-1800s, escaping and becoming naturalized by the 1950s. The Soil Conservation Service also planted pampas grass for erosion control. By the 1970s, Andean pampas grass had become naturalized in northern California on hundreds of acres near the redwoods. While Andean pampas grass is more aggressive, true pampas grass has also escaped domestic plantings and established itself in arid and semiarid areas.

True pampas grass plumes are held just above the leaves.

MANAGEMENT

Pampas grass removal is very difficult once the tussocks are established. Young plants can be hand pulled if the soil is moist or soft. Larger plants might require picks and shovels for removal. Using a rope, chain, or cable around the base of the tussock and pulling with a tractor or truck removes larger plants. Glyphosate can be sprayed in early summer or fall. Fluazifop, which only kills grasses, can be sprayed in fall.

FOR MORE INFORMATION

Chimera, C., and F. Starr, K. Martz, and L. Loope. 1999. *Pampas grass* (Cortaderia jubata & C. selloana): *An alien plant report.* U.S. Geological Survey Biological Resources Division in cooperation with American Water Works Association Research Foundation Maui County Board of Water Supply. http://hear.org/species/reports/corspp_fskm_awwa_report.pdf.

DiTomaso, J. M., J. J. Drewitz and G. B. Kyser. 2008. "Jubata grass (*Cortaderia jubata*) control using chemical and mechanical methods." *Invasive Plant Science and Management* 1:82–90.

Lambrinos, J. G. 2001. "The impact of the invasive alien grass *Cortaderia jubata* (Lemoine) Stapf on an endangered mediterranean-type shrubland in California." *Diversity and Distributions* 6:217–231.

———. 2001. "The expansion history of a sexual and asexual species of *Cortaderia* in California, U.S.A." *Journal of Ecology* 89:88–98.

Weeping lovegrass   *Eragrostis curvula*

NAME AND FAMILY
Weeping lovegrass, African lovegrass (*Eragrostis curvula* (Schrad.) Nees); grass family, (Poaceae).

IDENTIFYING CHARACTERISTICS
This perennial bunchgrass is named for the weeping form of its slender, arching grass blades. The long blades are 0.1 inch (3 mm) wide, with the edges sometimes rolled inwards. The whole grass clump can reach 2 to 4 feet (0.8 to 1.2 m) tall. A short, flattened group of stems with hairy sheaths may be visible in winter at the base of the plant. Nodding lavender-grey clusters of flowers and seeds 8 to 10 inches (20 to 25 cm) long wave on stems up to 6 feet (1.9 m) tall from early summer into fall. Weeping lovegrass may remain green all winter in warmer climates.

Left: *An invasion of weeping lovegrass in winter.*
Right: *Flower clusters are held on tall stems above the slender leaves.*

HABITAT AND RANGE

Scattered populations grow in all but the north-central United States in a wide range of habitats. Cold temperatures limit its northward expansion, but it may grow as an annual in cold climates. It grows in desert shrublands and grasslands and in open pine and mixed hardwood forests. It particularly likes open, sunny areas with sandy soils. Weeping lovegrass is tolerant of fire but not of flooding or salt.

WHAT IT DOES IN THE ECOSYSTEM

Weeping lovegrass often lowers plant diversity, but in the southeast many rare insect species were found in areas long dominated by weeping lovegrass. It changes bird and mammal habitat, favoring some species over others. It reproduces through self-fertilized seeds that have high germination rates.

HOW IT CAME TO NORTH AMERICA

Native to Africa. The first ecotype introduced to the U.S. came from South Africa in 1927. A second ecotype was introduced from Tanzania in 1935 to Stillwater, Oklahoma. Other ecotypes from South Africa were introduced later. Between 1940 and 1980, it was widely used for erosion control due to its dense, fibrous roots and was planted along highways in the southwest and southeastern U.S. It is also sold as an ornamental plant.

MANAGEMENT

Any control measures should take into account that loss of weeping lovegrass could result in increased erosion until native plants reestablish. Plants can be dug out. Herbicides containing glyphosate can be applied to the foliage in spring before the plants flower. Seeds germinate best if exposed to light, so mulching and minimizing soil disturbance will reduce establishment.

FOR MORE INFORMATION

Gucker, Corey L. 2009. *"Eragrostis curvula." Fire Effects Information System.* U.S. Department of Agriculture, Forest Service, Rocky Mountain Research Station, Fire Sciences Laboratory. http://www.fs.fed.us/database/feis.

Miller, J. H., E. B. Chambliss, N. J. Loewenstein. 2010. *A field guide for the identification of invasive plants in southern forests.* General Technical Report SRS-119. Asheville, NC: U.S. Department of Agriculture, Forest Service, Southern Research Station. http://www.srs.fs.fed.us/pubs/gtr/gtr_srs119.pdf.

Cogon Grass   *Imperata cylindrica*

NAME AND FAMILY
Cogon grass, satintail, spear grass (*Imperata cylindrica* [L.] P. Beauv.); grass family (Poaceae). Considered one of the world's worst weeds and listed as a noxious weed in the United States. Can look similar to Johnson grass (*Sorghum halepense*, p. 411) before flowering and looks very similar to Brazilian satintail (*I. brasiliensis* Trin.), an agricultural weed.

IDENTIFYING CHARACTERISTICS
Growing in loose bunches, cogon grass leaves originate near the ground and grow to be 1 to 4 ft. (0.3 to 1.2 m) long. The stiff leaf blades with an off-center white midrib are 0.5 to 0.75 in. (1.3 to 1.9 cm) wide, ending in a sharp point. They are generally light green, turning red brown in cold weather. The edges of the leaves are finely serrated and razor sharp. Cogon grass flowers in late fall or winter and in spring, and fluffy white seed heads (panicles) form that are about 2 to 11 in. (5 to 28 cm) long and 1.5 in. (3.7 cm) wide. Seeds are attached to a plume of long silky hairs that are carried off on the wind or by animals. Plants mainly spread by pointy, white, branching rhizomes, and can produce up to 3 tons of rhizomes per acre. Johnson grass has a more prominent stem and shorter leaves. Brazilian satintail flowers have only one stamen, whereas cogon grass flowers have two stamens.

Cogon grass flowers are held in a dense spike flower.

Leaf blades have an off-center white midrib.

Top: *In spring, fluffy seed heads open up above the leaves.* Left: *Cogon grass can form extensive stands with dense cover.*

HABITAT AND RANGE
Not tolerant of cold weather, cogon grass grows from South Carolina south to Florida and west to Texas. It grows along roadsides, in forests, in pine savannahs, along streams, on sand dunes, and in pastures and fields. It can tolerate drought, salinity, and shade.

WHAT IT DOES IN THE ECOSYSTEM
Fast-spreading, strong rhizomes allow quick occupation of new territory, where the grass often forms extensive stands. The leaves form a dense mat on the ground that prevents other plants from establishing and that can cause more frequent and intense fires. Species endemic to longleaf pine habitats decline when cogongrass invades, and loblolly pine seedlings show reduced survival and productivity. The roots of cogon grass may release chemicals toxic to other plants in addition to taking up nitrogen. The dense cover reduces nesting by ground nesting birds and other animals.

HOW IT CAME TO NORTH AMERICA
In 1912, cogon grass arrived as packing material in a crate of Satsuma oranges destined for Alabama. It is native to southeastern Asia. In the 1920s and 1930s it was introduced to other states as forage in pastures and for soil stabilization, but it is only useful for forage when very young, before the serrated leaf edges develop.

A variety of cogon grass known as Japanese blood grass (*Imperata cylindrica* var. *rubra*) is commonly sold as an ornamental and although it rarely sets seed in gardens there is concern about its escape or hybridization with other strains.

MANAGEMENT

A combination of mowing or burning, followed by plowing or disking and herbicide application, can control stands of cogon grass. Glyphosate can be applied in spring from when grasses begin to green until the appearance of flowers to eliminate seed production. Glyphosate can also be sprayed in fall to kill plants. Multiple applications will be needed to control well-established stands. Imazapyr is also used to control cogongrass but it can harm surrounding plants and remains active in soil for long periods of time.

FOR MORE INFORMATION

Cogongrass.org. http://www.cogongrass.org.

Johnson, E. R. R. L. and D. G. Shilling. 2009. "Cogon grass." Plant Conservation Alliance Alien Plant Working Group. http://www.nps.gov/plants/alien/fact/imcy1.htm.

Lippincott, C. L. 2000. "Effects of *Imperata cylindrica* [L.] Beauv. (Cogongrass) invasion on fire regime in Florida sandhill (U.S.A.)." *Natural Areas Journal* 20:140–149.

MacDonald, G. E., D. G. Shilling, B. J. Brecke, J. F. Gaffney, K. A. Langeland, and J. T. Ducar. 2006. "Weeds in the sunshine: Cogongrass (*Imperata cylindrica* [L.] Beauv.) biology, ecology, and management in Florida." SS-AGR-52. Agronomy Department, Florida Cooperative Extension Service, Institute of Food and Agricultural Sciences, University of Florida. http://edis.ifas.ufl.edu/WG202.

Chinese Silver Grass   *Miscanthus sinensis*

NAME AND FAMILY

Chinese or Japanese silver grass, miscanthus, eulalia, maiden grass (*Miscanthus sinensis* Anderss.); grass family (Poaceae).

IDENTIFYING CHARACTERISTICS

Tall reedlike stems of this clumping grass grow to 4 to 12 ft. (1.2 to 3.6 m) high and hold feathery flower heads. One inch (2.5 cm) wide green leaves whose blades can be 18 in. (45 cm) long have a distinct white midrib. Leaves rise from the base to 6 ft. (1.8

Seeds are held in a fan of spikes.

m) and droop over toward the ends. In fall, leaves turn tan to yellow. In late summer, fan-shaped flower heads up to 1 ft. (0.3 m) long bloom red to maroon, then pinkish, before maturing to silvery and drying to tan. Seed heads endure well into winter. Dozens of cultivars have been developed for leaf color, size, and biomass, as well as some hybrids, including some that are a hybrid with sugar cane.

HABITAT AND RANGE
Prefers sunny locations with rich, moist, but not soggy, soils. Cold tolerant but not as well adapted to hot southern summers. It grows in most of the United States, except the driest states, and in some areas of southern Canada.

WHAT IT DOES IN THE ECOSYSTEM
Chinese silver grass will take over roadsides and burned pastures or brushland, spreading by rhizomes and seeds. The fast-growing grass will displace native grasses with its thick bunching growth. Where it grows densely, its thick, dry clumps increase fire hazards. While cattle and other domestic grazers will eat this grass, some reports say that wild grazers do not browse it. In Asia, its sturdy reeds have been used for roof thatching and in both Europe and Asia, certain varieties are grown for energy generation and paper pulp.

Flowering stalks are held above the fountain of leaves.

HOW IT CAME TO NORTH AMERICA
The first varieties were apparently introduced at the end of the nineteenth century for ornamental plantings. New varieties continued to be developed and many are actively sold by landscaping retailers today. Many new varieties do not set seed. Chinese silver grass is native to China, Korea, and Japan.

MANAGEMENT
Very small pieces of rhizome can propagate new clumps, so digging or hand pulling must be very thorough.

Fluffy seeds disperse by wind.

Disking or harrowing is usually counterproductive because it breaks up the rhizomes. Mowing or grazing many times each growing season can reduce or eliminate stands. Mowing is best done before formation of flower heads and repeated to prevent flower and seed formation. Burning or mowing in fall stimulates new growth. The best time for applying herbicides with glyphosate is in fall when plants are sending nutrients down into the root system. Spraying may have to be repeated until no new shoots appear.

FOR MORE INFORMATION

Meyer, M. H. "*Miscanthus:* Ornamental and invasive grass." University of Minnesota. http://miscanthus.cfans.umn.edu.

Miller, J. H., E. B. Chambliss, and N. J. Loewenstein. 2010. *A field guide for the identification of invasive plants in southern forests.* GTR SRS-119. USDA Forest Service Southern Research Station. http://www.srs.fs.fed.us/pubs/gtr/gtr_srs119.pdf.

Scally, L. T. Hodkinson, and M. B. Jones. 2001. "Origins and taxonomy of *Miscanthus.*" In Miscanthus *for energy and fibre,* edited by M. B. Jones and M. Walsh, 1–9. London: James and James.

Watson, L., and M. J. Dallwitz. 1992 onward. "Grass genera of the world." Version 28th November 2005. http://delta-intkey.com/grass/.

African Fountain Grass and Buffel Grass   *Pennisetum* spp.

NAME AND FAMILY
African fountain grass, tender fountain grass, fountain grass, purple fountain grass (*Pennisetum setaceum* [Forssk.] Chiov.) and buffel grass, African foxtail grass (*Cenchrus ciliaris* L., formerly *P. ciliare* [L.] Link.); grass family (Poaceae). Other nonnative *Pennisetum* species include yellow foxtail (*P. glaucum*), elephant or Napier grass (*P. purpureum*), and kikuyu grass (*P. clandestinum*).

IDENTIFYING CHARACTERISTICS
African fountain grass features 6 to 15 in. (15 to 38 cm) long, nodding, pink to purple flower heads in late summer to early fall. Seeds have long bristles extending outward from the flower head and the whole flower head looks like a bristly tail. The plants grow in clumps or tussocks and rise 1.5 to 5 ft. (0.46 to 1.5 m) tall. Green leaves are very narrow and up to 3 ft. (1 m) long. Several different cultivars have produced leaves that range from green to purple, but most cultivars do not produce viable seed. Buffel grass is a close relative, and has a gray to purple to yellowish cylindrical seed head, 1 to 5 in. (2.5 to 13 cm) long, held upright. Its bluish-green leaves are covered with soft hairs on the upper side and are 2 to 12 in. (5 to 31 cm) long and less than 0.5 in. (1.3 cm) wide. Buffel grass forms tussocks to 3 ft. (1 m) tall or mats.

African fountain grass has narrow leaves and ornamental seed heads.

HABITAT AND RANGE

Fountain grass and buffel grass share the same habitats, although fountain grass has been favored for landscaping and buffel grass for grazing. Most common in the Southwest, they are also found across the southern states to California and as far north as Tennessee and Nebraska. These grasses prefer arid and semiarid open spaces, including deserts and lava beds. They are often found along roadsides and in fields.

Seeds of African fountain grass are held in a bottlebrush-like arrangement.

WHAT IT DOES IN THE ECOSYSTEM

Where the *Pennisetum* species have established themselves in large numbers they have usually made the area much more vulnerable to fire. While the fires may devastate other plants and ground nesting animals, the varieties of *Pennisetum* are well adapted to fire, and after fires, they thrive with the decreased competition. African fountain grass can reproduce from both fertilized and unfertilized seed (it is "apomictic") and seeds remain viable for up to seven years. Buffel grass is a good forage grass in dry pastures, and just south of the U.S. border it has taken over more than a million acres of native rangeland. The California Exotic Pest Plant Council calls buffel grass one of the "most invasive widespread wildland pest plants." In the Sonoran desert the fires it fuels threaten the survival of the saguaro cactus and other fire-sensitive desert plants.

HOW IT CAME TO NORTH AMERICA

African fountain grass and buffel grass are natives of Africa and the Middle East. Fountain grass came to the southwestern United States in the 1940s as an ornamental while buffel grass was brought in during the 1940s by the U.S. Department of Agriculture to improve cattle production by increasing forage on dry rangelands.

MANAGEMENT

The *Pennisetum* species are very tenacious grasses with long-lived seeds that require years of monitoring for removal to be permanent. Where populations are widespread, control should begin with outlying scattered plants and progress to the main source of seeds. Cutting, hand pulling, and digging can be effective on small areas, especially if repeated. Burning is not recommended since it can stimulate new growth. Systemic herbicides are the most effective chemical controls. Those containing hexazinone can be used preemergent or postemergent but should not be used near trees or water. Glyphosate is effective at controlling buffel grass if used when at least half of the plant is green.

FOR MORE INFORMATION

James, D. 1995. "The threat of exotic grasses to the biodiversity of semi-arid ecosystems." *The Arid Lands Newsletter* 37(3).

Southern Arizona Buffelgrass Coordination Center. http://www.buffelgrass.org.

Tunison, J. T. 1992. "Fountain grass control in Hawaii Volcanoes National Park: Management considerations and strategies." In *Alien plant invasions in native ecosystems of Hawaii: Management and research,* edited by C. P. Stone, C. W. Smith, and J. T. Tunison. University of Hawaii Cooperative National Park Resources Studies Unit. Honolulu: University of Hawaii Press.

Williams, D. G., and B. Zdravko. 2000. "African grass invasion in the Americas: Ecosystem consequences and the role of ecophysiology." *Biological Invasions* 2(2):123–140.

Terrestrial Plants—Grasses and Sedges— Leaves Not Angular (Grasses)—Plants Not in Distinct Clumps or Growing along Creeping Stems, over 4 ft. (1.2 m) Tall

Reed sweetgrass   *Glyceria maxima*

NAME AND FAMILY
Reed sweetgrass, reed mannagrass, reed meadow grass (*Glyceria maxima* [C. Hartm.] Holmb.); grass family, (Poaceae). The native American manna grass (*G. grandis*) is shorter.

IDENTIFYING CHARACTERISTICS
This perennial wetland grass grows to be 8 feet (2.5 m) tall and spreads by rhizomes to form dense stands. Flat leaf blades reach 12 to 24 inches (30 to 60 cm) long and 0.2 to 0.75 inches (0.6 to 2 cm) wide with a distinct midrib. The edges of the leaves have short, stiff hairs that feel rough. The sheath where the leaf meets the stem feels rough and has a reddish-brown band. The flowers, which bloom in summer, are held in a branched cluster, 6 to 12 inches (15 to 30 cm) long that may be either open or held tightly. The flower branches also have stiff hairs like the leaf margins. American manna grass is usually less than 5 feet (1.5 m) tall and has smooth sheaths.

HABITAT AND RANGE
Usually found in wet, nutrient-rich wetlands and seasonally flooded areas. Found in wetlands in Alberta and Ontario, Canada, as well as in Wisconsin and Massachusetts. Small infestations have also been reported in Washington and Illinois.

WHAT IT DOES IN THE ECOSYSTEM
Forms dense stands that exclude native wetland species. Reduces seed availability for birds and provides poor nesting habi-

Reed sweetgrass growing at the edge of a wetland.

The seed head forms an open panicle.

tat. Despite being introduced as forage for livestock, it is considered poor forage for most animals because the young shoots contain cyanide. Studies in Australia show that it converts fast-flowing streams to poorly aerated marsh. Although seeds spread by water or in soil carried by equipment, most spread probably occurs when rhizomes are moved to new places, since seed germination rates are thought to be low in North America.

HOW IT CAME TO NORTH AMERICA
Native to Europe and colder regions of Asia, reed sweetgrass was introduced in the 1940s as a forage plant that could be planted in wet pasture areas for cattle. It was found growing in Wisconsin in the 1970s and in Massachusetts in the 1990s. A variegated variety is sold as an ornamental plant.

MANAGEMENT
Small infestations can be hand pulled or smothered using black plastic. Cutting plants and flooding the stubble may drown plants. Herbicides containing glyphosate are used to treat reed sweetgrass in summer.

FOR MORE INFORMATION
Howard, V. M. 2007. "*Glyceria maxima*." USGS Nonindigenous Aquatic Species Database
    http://nas.er.usgs.gov/queries/factsheet.aspx?SpeciesID=1120.
Martin, T. 2000. "Weed Alert! *Glyceria maxima* (C. Hartm.) Holmb." The Nature Conservancy.
    http://www.invasive.org/gist/alert/alrtglyc.html.

Reed Canarygrass   *Phalaris arundinacea*

NAME AND FAMILY
Reed canarygrass, ribbon grass, gardener's garters (*Phalaris arundinacea* L.); grass family (Poaceae). Harding grass (*P. aquatica*) and orchard grass (*Dactylis glomerata*) look similar to reed canarygrass and are also considered invasive in parts of North America. The native bluejoint (*Calamagrostis canadensis*) can also be confused with reed canarygrass.

IDENTIFYING CHARACTERISTICS
This perennial grass begins to grow very early in spring. It produces 1 to 6 ft. (0.3 to 1.8 m) tall, smooth, sturdy, sometimes hollow, stems from underground roots. The leaves are flat, 0.25 to 0.75 in. (0.6 to 1.9 cm) wide, and generally without hairs. The early-summer flowers are held above the leaves in dense, narrow plumes 3 to 8 in. (7 to 20 cm) long. The amount of branching, the leaf color, and the size, shape, and density of flowers can vary considerably. To distinguish the native bluejoint reed grass, look at the ligule (where the leaf blade meets the stem). Reed canarygrass has a distinctive transparent ligule. Harding grass stems are swollen at the base and the plumes are shorter and more compact. Orchard grass has narrower leaf blades, 0.1 to 0.13 in. (0.3 to 0.6 mm) wide.

HABITAT AND RANGE
Reed canarygrass occurs around the globe in boreal regions. There are probably native strains across Canada and the northern United States, but scientists think the invasive strains originated in Eurasia or resulted from crosses between Eurasian and native plants. It occurs across Canada and in all but the Gulf coast

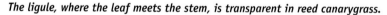

The ligule, where the leaf meets the stem, is transparent in reed canarygrass.

states (Alabama excepted) in the United States. It prefers wet soils but does not like to sit in standing water unless the stand is well established. It is found most often in shallow marshes and meadows and along drainage ditches.

WHAT IT DOES IN THE ECOSYSTEM
Reed canarygrass forms dense colonies that exclude other plants and alter animal habitat. It can change the hydrology of an area because it traps silt, or sometimes water undercuts the dense root mats. Plants spread by seed and through the extension of underground stems. Nitrogen enrichment of wetlands, including pollutants, increases reed canarygrass's ability to suppress native species growth. Used for making mats and hats by some Pacific Northwest tribes.

HOW IT CAME TO NORTH AMERICA
Agronomists began developing strains of reed canarygrass for forage and hay in New England in the 1830s and continued to develop strains adapted to other regions and to drier conditions. Invasive strains may be from Eurasian plants or from crosses between Eurasian and native strains. Nurseries sell a variety of

Left: *Mature seed heads form an open panicle.* **Right:** *Reed canarygrass forms dense stands along waterways.*

Seed heads gradually open as they mature.

reed canarygrass called ribbon grass (*P. arundinacea* var. *picta* L.) because of its variegated leaves.

MANAGEMENT

Because of the extensive underground root system, reed canarygrass is difficult to control and will often require a combination of approaches. Repeated cutting during the growing season can eliminate small patches. In areas with fire-adapted native species, prescribed burns can be successful. In areas where water level can be controlled, prolonged inundation can kill reed canarygrass. Glyphosate or fluazifop may control reed canary grass when applied when grass is actively growing, but treatments will often need to be repeated.

FOR MORE INFORMATION

Hutchison, Max. 1992. "Vegetation management guideline: reed canary grass (*Phalaris arundinacea* L.)." *Natural Areas Journal.* 12(3): 159.

Waggy, Melissa, A. 2010. "*Phalaris arundinacea*." *Fire Effects Information System.* U.S. Department of Agriculture, Forest Service, Rocky Mountain Research Station, Fire Sciences Laboratory. http://www.fs.fed.us/database/feis.

Merigliano, M. F., and P. Lesica. 1998. "The native status of reed canarygrass (*Phalaris arundinacea* L.) in the inland Northwest, U.S.A." *Natural Areas Journal* 18:223–230.

Spyreas, G., B. W. Wilm, A. E. Plochner, D. M. Ketzner, J. W. Matthews, J. L. Ellis, and E. J. Heske. 2010. "Biological consequences of invasion by reed canarygrass *(Phalaris arundinacea)*." *Biological Invasions* 12:1253–1267.

Phragmites   *Phragmites australis*

NAME AND FAMILY
Phragmites or common reed (*Phragmites australis* [Cav.] Trin. ex Steud.); grass family (Poaceae). *P. australis* includes more than 20 genetic strains or lineages, 11 of which are considered native to the United States. Some researchers consider the native phragmites to be rare and worth special efforts to preserve.

IDENTIFYING CHARACTERISTICS
Phragmites is a tall (up to 20 ft. [6 m]) wetland grass with leaves that stick out from the stems in the fashion of a corn plant. Leaves are 50 to 100 in. (20 to 40 cm) long and 0.4 to 1.75 in. (1 to 4 cm) wide. In summer it produces fluffy plumes of flowers held above the stems. Distinguishing native and exotic

phragmites is one of the more difficult field challenges, but by checking field observations against genetic analysis, scientists note several differences that can be seen in the field. Natives are generally more scattered while exotic phragmites forms dense, dome-shaped masses. Natives usually have more gray-green leaves while exotics have more yellow-green leaves. Peel the leaf back from the stem, and the native will tend to have smooth, shiny stems and a reddish color in spring and summer, whereas the exotic will have finely ribbed and dull stems with a tan color in spring and summer.

HABITAT AND RANGE
Exotic strains of phragmites have invaded all of the lower 48 states and the southern tier of Canada. Native strains occur in scattered populations in the South, Mid-Atlantic, intermountain West, and Midwest. Phragmites grows in wet areas—marshes, floodplains, drainage ditches, lake edges, and wet meadows and prairies. It tolerates brackish water.

Phragmites has very dense, fluffy seed heads.

Phragmites forms dense stands in marshes, displacing native vegetation.

WHAT IT DOES IN THE ECOSYSTEM

Phragmites invasions begin in wet areas, and stands become so aggressive that they can consume shallow ponds in a few years, yet the quick-growing stands can also stabilize eroding banks and shores. They clog waterways and their tall stems and dense growth shade out native aquatic or marsh plants. By invading wetlands that have been newly made to offset previous losses (part of the government policy of "no net loss of wetlands"), phragmites may actually decrease the extent of wetlands. Stalks often break near the midpoint but don't fall to the ground. This results in both denial of light to lower-growing plants and animals and a fire hazard. Stands shelter several species of small animals and are favored by red-winged blackbirds, though they eliminate favored habitat of other birds.

HOW IT CAME TO NORTH AMERICA

Archeological records show *P. australis* has been present on some sites for almost 3,000 years, but its rapid spread began after European colonization, and became even faster in the era of the automobile and motorboat. It spreads mainly when water carries root fragments from one wetland to another. Phragmites has been used throughout history for many purposes, from mats to cigarette paper. Young shoots are often grazed by wild and domestic animals. Phragmites rhizomes (underground stems) can be roasted and the seeds can be eaten as a porridge. Reeds are used for thatching roofs and dense stands remove pollutants from the ground.

Left: *Gray-green leaves stick out from the stems.* **Right:** *Phragmites seed heads remain visible in winter.*

MANAGEMENT

Once an invasion takes root, extermination becomes very difficult. Small stands can be controlled through repeated cutting or by cutting and dripping glyphosate formulated for use near water into the cut stems in late summer. Controls for large stands include burning, flooding, grazing, disking, and aerial spraying of herbicides. Burning is ineffective if water level is above the ground. Special permits will often be required to treat phragmites growing along waterways.

FOR MORE INFORMATION

Cornell University Ecology and Management of Invasive Plants Program. http://www.invasive plants.net/phragmites.

Gucker, Corey L. 2008. *"Phragmites australis."* Fire Effects Information System. U.S. Department of Agriculture, Forest Service, Rocky Mountain Research Station, Fire Sciences Laboratory. http://www.fs.fed.us/database/feis.

Saltonstall, K. 2002. "Cryptic invasion by non-native genotypes of the common reed, *Phragmites australis,* into North America." *Proceedings of the National Academy of Sciences* 99:2445–2449.

Golden Bamboo   *Phyllostachys aurea*

NAME AND FAMILY
Golden bamboo, fish pole bamboo (*Phyllostachys aurea* Carr. ex A. & C. Riv-
ière); grass family (Poaceae). Looks similar to the native switchcane (*Arundi-
naria gigantea*).

IDENTIFYING CHARACTERISTICS
Golden bamboo grows in dense evergreen thickets and has hollow stems, or
culms, up to 30 ft. (9 m) high and 1 to 6 in. (2.5 to 15 cm) in diameter.
Golden bamboo is distinguished by its inflated and often contorted internodes
near the base of the stem. The green-gold stems and twigs bear leaves 0.25 to
0.75 in. (0.6 to 1.9 cm) wide and 3 to 10 in. (7.6 to 25.4 cm) long, generally
growing alternately or in fanlike clusters pointing upward. The solid nodes are
often darker than the rest of the stem. Lower shoots and branches have papery
sheaths that quickly form a carpet on the ground. The plant spreads by shallow
underground stems (rhizomes) that send up new stems from alternate nodes.
This bamboo may flower only once every 7 to 12 years. The southeastern
native switchcane is distinguished from golden bamboo by having one flat side
in its otherwise rounded stem; it also has persistent papery sheaths on the
stems and only grows to 8 ft. (2.4 m).

Left: *Golden bamboo spreads quickly via tough underground rhizomes.*
Right: *The stalks are tough and woody, with darkened solid nodes.*

Golden bamboo is named for its yellow-green stems.

HABITAT AND RANGE
Found from New York south to Florida, west to Texas and also in Oregon and California. Golden bamboo prefers open sunlight and warm climates, though it tolerates winter temperatures to 0° F (–18°C). It also grows in partially wooded areas. The optimum soils are light and moist or southeastern clays that hold moisture.

WHAT IT DOES IN THE ECOSYSTEM
In open or disturbed areas, golden bamboo spreads very rapidly by rhizomes. It has been used as privacy screening because it grows so densely. The papery sheaths that fall to the ground become a moisture-retaining mulch over its runners and, with the shade of the dense, high stems and spreading leaves, they suppress the growth of native plants. Because of its length and strength, it has been a favorite for cane fishing poles. Shoots are edible but seldom eaten by Americans.

HOW IT CAME TO NORTH AMERICA
Several hundred bamboo species have been brought to North America for ornamental plantings. Golden bamboo, the most invasive, was introduced in Alabama in 1882 from Asia.

MANAGEMENT
Small stands can be controlled and contained by mowing repeatedly to kill shoots sent up by rhizomes. Herbicides can be applied on regrowth of bamboo shoots several weeks after cutting. Glyphosate and imazapyr are most effective, but imazapyr should not be used if there are hardwood trees, shrubs, or other

Plants flower in late spring to early summer. Each flower is held in a structure called a spike. The spikes are held on a zigzag stem in a bunch similar to the grains of wheat. Three stiff, 1 in. (2.5 cm) long, barbed bristles called awns stick out above the spike. The spikes disperse as a group but eventually break apart into individual spikes. Young spikes can be reddish in color.

HABITAT AND RANGE
Barbed goat grass grows principally in California, but is also found in southern Oregon and some Mid-Atlantic states. It grows in grasslands, oak woodlands, rangelands, and field edges. It is of particular concern where it grows in serpentine grasslands in California.

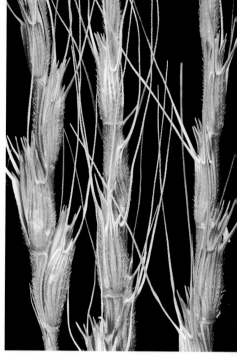

Seed-holding spikelets zigzag up the stem.

WHAT IT DOES IN THE ECOSYSTEM
Because barbed goat grass is a close relative of wheat, it crosses with wheat to produce sterile seeds. On rangeland, it reduces forage for cattle, sheep, and wildlife. It uses more moisture than other plants and can change soil microbial communities, making it more difficult for native plants to establish. Because it has a relatively high silica content, it stands in dry condition longer than other grasses and can change fire frequencies. In serpentine grasslands, it changes nutrient cycling rates in the soil. On rangeland, since grazers avoid barbed goat grass, it tends to survive and thrive compared with the preferred forage. The long barbed awns surrounding the seeds catch on fur, wool, clothes, and shoes, transporting seeds to new areas. A fungus aids germination of barbed goat grass by weakening the hard seed head. Gophers reduce establishment by burying patches of seedlings. Plants that survive being buried produce seeds uninfected with the fungus, further reducing seed germination, so pastures with high gopher activity have lower densities of goat grass.

HOW IT CAME TO NORTH AMERICA
Native to southern Europe and western Asia, it was probably introduced to California around 1915. It may also have arrived accidentally as a contaminant in seed. Goat grasses are sometimes hybridized with wheat to increase cold tolerance and disease resistance, but they harbor pests of winter wheat.

Left: *The upper edge of the ligule is finely fringed.* Right: *Long barbs aid in seed dispersal.*

MANAGEMENT

Goat grass is difficult to control once established, so prevention of new infestations is key. Remove grazing animals from infested rangelands to prevent spread. Heavy grazing and mowing can encourage the grass to spread and produce more seeds. For small areas, hand pulling and hoeing plants works, letting roots air dry after removal. Burning in early summer, before goat grass seeds mature but after native annual seeds have matured, reduces goat grass densities. Glyphosate applied before plants flower will kill them, but the grass-specific herbicides fluazifop and clethodim are also effective when applied to plants as they are beginning to flower.

FOR MORE INFORMATION

Aigner, P. A. and R. J. Woerly. 20101. "Herbicides and mowing to control barb goatgrass (*Aegilops triuncialis*) and restore native plants in serpentine grasslands." *Invasive Plant Science and Management* 4:448–457.

California Department of Food and Agriculture. http://www.cdfa.ca.gov/phpps/ipc/weedinfo/aegilops.htm.

DiTomaso, J. M. 2004. *Aegilops triuncialis* plant assessment form. California Invasive Plant Council. http://www.cal-ipc.org/ip/inventory/PAF/Aegilops%20triuncialis.pdf.

Eviner, V. T., and F. S. Chapin. 2003. "Gopher-plant-fungal interactions affect the establishment of an invasive grass." *Ecology* 84:120–128. http://www.ecostudies.org/reprints/Eviner_Gopher-plant-fungal.pdf.

Oregon Department of Agriculture. "Barbed Goat Grass" http://www.oregon.gov/ODA/PLANT/WEEDS/profile_barbedgoatgrass.shtml.

Creeping Bentgrass   *Agrostis stolonifera*

NAME AND FAMILY
Creeping bentgrass, redtop (*Agrostis stolonifera* L.); grass family (Poaceae). Often called *A. palustris* in the turfgrass industry. Many varieties and cultivars of this species exist. Creeping bentgrass hybridizes with at least 13 species of native and nonnative *Agrostis* and *Polypogon* spp., including nonnative redtop (*A. gigantea* Roth) and colonial bentgrass (*A. capillaris* L.). Although native populations of creeping bentgrass may exist in the northernmost United States and Canada along salt marshes and lake edges, in most areas creeping bentgrass populations are European in origin.

IDENTIFYING CHARACTERISTICS
A perennial grass with creeping belowground and aboveground stems (rhizomes and stolons). Multiple slender stems in tufts grow 1 to 3 ft. (0.3 to 1 m) tall. Leaves are smooth, less than 0.5 in. (1.3 cm) wide and 0.5 to 6 in. (1.3 to 15 cm) long, and come to a fine point. The flower and seed heads occur in a short plume 1 to 8 in. (2.5 to 20 cm) long and 0.5 to 5 in. (1.3 to 13 cm) wide. When seeds form, the seed heads tighten up. Flowers in summer. Leaves remain green throughout the summer.

Redtop spreads mainly by underground rhizomes instead of aboveground stolons and the seed heads remain open, whereas creeping bentgrass's seed

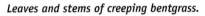

Leaves and stems of creeping bentgrass.

Flowers and seeds are arranged in a short plume that gradually opens as it matures.

heads close up. Colonial bentgrass is smaller than creeping bentgrass and its seed heads also remain open rather than closing up. It spreads by both rhizomes and stolons.

HABITAT AND RANGE
Found along stream banks and in wet meadows throughout most of Canada and the United States except in the Southeast. It is tolerant of a wide range of soil types, and although it prefers moist soils, it is drought tolerant. Frequently establishes after disturbances such as fire or land clearing.

WHAT IT DOES IN THE ECOSYSTEM
Creeping bentgrass can form dense patches that outcompete native plants and prevent establishment of shrubs. In New Zealand it has been shown to change the types of insects occurring in a community. Seeds are dispersed by wind and water, and by grazers consuming seeds as they feed on the plants. In North America, it is not generally considered a weed of major importance and most reports of invasiveness are from the western United States. Because of its popularity as a turfgrass, Monsanto and Scott Seed developed a strain that is glyphosate resistant. There is evidence, however, that glyphosate resistance can spread to uncultivated populations and to other species that hybridize with creeping bentgrass. The resistant variety is not being marketed yet (as of October 2012).

HOW IT CAME TO NORTH AMERICA
Native to Europe and Northern Africa, creeping bentgrass was planted as a pasture grass and turfgrass and probably arrived accidentally in hay or seeds before the 1750s. It is one of the most commonly used grasses on golf course putting greens in the northern United States and Canada.

MANAGEMENT
Small areas can be hand pulled, but plants will resprout from roots left in the ground. On larger infestations, herbicides such as glyphosate are effective.

FOR MORE INFORMATION

APHIS. 2006. *White Paper: Perspective on Creeping Bentgrass* Agrostis stolonifera *L.* http://www
.aphis.usda.gov/peer_review/downloads/cbg_reviewers_charge.pdf.

Esser, Lora L. 1994. *"Agrostis stolonifera." Fire Effects Information System.* U.S. Department
of Agriculture, Forest Service, Rocky Mountain Research Station, Fire Sciences Laboratory.
www.fs.fed.us/database/feis.

Levine, J. M. 2001. "Local interactions, dispersal, and native and exotic plant diversity along
a California Stream." *Oikos* 95:397–408.

Zapiola, M. L., C. K. Campbell, M. D. Butler, and C. A. Mallory-Smith. 2008. "Escape and establishment of transgenic glyphosate-resistant creeping bentgrass *(Agrostis stolonifera)* in
Oregon, U.S.A.: A 4-year study." *Journal of Applied Ecology* 45:486–494.

Small Carpgrass   *Arthraxon hispidus*

NAME AND FAMILY

Small carpgrass, hairy jointgrass, jointhead arthraxon (*Arthraxon hispidus*
[Thunb.] Makino); grass family (Poaceae). Can be confused with deer-tongue
grass (*Dicanthelium clandestinum*), wavyleaf basketgrass (*Oplismenus hirtellus* ssp.
undulatifolius, p. 434), and sometimes with Japanese stilt grass (*Microstegium
vimineum*).

IDENTIFYING CHARACTERISTICS

This annual grass grows to 1.5 ft. (0.46 m) tall, but its wiry stems can creep
along the ground. The heart-shaped bases of the leaf blades encircle the stems.

Left: *Hairs protrude from the edges of the leaf and the leaf blade clasps the stem.*
Right: *Short, wide leaves and a low growing habit characterize small carpgrass.*

Leaf blades are oval to lance shaped, 1 to 3 in. (2.5 to 7.6 cm) long and 0.2 to 0.6 in. (0.5 to 0.6 cm) wide. Hairs stick out from the margins of the leaf blades. Flowers are held on one to several 1 to 3 in. (2.5 to 7.6 cm) long spikes blooming in early fall. Deer-tongue grass has much more open, branched clusters of flowers (as opposed to the spikes of carpgrass), is covered in stiff sparse hairs, and grows larger and in drier sites. Wavyleaf basketgrass has scattered hairs on the leaves and ripples across the entire leaf blade. Japanese stilt grass leaves are not heart shaped where they join the stem.

HABITAT AND RANGE
Although most common in the southeastern United States, small carpgrass grows from Massachusetts south to Florida, west to Illinois, Texas, and into Mexico, and is reported in Oregon. Preferring sunny, moist habitats, small carpgrass inhabits floodplains, pastures, shorelines, and stream banks.

WHAT IT DOES IN THE ECOSYSTEM
Small carpgrass can form dense stands, particularly along streams, which can inhibit the growth of native plants. Stems creep along the ground and can root at the nodes.

HOW IT CAME TO NORTH AMERICA
First noticed in the United States in the 1930s in Virginia, small carpgrass may have been an accidental introduction or may have been brought by immigrants. It is sometimes used in Asia as a medicinal plant for relieving cough and asthma. Native to Japan and eastern Asia.

MANAGEMENT
Small areas can be hand pulled. Because the grass generally occurs near wetlands, systemic herbicides formulated for use near water could be used on it.

FOR MORE INFORMATION
"*Arthraxon hispidus.*" *Invasive Plant Atlas of New England.* http://nbii-nin.ciesin.columbia.edu/
  ipane/icat/browse.do?specieId=40.

False Brome   *Brachypodium sylvaticum*

NAME AND FAMILY
False brome, slender false brome, (*Brachypodium sylvaticum* [Huds.] P. Beauv). grass family (Poaceae). Often confused with *Bromus* species.

IDENTIFYING CHARACTERISTICS
This perennial grass can grow to 2.5 feet (0.7 m) high but its thin, flat, 0.25 to 0.3 in. (4 to 10 mm) wide, bright green leaf blades arch gracefully. Hold a leaf

Left: *Hairy stems and leaf blades.* Right: *A plant showing the bright green, arching leaf blades.*

up to the sky and the fine hairs along the edge are clearly visible. The flower heads that appear in July and August have 5 to 10 spikelets drooping on very short stalks (pedicels) off the flowering stem. The straight awns projecting from the ends of the spikelets are 0.24 to 0.7 in. (6 to 18 mm) long. The leaf joins a hollow, hairy stem (culm) and the leaf sheath is open. Other *Bromus* species have closed leaf sheaths and spikelets on long pedicels.

HABITAT AND RANGE
Now a Class A noxious weed in Washington state and spreading extensively in Oregon, particularly in the Willamette Valley south from Portland. It establishes stands in both shady woods and open prairies and has spread into higher areas and to some coastal counties. Also present in British Columbia, northern California, New York, and Virginia.

WHAT IT DOES IN THE ECOSYSTEM
Individual plants can multiply rapidly to form mats that dominate a location, outcompeting native grass species and other plants. Also said to retard or prevent oak seed germination. It also shelters rodents that attack young trees and it eliminates plants that support rare butterflies. The plant itself is fairly fire resistant and regrows rapidly after burns, but its dead thatch may increase fire

danger. (In Willamette Valley oak savannahs, however, fire is necessary to suppress evergreen competition for the white oaks.)

HOW IT CAME TO NORTH AMERICA

This native of North Africa and Eurasia was noticed only recently, in the late 1980s in Oregon's Willamette Valley, but it may have been among the species planted experimentally in the 1930s to improve open rangeland leased to cattle farmers. Reported from Eugene, Oregon, in 1939. Probably spread by seeds carried on clothing, vehicles, and footwear. It is occasionally sold as an ornamental plant.

MANAGEMENT

Individual plants or small clusters can be dug out, preferably before dropping seed. Larger stands yield to glyphosate or herbicides designed for killing grasses only. The best time to spray is from midsummer until fall rains. Frequent mowing before seeds set also reduces populations.

FOR MORE INFORMATION

Daniel, S. and D. Werier. 2010. "Slender false brome *(Brachypodium sylvaticum* ssp. *sylvaticum)* a new invasive plant in New York." *New York Flora Association* 21(1), Winter 2010. http://www.nyflora.org/download_file/view/40/74.

Institute for Applied Ecology, False Brome Working Group. http://appliedeco.org/invasive-species-resources/FBWG.

King County Washington Noxious Weeds. "False brome." http://www.kingcounty.gov/environment/animalsAndPlants/noxious-weeds/weed-identification/false-brome.aspx.

Smooth Brome   *Bromus inermis*

NAME AND FAMILY

Smooth brome, awnless brome (*Bromus inermis* Leyss. ssp. inermis); grass family (Poaceae). A native subspecies, Pumpelly's brome (*B. inermis* ssp. *pumpellianus*) occurs in the United States, but the widespread smooth brome subspecies was introduced from Europe. The species is quite variable, with more than 30 known varieties.

IDENTIFYING CHARACTERISTICS

This perennial grass has smooth stems 1 to 3 ft. (0.3 to 1 m) tall that begin growth in very early spring. Wide-spreading rhizomes are dark colored and jointed, with the joints covered by scaly, dark-brown sheaths. Flat leaf blades are 0.25 to 0.75 in. (0.6 to 1.9 cm) wide and 6 to 15 in. (15.2 to 38 cm) long. The leaf wraps around the stem with a V-shaped notch and the ligule, less than

Top: *Seed heads of smooth brome droop slightly.* **Right:** *Flower panicles have one to four stiff branches at each node that stick straight out or slightly up.*

0.06 in. (0.15cm) long, is brownish at the base. A nodding open cluster of flowers blooms in early summer. Each cluster has one to four stiff branches per node that stick straight out or slightly up, distinguishing smooth brome from similar native bromes. Each branch has several purplish cylindrical spikelets holding 7 to 10 flowers. The bract enclosing the flower (the lemma) does not have a bristle (awn), further distinguishing smooth brome from native Pumpelly's brome and from the invasive cheatgrass (*B. tectorum*, p. 426).

HABITAT AND RANGE
Occurs throughout Canada and in all states in the United States except for Florida and Alabama. It prefers sun along riverbanks, roadsides, field edges, prairies, and pastures.

WHAT IT DOES IN THE ECOSYSTEM
Smooth brome persists in rangelands and fields and its dense mat of rhizomes prevents establishment of native species. The plants compete with native grassland plants for water and light, reducing native plant diversity. Rhizomes start to form as early as three weeks after seeds germinate. It hybridizes with Pumpelly's brome in the Rocky Mountains. Serves as an alternate host for several crop viruses but is a good drought-tolerant pasture grass. It is also used for erosion control because of the interlocking roots.

HOW IT CAME TO NORTH AMERICA
Some smooth brome may have been introduced as early as 1875, but is also known to have arrived at the California Experiment Station in 1884 as a forage

grass. There are northern and southern strains; the northern strains were developed from Siberian plants and the southern strains from Hungarian plants.

MANAGEMENT

Because smooth brome occurs in so many different habitat types, management options vary. Burning in late spring before flowering stems emerge achieves good control. Repeated mowing can reduce smooth brome populations. Glyphosate can be sprayed on plants before flowering or in fall.

FOR MORE INFORMATION

Fink, K. A. and S. D. Wilson. 2011. "*Bromus inermis* invasion of a native grassland: Diversity and resouce reduction." *Botany* 89:157–164.

Royer, F., and R. Dickinson. 2004. *Weeds of the Northern U.S. and Canada*. Renton, WA, and Alberta: Lone Pine Publishing and University of Alberta Press.

Salesman, J. B. and M. Thomsen. "Smooth brome *(Bromus inermis)* in tallgrass prairies: A review of control methods and future research directions." *Ecological Resotration:* 29:374–381.

Cheatgrass   *Bromus tectorum*

NAME AND FAMILY

Cheatgrass, downy brome, early chess, bronco grass, thatch bromegrass, and military grass (*Bromus tectorum* L.); grass family (Poaceae). Several other species of Bromus, including red brome or foxtail brome (*B. madritensis* ssp. *rubens*) and Japanese brome (*B. japonicus*), are also considered invasive in North America and can co-occur with cheatgrass, but cheatgrass is the most widespread.

Cheatgrass dries out in summer, creating a fire hazard.

IDENTIFYING CHARACTERISTICS

Bright-green seedlings grow to 2 in. (5.1 cm) in fall to early winter, then up to 18 in. (4.5 cm) by early summer, bearing tiny flowers and distinctly purplish seeds that give a stand of cheatgrass a rosy to purplish color. As the flowers and seed heads develop, the top of the plant arches over—dangling its seed-containing spikelets in curtainlike fashion. Spikelets have several long, sharp bristles extending beyond the seed (awns). This is a highly adaptive annual grass that can mature and set seed at anywhere from 1 in. (2.5 cm) to 2 ft. (0.6 m) high. The main stem is thin and bears nar-

Seed heads hang down from the flowering stalks.

row, long, hairy or downy leaves at intervals of several inches. Leaves are not always downy.

Japanese brome can only be distinguished by looking at the number of nerves on the first "glume," a bract at the base of the flower head or seed head. Cheatgrass's glume has one nerve, whereas Japanese brome's has three nerves. Red brome's seed heads stick upright in tufts like bristle brushes and have a purplish tinge.

HABITAT AND RANGE

Cheatgrass grows in all 50 states, throughout Canada, and even in Greenland. In the East it favors disturbed, dry roadsides and meadows while in the West it has become the dominant small plant in many dry rangelands and pastures, being most invasive where rainfall is between 10 and 20 in. (25 to 50 cm) and where rains fall largely in the fall and winter, during its germination and early growth. It grows well in both open range and in relatively open western pine and fir forests.

WHAT IT DOES IN THE ECOSYSTEM

Cheatgrass's best opportunities include soil disturbance (such as in overgrazing, logging, or cultivation) and fire. Since its seeds germinate in the fall and it tolerates cold weather and utilizes carbohydrates efficiently, it gets a jump on natives that have a longer dormancy. It tends to displace natives like Idaho fescue, wheatgrass, and even sagebrush. Seeds are blown by the wind, or their bristles hitch rides on fur and clothing. In early spring, while cheatgrass is tender, grazing animals, including bighorn sheep, antelope, mule deer, and rabbits, feed on it. Birds and insects feed on the seeds. U.S. Geological Survey scien-

Each spikelet has long bristles extending past the seed.

tists estimated that in the late 1990s cheatgrass dominated over 17 million acres of federal lands in the West and up to 85 percent of vascular plants in these areas were cheatgrass. The results of invasions include more frequent wildfires and less diversity of plants and animals.

HOW IT CAME TO NORTH AMERICA

Cheatgrass is a Eurasian or Mediterranean native that evolved with steppe grazers and survives heavy use by both wild and domestic animals. Scientists identified it in New York and Pennsylvania in 1861, though its means of arrival are uncertain. In the 1880s, it was found in the wheat-growing districts of Ontario, British Columbia, Washington, and Utah. Scientists in Washington State imported cheatgrass for an experiment in 1898, but most likely it also spread through contaminated seed. By the early 1900s it had established itself in both the East and the West and by the mid-1900s it grew in all states and bordering Canadian provinces.

MANAGEMENT

Controlling cheatgrass requires a combination of four tactics: Killing live plants, preventing seed formation, stopping germination and new growth, and substituting effective native competitors. Burning and spraying are the main tools for controlling larger stands of cheatgrass, while repeated mowing in spring and summer can prevent the formation of seeds, though it will not kill the grass any more than mowing kills a lawn. After burning in fall, the planting of native perennials like Idaho fescue helps to replace cheatgrass. In some areas managers burn or spray cheatgrass in the fall, seed competitors in the spring, and simultaneously release grazing animals that eat new cheatgrass while walking the competing seeds into the soil. Grazing beyond cheatgrass's period of seed head formation can be counterproductive, as grazers also eat native competitors. Herbicides containing glyphosate, quizalofop, fluazifop, or sethoxidym are commonly used to control cheatgrass in natural areas, and are applied before the seedheads form.

FOR MORE INFORMATION

Belnap, J., M. Reheis, and R. Reynolds. 2000. "Conditions favoring and retarding cheatgrass invasion of arid lands in the southwestern U.S." http://esp.cr.usgs.gov/info/sw/cheatgrass.

Carpenter, A., and T. Murray. 2004. "Element stewardship abstract for *Bromus tectorum* L. (*Anisantha tectorum* [L.] Nevski)." The Nature Conservancy. http://wiki.bugwood.org/Bromus_tectorum.

Mack, R. N. 2010. "Fifty years of 'Waging war on cheatgrass': Research advances, while meaningful control languishes." In *Fifty Years of Invasion Ecology: The Legacy of Charles Elton*, edited by D. M. Richardson. Oxford, UK: Wiley-Blackwell.

Young, J. A. and C. D. Clements. 2009. *Cheatgrass: Fire and Forage on the Range.* Reno, NV: University of Nevada Press.

Bermuda Grass   *Cynodon dactylon*

NAME AND FAMILY
Bermuda grass, devil's grass (*Cynodon dactylon* [L.] Pers.); grass family (Poaceae). Sometimes confused with large crabgrass (*Digitaria sanguinalis* [L.] Scop.).

IDENTIFYING CHARACTERISTICS
This perennial grass spreads by long underground and aboveground runners. Aboveground runners root where they touch the ground. Underground rhi-

Left: *Aboveground runners can root and send up leaves from the nodes.*
Right: *The ligule has a ring of white hairs.*

Above and below the ground, Bermuda grass spreads rapidly.

The rhizome's pointed tip allows it to come up through heavy mulch or newspaper.

zomes are usually whitish in color. The foliage is gray green and relatively fine, leading to its popularity as a turfgrass. The ligule, where the leaf meets the stem, has a ring of white hairs. Bermuda grass flowering stalks grow in late summer and are topped by a whorl of three to seven, 0.5 in. (1.3 cm) long branches holding the flowers densely along each branch. Crabgrass flowering stalks look similar to those of Bermuda grass, but crabgrass does not have stout runners.

HABITAT AND RANGE

Prefers warmer climates from Florida to California, but will grow as far north as Massachusetts, Michigan, and Utah. Grows in fields, waste places, lawns, gardens, and roadsides where there is high light.

WHAT IT DOES IN THE ECOSYSTEM

Bermuda grass is principally a problem in disturbed areas. It can outcompete many other early successional plants and spreads rapidly. It also releases chemicals from both living and dead plants that may inhibit the growth of other plants. It spreads mainly when rhizomes are fragmented and carried to new areas. It has the ability to push its way through heavy layers of mulch. Its high pollen production makes it a significant contributor to allergies in some areas.

HOW IT CAME TO NORTH AMERICA
Introduced in the mid-1800s as a pasture grass because it stays green in summer and grows under drought conditions. Probably originally from eastern Africa. Commonly used in southern states as a turfgrass.

MANAGEMENT
Bermuda grass is difficult to remove because of its extensive, fine root system. Shading will reduce and eventually eliminate the grass. Since Bermuda grass can grow through more than a foot of loose mulch, black plastic applied during summer will smother plants but some deep rhizomes may survive. Glyphosate or a grass-specific herbicide containing fluazifop, sethoxydim, or clethodim applied when rhizomes are actively growing can be effective. Clipping or mowing weekly, particularly during summer droughts, will reduce spread.

FOR MORE INFORMATION
Guglielmini, A. C., and E. H. Satorre. 2002. "Shading effects on spatial growth and biomass partitioning of *Cynodon dactylon*." *Weed Research* 42:124–134.

Newman, D. 1992. "Element stewardship abstract for *Cynodon dactylon*." The Nature Conservancy. http://wiki.bugwood.org/Cynodon_dactylon.

Japanese Stilt Grass   *Microstegium vimineum*

NAME AND FAMILY
Japanese stilt grass, Nepalese browntop, eulalia, Chinese packing grass (*Microstegium vimineum* [Trin.] Camus); grass family (Poaceae).

IDENTIFYING CHARACTERISTICS
At maturity, the clumps of annual stilt grass are more stem than leaves. The slender stems rise to 3.5 ft. (1.1 m), but often bend over. They bear lime-green leaves with a distinct, off-center, shiny (due to minute silvery hairs) midrib. The alternate, well-spaced leaves are lance shaped, pointed fore and aft, 1 to 4 in. (2.5 to 10 cm) long, and 0.5 in. (1.3 cm) wide. In late summer, thin flower stalks appear at the leaf axils or ends of the stems. Flower heads consist of one to three thin spikes with flowers or seeds closely clustered along the spike, somewhat resembling crabgrass flower heads. By late fall, after the plants have shed hundreds of yellow to reddish, elliptical seeds, their leaves darken, then the plants die. Japanese stilt grass resembles some other grasses, but can be easily distinguished by the silver stripe along the leaf and the leaf shape.

HABITAT AND RANGE

Japanese stilt grass has naturalized from southern New England south to the Carolinas and west to Texas and Illinois. Its preferred habitat is acidic to neutral, highly organic soils in floodplain forests and streambanks, but it also grows in fields, brushy areas, along roads, and under utility lines. It can also grow in slightly alkaline soils and in fairly heavy shade.

WHAT IT DOES IN THE ECOSYSTEM

By itself, Japanese stilt grass does not readily take over established natural plant communities, but given a start by some land disturbance (grazing, burning, mowing, or logging), it can monopolize the ground-level plant community within five years. Large patches of Japanese stilt grass, spreading by seeds, can outcompete native plants and rob the lower ones of sunlight. It roots at stem nodes that touch the ground but, as an annual, produces new plants only by seed. The decaying plants increase organic matter in the soil and increase pH. Nutrient cycling is also altered in Japanese stilt grass stands. Few animals browse Japanese stilt grass, but rats will live in it. They prey on the eggs of ground-nesting birds. Deer will not readily eat Japanese stilt grass, but unchecked populations of deer that reduce natural vegetation and cause soil disturbance can encourage the spread of the plant.

Japanese stilt grass flowers and seeds are held closely clustered on thin spikes.

Left: *Large patches of Japanese stilt grass outcompete native plants and change nutrient cycling in the soil.* Right: *The silver stripe in the middle of the leaf helps identify Japanese stilt grass.*

HOW IT CAME TO NORTH AMERICA

The tradition of packing oriental porcelain in dry Japanese stilt grass (also called packing grass) probably accounts for its appearance in Tennessee around 1919 and elsewhere later on. It is native to temperate regions of Southeast Asia.

MANAGEMENT

This shallow-rooted grass is relatively easily pulled by hand when the ground is moist and the plants are tall enough to grab firmly at the base, but large colonies require many hours of work and plants will have to be pulled for several years to exhaust the seed bank. Mowing and weed whacking late in the season before seeds set can also be useful, but mowing too early can give the plants time to produce new flower spikelets at the axils. Systemic herbicides like glyphosate kill Japanese stilt grass but also its competitors. However, herbicides containing imazameth, fluazifop, or sethoxydim will kill Japanese stilt grass but not most competing natives like asters, legumes, and sedges. Prevent soil disturbances that will create a place for stilt grass seeds to germinate.

FOR MORE INFORMATION

Adams, S. N. and K. A. M. Englehardt. 2009. "Diversity declines in *Microstegium vimineum* (Japanese stiltgrass) patches." *Biological Conservation* 142:1003–1010.

Fairbrothers, D. E., and J. R. Gray. 1972. "*Microstegium vimineum* (Trin.) A. Camus (Gramineae) in the United States." *Bulletin of the Torrey Botanical Club* 99:97–100.

Fryer, Janet L. 2011. "*Microstegium vimineum.*" *Fire Effects Information System.* U.S. Department of Agriculture, Forest Service, Rocky Mountain Research Station, Fire Sciences Laboratory. http://www.fs.fed.us/database/feis.

Kourtev P. S., J. G. Ehrenfeld, and M. Haggblom. 2003. "Experimental analysis of the effect of exotic and native plant species on the structure and function of soil microbial communities." *Soil Biology and Biochemistry* 35:895–905.

Wavyleaf basketgrass   *Oplismenus hirtellus* ssp. *undulatifolius*

NAME AND FAMILY
Wavyleaf basketgrass (*Oplismenus hirtellus* ssp. *undulatifolius* (Ard.) U. Scholz); grass family (Poaceae). Similar in appearance to native basketgrass species and to nonnative small carpgrass (*Arthraxon hispidus*, p. 421).

IDENTIFYING CHARACTERISTICS
Wavyleaf basketgrass's rippled leaves set it apart from other grasses. The leaves are about 0.5 in. (1.5 to 2 mm) wide and 1.5 to 4 in. (4 to 8 cm) long and are borne on stems that reach 8 to 11 in. (20 to 30 cm) tall. The stem and leaf sheath are covered in short hairs. Flowering stalks appear in late summer with three to seven spikelets arranged alternately on the stalk. The seeds mature in fall and feature a long pointed lower bract (awn) that secretes a sticky substance that causes seeds to sticks to fur, clothes, shoes, and tires. This perennial grass's delicate stems trail on the ground and root from the nodes. Two native subspecies of basketgrass grow farther south but will only have a few hairs on the stems, if any. Small carpgrass leaves have stiff hairs along the leaf edges and are not as wavy.

HABITAT AND RANGE
Occurs in forests and forested floodplains in Maryland and Virginia. Intolerant of full sun.

WHAT IT DOES IN THE ECOSYSTEM
This grass has not been present long enough for extensive ecological studies to be conducted, but because it spreads rapidly and forms dense stands in relatively undisturbed forests, land managers are concerned that it is likely to compete with native plants and possibly change soil characteristics like Japanese stilt grass does.

Left: *Leaves have a wavy appearance.* **Right:** *Plants carpet the ground in shady areas.*

HOW IT CAME TO NORTH AMERICA
Found along the Patapsco River in Maryland in 1996, this grass has been spreading rapidly in forests in Maryland and Virginia. Possibly introduced as an ornamental grass from its native range in southern Europe and southeast Asia.

MANAGEMENT
Because this species is a recent invader with a limited distribution, report any suspected infestations to university extension offices or other appropriate state agencies. This shallowly rooted grass is easy to hand pull. Herbicides containing glyphosate or grass-specific herbicides can be used to treat large infestations.

FOR MORE INFORMATION
Global Invasive Species Database. 2010. *"Osplimenus hirtellus* ssp. *undulatifolius."* http://www.issg
.org/database/species/ecology.asp?si=1557&fr=1&sts=&lang=EN.
Swearingen, J., B. Slattery, K. Reshetiloff, and S. Zwiker. 2010. *Plant invaders of Mid-Atlantic natural areas,* 4th ed. Washington, DC: National Park Service and U.S. Fish and Wildlife Service. http://www.nps.gov/plants/alien/pubs/midatlantic/ophiu.htm.

Torpedo Grass   *Panicum repens*

NAME AND FAMILY
Torpedo grass, bullet grass, quack grass (*Panicum repens* L.); grass family (Poaceae). Torpedo grass looks similar to many native panic grasses.

Extensive rhizomes allow torpedo grass to spread rapidly.

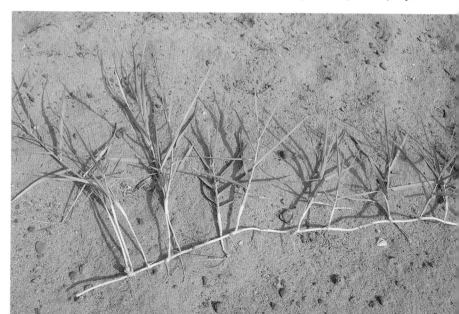

Left: *Torpedo grass is particularly problematic in shallow marshes in the southern United States.* Right: *The flowers and seeds occur on an inflorescence with upward-pointing branches.*

IDENTIFYING CHARACTERISTICS

This perennial grass is known for its extensive rhizomes, reaching 18 ft. (5.6 m) or longer. The pointed tips of the rhizomes are torpedo shaped, giving the grass its name. Rhizomes have brown or white scales and often look knotty. Stems grow along the rhizome, forming mats of vegetation. Stiff stems grow to 3 ft. (1 m) tall with leaves 2 to 10 in. (2.5 to 25.4 cm) long and only 0.06 to 0.25 in. (0.15 to 0.63 cm) wide. The upper surfaces of the leaves are usually hairy or have a whitish bloom, and the edges are rolled under. Flowers and seeds are held in a branched inflorescence, 3 to 9 in. (7.6 to 22.9 cm) long, with the branches pointed upward. Plants, which flower almost year-round, produce whitish seeds, but seed viability is generally low. Torpedo grass can generally be distinguished from other panic grasses by the hairiness of the leaves and by the torpedo-shaped rhizome tips.

HABITAT AND RANGE

This adaptable grass usually grows in damp soils along swamps, lakes, and canals—often with mats extending out over the water—but it can also grow on sand dunes and in pastures. Can tolerate some shade and is salt tolerant. Grows in coastal areas along the Atlantic from North Carolina to Florida, and in Gulf Coast states as far west as Texas. Also occurs in California.

WHAT IT DOES IN THE ECOSYSTEM

Torpedo grass grows quickly into monoculture stands that dominate and displace native plants, especially in shallow marshes. In Florida's Lake Okeechobee alone it covered nearly 14,000 acres (5,600 ha) in 1992. Plants mainly spread by fragmentation of rhizomes, but seeds also add to its spread. It blocks flood control systems and is weedy in agricultural fields and lawns, outcompeting even Bermuda grass (*Cynodon dactylon*).

HOW IT CAME TO NORTH AMERICA

Torpedo grass was first collected near Mobile, Alabama, in 1876, and starting in 1926 it began to be widely planted as a forage grass in the southeastern United States. Torpedo grass came to North America from southern Europe or tropical to subtropical Africa or Asia; some authorities believe it is native to Australia.

MANAGEMENT

Small patches can be dug out, taking care to remove all the rhizomes. Grasses can be controlled using nonselective herbicides like glyphosate or imazapyr. Due to the extensive rhizomes, more than one treatment may be necessary. If water levels can be manipulated, grasses can be drowned out.

FOR MORE INFORMATION

Hanlon, C. G., and K. Langeland. 2001. "Comparison of experimental strategies to control torpedo-grass." *Journal of Aquatic Plant Management* 38:40–47.

Langeland, K. A., J. A. Ferrell, B. Sellers, G. E. MacDonald, and R. K. Stocker. 2011. "Integrated Management of Nonnative Plants in Natural Areas of Florida." SP 242. Gainesville, FL: University of Florida IFAS. http://edis.ifas.ufl.edu/wg209.

Mediterranean Grass   *Schismus arabicus*

NAME AND FAMILY

Mediterranean grass, Arabian schismus (*Schismus arabicus* Nees), and common Mediterranean grass (*S. barbatus* [Loefl. ex L.] Thellung); grass family (Poaceae). These two grasses are very closely related and occur in similar habitats across a similar range.

IDENTIFYING CHARACTERISTICS

These annual grasses grow with tufted stems 5 to 8 in. (12.7 to 20 cm) tall. Narrow (0.06 in. [0.15 cm]), in-rolled leaves alternate along the stems and are 2 to 4 in. (5 to 10 cm) long. Flowers grow in tight, elliptical, green and purple clusters, 1.5 to 2 in. (3.7 to 5 cm) long, in late winter and early spring. The two species can be distinguished only by looking at details of the spikelets (flowers), *S. barbatus* having slightly shorter and more rounded spikelets.

HABITAT AND RANGE
Mediterranean grass grows in the southwestern United States from southern California to New Mexico. Found in spring in the Mojave and Sonoran deserts and common along roadsides, dry riverbeds, and waste places.

WHAT IT DOES IN THE ECOSYSTEM
Mediterranean grass is of greatest concern in deserts, where it creates a green carpet between shrubs in winter and early spring. It can go from a seedling to seed production in as little as two weeks. It competes for nutrients and water with the native annual grass six-weeks fescue (*Vulpia octoflora*) and other native annual herbaceous plants. It may have contributed to the increased extent of fires in scrublands because the dead stems carry fire from one shrub to another. Seeds are dispersed by wind as spikelets tumble across the desert. The seeds are sand-grain size and only some seeds germinate each year, leaving a reserve in case of poor growing conditions in any one year. The benefit of forage it provides to desert tortoises may be offset by the increased fire threat to the tortoise.

Mediterranean grass can form a green carpet in deserts of the southwestern United States.

HOW IT CAME TO NORTH AMERICA

Mediterranean grass apparently spread westward from Arizona into California by grazers or movement of vehicles and other equipment. First reported in Arizona in 1926 and in California in 1936. It was probably introduced accidentally from Eurasia.

MANAGEMENT

Arabian schismus is difficult to control. Plants are very small and have extensive root mats, making it hard to pull them up. Plants are susceptible to glyphosate, but because of the small amount of leaf area, it is difficult to effectively apply herbicide. Preemergent herbicides can prevent seeds from germinating.

Flowers grow in a tight purplish cluster in late winter and early spring.

FOR MORE INFORMATION

Brooks, M. *"Schismus* spp." California Invasive Plant Council. http://www.cal-ipc.org/ip/
management/ipcw/pages/detailreport.cfm@usernumber=73&surveynumber=182.php.
Halvorson, W. L. and P. Guertin. 2003. *Factsheet for* Schismus barbatus *(Loefl. ex L.) Thellung.*
USGS Weeds in the West Project: Status of Introduced Plants in Southern Arizona Parks.
Tucson, AZ: USGS Southwest Biological Science Center, University of Arizona. http://sdrsnet
.srnr.arizona.edu/data/sdrs/ww/docs/schibarb.pdf.

Tall Ryegrass   *Schedonorus phoenix*

NAME AND FAMILY

Tall ryegrass, tall fescue, Kentucky fescue (*Schedonorus phoenix* (Scop.) Holub, formerly *Lolium arundinaceum* [Schreb.] S. J. Darbyshire); grass family (Poaceae). Several other species of introduced fescues are also considered invasive in North America, including perennial ryegrass (*L. perenne* L.) and meadow ryegrass or tall fescue (*L. pratense* [Huds.] S. J. Darbyshire). If you have a lawn and live anywhere in North America from the upper South to southern Canada, chances are you are growing one of these or a related species.

IDENTIFYING CHARACTERISTICS

Tall ryegrass and its relatives are known as cool-season grasses because they stay green through winter and grow fastest in cool weather. Tall ryegrass produces small tufts of leaves and has short underground stems (rhizomes). From the tuft of leaves at the base, a stem grows 1 to 6 ft. (0.3 to 1.8 m) tall. Leaves are distinguished by their thickness, distinctly ridged veins on the upper surface without a distinct midvein, toothed edges, and glossy undersides. Leaves can be 2 to 24 in. (5 to 60 cm) long. The base of the leaf clasps the stem with little, hairy, ear-shaped projections (auricles). Meadow ryegrass and perennial ryegrass can be distinguished from tall ryegrass because their auricles are not hairy. Flowers and seeds are held in loosely branched, terminal clusters (panicles), 2 to 10 in. (5 to 25.4 cm) long, the lowest node of which has two to three branches. The panicle has a purplish color as it matures.

HABITAT AND RANGE
Found across the United States and southern Canada, both as cultivated specimens in lawns and pastures and escaped specimens in meadows and grasslands. It grows best on sunny, moist, nutrient-rich soils, but with adequate moisture will grow in many soil types. Plants can tolerate some drought.

WHAT IT DOES IN THE ECOSYSTEM
Tall ryegrass is aggressive and can displace native species in grasslands. Some populations are infected by an internal (endophytic) fungus (*Neotyphodium coenophialum*). Plants infected with the fungus have an increased competitive advantage against native plants, causing a decline in species diversity over time because infected plants germinate earlier, produce more biomass and more seeds, and spread by rhizomes to a greater extent. Infected plants also

Tall ryegrass lawns, if left uncut, will flower and set seed.

produce alkaloids that can cause wild and domesticated grazers to become ill. Songbirds and small mammals will consume seeds, but can become ill if seeds are infected with the fungus. The fungus spreads only by growing in seeds of infected plants, not by spores. Generally, tall ryegrass lowers the overall quality of wildlife habitat because it displaces favored native food sources. Tall ryegrass also releases chemicals from the roots that have been shown to hamper the growth of woody plant species. The coarse roots help loosen compacted soil and help control soil erosion.

Seed heads gradually open into a loosely branched panicle.

HOW IT CAME TO NORTH AMERICA

Tall ryegrass was imported from Europe in the late 1800s. The cultivar K31 (Kentucky 31) was found on a Kentucky farm in 1931 and promoted as a pasture grass. In the 1940s, tall ryegrass was promoted for improving pastures and reducing erosion and became very widespread. It is still promoted by numerous agricultural agencies.

MANAGEMENT

Small patches can be pulled or dug up, taking care to remove all roots. Burning grasslands infested with tall ryegrass in early spring can reduce its abundance. Herbicides (including glyphosate or imazapic) can be sprayed on plants in late fall or early spring, when most native grasses and plants are dormant.

FOR MORE INFORMATION

Applegate, R. D. 2009. "Tall Fescue." Plant Conservation Alliance Alien Plant Working Group. http://aggie-horticulture.tamu.edu/plantanswers/turf/publications/tallfesc.html.

Ball, D. M., J. F. Pedersen, and G. D. Lacefield. 1993. "The tall-fescue endophyte." *American Scientist* 81:370–379.

Batcher, M.S. 2000. "Element stewardship abstract for *Festuca arundinacea* (Schreb.)." The Nature Conservancy. http://tncweeds.ucdavis.edu/esadocs/documnts/festaru.html.

Clay, K., and J. Holah. 1999. "Fungal endophyte symbiosis and plant diversity in successional fields." *Science* 285:1742–1744.

Missouri Department of Conservation. "Tall Fescue Control." http://mdc.mo.gov/landwater-care/invasive-species-management/invasive-plant-management/tall-fescue-control.

Medusahead   *Taeniatherum caput-medusae*

NAME AND FAMILY
Medusahead, medusahead rye (*Taeniatherum caput-medusae* [L.] Nevski); grass family (Poaceae).

IDENTIFYING CHARACTERISTICS
The seed heads distinguish this annual grass. They are made up of several spikelets, each containing one seed. Each spikelet has two awns, 1 to 4 in. (2.5 to 10 cm) long, the longer of which has minute upright barbs that catch on fur and clothing. As the awns dry, they twist, giving the plant its name "medusahead." The flowers, which bloom in late spring, are usually self-pollinated, although some wind pollination occurs. Plants produce few leaves, instead putting their energy into producing seeds and tillers. Leaves are less than 0.13 in. (0.3 cm) wide and often have in-rolled edges. Plants germinate in fall and reach a height of 8 to 24 in. (20 to 60 cm). They are often noticeable because they stay green one to two weeks longer than many other annual grasses.

Long awns with hooked barbs twist to give the plant the name "medusahead."

HABITAT AND RANGE
Medusahead can be found throughout most of the western United States and Canada, from British Columbia south to California and west to Idaho. It is also sometimes found in eastern states. Often found with cheatgrass (*Bromus tectorum*, p. 426), medusahead thrives on dry, open lands like sagebrush and roadsides that face frequent disturbance from overgrazing or frequent fires.

WHAT IT DOES IN THE ECOSYSTEM
Medusahead displaces other plants, including cheatgrass in some sites, by reaching densities of up to 200 plants

per ft^2 (2,000 plants per m^2). After plants die, they form a dense litter layer that reduces the ability of other seeds to germinate and increases the risk of fire. Medusahead's own seeds are well adapted to germinating in the dense leaf litter. Few animals choose to graze or eat the seeds of medusahead. The foliage has a high silica content. The long barbs on the seeds can injure grazers by sticking into their eyes or noses. Seeds are dispersed by catching on fur or passing through the digestive tracts of animals that do manage to consume the seeds.

HOW IT CAME TO NORTH AMERICA
Native to the Mediterranean region, medusahead was probably accidentally introduced in the 1880s. The first botanical specimen was collected near Roseburg, Oregon, in 1887. It began to spread rapidly in the 1930s.

MANAGEMENT
Plowing in spring after medusahead seeds germinate can help control large infestations. Burning before seeds are dispersed can also largely eliminate seed production. Intensive mid-spring grazing by sheep significantly reduces medusahead populations. Glyphosate can be sprayed on plants before seeds form or imazapic can be used as a preemergent herbicide applied in fall. Reestablishment of perennial grasses can help reduce reinfestation by medusahead.

FOR MORE INFORMATION
Archer, A. J. 2001. "*Taeniatherum caput-medusae.*" *Fire Effects Information System*. U.S. Department of Agriculture, Forest Service, Rocky Mountain Research Station, Fire Sciences Laboratory. http://www.fs.fed.us/database/feis.

DiTomaso, J. M., G. B. Kyser, M. R. George, M. P. Doran, and E. A. Laca. 2008. "Control of medusahead (*Taeniatherum caput-medusae*) using timely sheep grazing." *Invasive Plant Science and Management* 1:241–247.

Miller, H. C., D. Clausnitzer, and M. M. Borman. 1999. "Medusahead." In *Biology and Management of Noxious Rangeland Weeds,* edited by R. L. Sheley and J. K. Petroff, 271–281. Corvallis, OR: Oregon State University Press.

Pollack, O., and T. Kan. "*Taeniatherum caput-medusae.*" California Invasive Plant Council. http://www.cal-ipc.org/ip/management/ipcw/pages/detailreport.cfm@usernumber=80&surveynumber=182.php.

Terrestrial Plants—Ferns

Climbing Ferns   *Lygodium* spp.

NAME AND FAMILY
Japanese climbing fern (*Lygodium japonicum* [Thunb.] Sw.) and Old World or small leaf climbing fern (*L. microphyllum* [Cav.] R. Br.); curly-grass family (Schizaeaceae). There is a native American climbing fern (*L. palmatum* [Bernh.] Sw.) that closely resembles Old World climbing fern but does not overlap in range.

IDENTIFYING CHARACTERISTICS
These are not your grandmother's garden ferns, but ferns whose fronds can grow 100 ft. (30 m) long (although the casual observer may not know that what appears as a stem is actually the central part of a large frond, the wiry rachis that bears many leaflets). Leaflets are on stalks. The many-branched fronds are attached to a main stem by a stalk only 4 to 8 in. (10 to 20 cm) long. In Japanese climbing fern, each frond is divided into pinnae, triangular in outline, which are further divided into two to three leaflets (twice compound). Leaflets are highly lobed, often palm shaped, on short stalks, and 2 to 3 in. (5 to 7.6 cm) long and 1 in. (2.5 cm) wide, and have curved hairs on the underside of

Old World climbing fern clambers over trees and shrubs in southern Florida.

Japanese climbing fern fronds have highly lobed leaflets on short stalks.

the leaflet. Compressed leaflets hold the sporangia, or spore-containing structures through which the ferns reproduce. The fronds of Old World climbing fern are divided once into oblong pinnae, 2 to 5 in. (5 to 13 cm) long by 1 to 2.5 in. (2.5 to 6.2 cm) wide, and are not hairy. The spore-bearing pinnae have lobed edges, which are often curled under. Spores are carried by wind and water. Aboveground stems die in winter where there is frost, but the black, wiry roots remain viable in the soil.

HABITAT AND RANGE
Climbing ferns prefer moist soils in sun or shade. Often found along edges of roadsides, lakes, marshes, creeks, and woods. Japanese climbing fern occurs from North Carolina south to Florida and west to Texas. Old World climbing fern does not yet grow beyond central and southern Florida but has invaded over 50,000 acres in that state.

WHAT IT DOES IN THE ECOSYSTEM
Ferns climb into trees, blocking light to vegetation below. The dead stems form a ladder for the next year's fronds to climb. The ladder of vines also allows fire to climb into the canopy and spread into flooded areas where fire would not normally travel well. Over the ground the ferns form dense mats, sometimes more than 4 ft. (1.2 m) thick, that prevent the germination and survival of other plants. Dense root mats can impede the flow of water in creeks and wetlands. Ferns spread extremely rapidly. In Loxahatchee National Wildlife Refuge, Old World climbing fern was undetected in 1990, yet by 1995 it covered 16,800 acres.

Left: *Spore pinnae of Old World climbing fern have lobed margins.* **Right:** *Leaflets of Old World climbing fern are oblong.*

HOW IT CAME TO NORTH AMERICA

Japanese climbing fern was introduced as an ornamental plant around 1900 and was first reported outside cultivation in Georgia in 1903. Old World climbing fern appeared outside of cultivation in Florida in 1960; also introduced as an ornamental plant. Japanese climbing fern is native to Eastern Asia and Australia. Old World climbing fern is native to tropical Asia, Africa, and Australia.

MANAGEMENT

Plants can be hand pulled or cut repeatedly. Ferns can also be cut near the ground and the cut stems sprayed with glyphosate. Aerial spraying of glyphosate and metsulfuron-methyl is used on large infestations. Two moths were introduced in Florida to control Old World climbing fern. One failed to establish, but the brown lygodium moth (*Neomusotima conspurcatalis*) has begun to spread. Other biological control agents are being investigated.

FOR MORE INFORMATION

Langeland, K. A., H. M. Cherry, C. M. McCormick, and K. A. Craddock Burks et al. 2008. *Identification and Biology of Nonnative Plants in Florida's Natural Areas,* 2nd ed. SP 257. Gainesville, FL: University of Florida IFAS.

Lott, M. S., J. C. Volin, R. W. Pemberton, and D. F. Austin. 2003. "The reproductive biology of the invasive ferns *Lygodium microphyllum* and *L. japonicum* (Schizaeaceae): Implications for invasive potential." *American Journal of Botany* 90:1144–1152.

Miller, J. H., E. B. Chambliss, N. J. Loewenstein. 2010. *A field guide for the identification of invasive plants in southern forests.* General Technical Report SRS-119. Asheville, NC: U.S. Department of Agriculture, Forest Service, Southern Research Station. http://www.srs.fs .fed.us/pubs/gtr/gtr_srs119.pdf.

Minogue, P. J., S. Jones, K. K. Bohn, and R. L. Williams. 2009. *Biology and control of Japanese climbing fern* (Lygodium japonicum). FOR218. University of Florida IFAS. https://edis.ifas .ufl.edu/pdffiles/FR/FR28000.pdf.

Sword Ferns 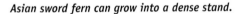 *Nephrolepis* spp.

NAME AND FAMILY
Erect sword fern, tuberous sword fern, fishbone fern (*Nephrolepis cordifolia* [L.] Presl.) and Asian sword fern (*N. multiflora* [Roxb.] Jarrett ex Morton); wood fern family (Dryopteridaceae). They can be easily confused with the native Boston sword fern (*N. exaltata* [L.] Schott), which is commonly used as a houseplant.

IDENTIFYING CHARACTERISTICS
The fronds of these ferns grow to 3 ft. (1 m) tall. Fronds are evergreen in areas without frost but will die back to the ground in colder areas. The frond is divided into 40 to 100 pairs of pinnae (leaflets), each oblong with a triangular point on the upper side of the pinna that overlaps the rachis (the frond stem). The edges of the pinnae can be smooth or slightly toothed. The rachis is covered on the upper side by two-toned brown scales. Some fronds will be fertile, with rounded sori (spore-containing structures) on the underside of the pinnae arranged along the edges of the pinnae. Wiry aboveground stems (stolons) are covered with tan scales, and they sometimes produce underground, rounded tubers. Rhizomes are also covered with scales. Asian sword fern has a line of short, upright hairs along the upper side of the midrib of the pinna. The native sword fern does not have

Asian sword fern can grow into a dense stand.

Erect sword fern (and Asian sword fern) fronds are divided into oblong leaflets with triangular points.

tubers and the scales on the rachis are only faintly two-toned if at all. The pinnae of the native do not overlap the rachis.

HABITAT AND RANGE

Sword ferns grow on rocks, on other plants, or on the ground in partly shady to shady, moist areas like hardwood hammocks (tree islands), pine rock lands, and marsh edges. They will also grow along roadsides and in other disturbed areas. They occur in natural areas throughout Florida and into southern Georgia.

WHAT IT DOES IN THE ECOSYSTEM

The ferns grow in dense stands that displace native plants, including the native sword fern. They spread by spores and by the dispersal of tubers, rhizomes, and aboveground stolons.

HOW IT CAME TO NORTH AMERICA

Asian sword fern is from tropical Asia, but the origin of erect sword fern is unclear. Sword ferns are commonly used as landscape plants in tropical areas. The most common houseplant fern is usually the native Boston fern, but can be one of several species of *Nephrolepis*.

MANAGEMENT

Ferns can be hand pulled or dug up, taking care to remove all rhizomes, tubers, and stolons. Stands can also be sprayed with glyphosate.

FOR MORE INFORMATION

Langeland, K. A., H. M. Cherry, C. M. McCormick, and K. A. Craddock Burks et al. 2008. *Identification and Biology of Nonnative Plants in Florida's Natural Areas,* 2nd ed. SP 257. Gainesville, FL: University of Florida IFAS.

Langeland, K. A. 2011. "Natural area weeds: distinguishing native and non-native 'Boston ferns' and 'sword ferns' (*Nephrolepis* spp.)." SS-AGR-22. University of Florida IFAS. http://edis.ifas.ufl.edu/ag120.

Aquatic Plants—Rosette only

Water Hyacinth   *Eichornia crassipes*

NAME AND FAMILY
Water hyacinth (*Eichhornia crassipes* [Mart.] Solms); water hyacinth family (Pontederiaceae). Sometimes confused with frog-bit (*Limnobium spongia*).

IDENTIFYING CHARACTERISTICS
This floating plant is often distinguished by a large spike of 15 to 18 lavender to pinkish-blue flowers some 2 in. (5 cm) wide. Each flower has six petals, with the uppermost petal having a yellow splotch bordered in blue. The next most distinctive feature is the bulbous, spongy leafstalk. The thick leaves, up to 6 in. (15 cm) wide, are generally oval shaped to rounded, have dense veins, and curve inward at the edges. Plants grow to 3 ft. (1 m) high. A three-segment fruit contains ribbed seeds. Water hyacinth is easily distinguished from frog's-bit, which does not have the inflated or spongy stem and has small, white, single flowers in leaf axils.

Left: *Bulbous leafstalks hold spoon-shaped leaves.* **Right:** *Each lavender flower has six petals, with the uppermost petal having a yellow splotch bordered in blue.*

Flowers are held on stalks just above the leaves.

Plants grow very densely on the surface of the water.

HABITAT AND RANGE

Water hyacinth is a warm-climate plant found across the southern United States in still and shallow waters where it can both float freely and put down roots in sediments. Water hyacinth has made cameo appearances as far north as New York and Illinois.

WHAT IT DOES IN THE ECOSYSTEM

Slow-moving or still southern waterways sometimes appear to be carpeted from shore to shore with the bright green leaves and blue flowers of water hyacinth. The heavy shadow of this carpet prevents native water plants from germinating and growing, reducing habitat for most native fish and other aquatic animals, although it increases the productivity of some insects, including mosquitoes. One acre of hyacinth can turn into 500 tons (1,200 tons/ha) of rotting plant materials that uses dissolved oxygen. The plant spreads from underwater stems (stolons) as well as by seed. Populations can double in size in two weeks, one of the highest growth rates known in the world of vascular plants. Southern states spend tens of millions of dollars annually for control on tens of thousands of acres of waterways important to wildlife, boaters, swimmers, and fishermen. Manatees do feed on water hyacinth and Southeast Asian farmers boil it into a paste and feed it, mixed with other ingredients, to pigs.

HOW IT CAME TO NORTH AMERICA

This native of northern South America was brought to the United States for the 1884 exposition in New Orleans. By 1890, it had reached Florida. By the 1950s, it had taken over more than 125,000 acres (50,500 ha) of Florida waterways. The U.S. Corps of Engineers estimates the Florida populations have been reduced to several thousand acres. Pieces of the plant are transported to new places by wind and water, and accidentally on boats or other equipment.

MANAGEMENT

Prevention is far more economical than the cures. Small populations can be stopped by hand pulling and careful raking with a pond rake. While nurseries sell the plant for small ponds, if a property owner can't resist, then at least he or she should be sure to dispose of unwanted or dead plants in dry places. For larger populations, herbicides like 2,4-D, triclopyr, and glyphosate are effective but can also kill native plants and animals. Another problem with herbicide control is that dead plants sink and use up large amounts of oxygen that is needed by other aquatic life. Mechanical harvesting is done with a Swamp Devil, a dredge-like boat with a powerful engine that draws the plants out of the water and shreds them for disposal on dry land.

FOR MORE INFORMATION

Batcher, M. S. 2000. "Element stewardship abstract for *Eichornia crassipes.*" The Nature Conservancy. http://wiki.bugwood.org/Eichhornia_crassipes (accessed January 25, 2012).

Langeland, K. A., H. M. Cherry, C. M. McCormick, and K. A. Craddock Burks et al. 2008. *Identification and Biology of Nonnative Plants in Florida's Natural Areas,* 2nd ed. SP 257. Gainesville, FL: University of Florida IFAS.

May, M., C. Grosso and J. Collins. 2003. "Water Hyacinth." In: *Practical Guidebook for the Identification and Control of Invasive Aquatic and Wetland Plantsin the San Francisco Bay-Delta Region.* San Francisco Estuary Institute. http://www.sfei.org/nis/hyacinth.html.

McCann, J. A. et al. 1996. *Nonindigenous aquatic and selected terrestrial species of Florida: Status, pathway, and time of introduction, present distribution, and significant ecological and economic effects.* Gainesville: Southeastern Biological Science Center.

European frog-bit   *Hydrocharis morsus-ranae*

NAME AND FAMILY
European frog-bit (*Hydrocharis morsus-ranae* L.); frog-bit family, (Hydrocharitaceae). Looks similar to American frog-bit, *Limnobium spongia*.

IDENTIFYING CHARACTERISTICS
Mats of European frog-bit float in wetlands, held together by tangled runners. Individual plants grow off the runners and have heart-shaped, leathery leaves 0.5 to 2.5 in. (1.2 to 6.3 cm) in diameter with smooth edges. The leaves resemble miniature lily pads. The undersides of the leaves are purple. Male and female flowers are found on separate plants. The three-petaled, small (0.5 in. [1 cm]) white flowers are held on stalks above the leaves. Male flowers occur in clusters of two to five flowers on stalks up to 1.5 in. (4 cm) long. Female flowers occur singly on stalks up to 3.5 in. (9 cm) long. Female flowers occasionally produce a single round fruit containing many seeds. In fall, buds called turions fall to the bottom of the wetland and remain dormant until spring. American frog-bit mainly occurs in the southeastern U.S. but can be found as far north as New York and Illinois. American frog-bit leaves have a thicker coating of spongy cells under the leaves and are held on long, ridged stalks.

HABITAT AND RANGE
European frog-bit can be found in Lake Erie, Lake Ontario, Lake Champlain, Lake St. Clair, and the Detroit River, as well as in other nearby ponds, lakes,

Flowers are held above small lily pad–like leaves.

wetlands, and drainage ditches. It prefers quiet, open waters. It is also found in Washington State near Meadow Lake.

WHAT IT DOES IN THE ECOSYSTEM
The dense mats restrict light to submerged aquatic plants. They can also impede boat traffic and clog water intakes. A single plant can produce 100 turions that will produce new plants in spring. The plantlets and turions are carried by boats and waterfowl and through interconnected waterways to new areas. Some waterbirds and insects feed on frog-bit.

HOW IT CAME TO NORTH AMERICA
European frog-bit was planted in a pond as an ornamental plant at the Central Experimental Farm Arboretum outside Ottawa, Canada, in 1932. It apparently escaped from there into nearby waterways. It is a popular plant for water gardens and aquaria. European frog-bit is native to Europe and temperate Asia and is considered threatened or endangered in parts of Europe.

MANAGEMENT
Small populations can be scooped out of waterways. Mechanical harvesters can be used to remove larger populations.

FOR MORE INFORMATION
Jacono, C.C. 2011. "*Hydrocharis morsus-ranae.*" USGS Nonindigenous Aquatic Species Database. http://nas.er.usgs.gov/queries/factsheet.aspx?SpeciesID=1110.

Nault, M. E. and A. Mikulyuk. 2009. "European frog-bit *(Hydrocharis morsus-ranae):* A technical review of distribution, ecology, impacts and management." PUB-SS-1048. Wisconsin Department of Natural Resouces Bureau of Science Services. http://wiatri.net/ecoatlas/ReportFiles/Reports2/1712AqInvasivesReport2.pdf.

O'Neill, C.R. Jr. 2007. "European frog-bit *(Hydrocharis morsus-ranae)*—floating invader of Great Lakes Basin waters." New York Sea Grant Invasive Species Factsheet Series: 07-1. http://www.seagrant.sunysb.edu/ais/pdfs/Frog-bitFactsheet.pdf.

Water Lettuce   *Pistia stratiotes*

NAME AND FAMILY
Water lettuce, water bonnets (*Pistia stratiotes* L.); arum family (Araceae). Water lettuce occurs around the world on all continents except Europe and Antarctica.

IDENTIFYING CHARACTERISTICS
Appropriately named, this aquatic plant looks like an open head of lettuce floating on the water. The light green leaves are thick and softly hairy, with distinct parallel ridges and a scalloped edge at the top of the leaf. They form a

rosette with the largest leaves about 6 in. (15 cm) long. Rosettes are connected by underwater stems. Feathery roots hang below the rosettes. The flowers are difficult to see, hidden among the leaves. A single female flower sits at the base of a short stalk topped by a whorl of male flowers. The fruit is a small green berry.

HABITAT AND RANGE
Water lettuce grows in lakes and slow-moving streams. It is most common in the Southeast, but does grow as far north as Rhode Island and as far west as California. It has even been reported in Ontario, Canada, but researchers are not sure if these plants overwinter. Its range is limited by severe cold. It can withstand water temperatures as low as 59°F (15°C).

WHAT IT DOES IN THE ECOSYSTEM
Water lettuce can form dense mats that block sunlight to submerged aquatic vegetation and push out other floating vegetation. It lowers oxygen content in the water by blocking the air-water interface and through decay, degrading habitat for aquatic organisms, including fish; it can also block access to water for terrestrial animals. The plants serve as a breeding ground for mosquitoes. The mats of plants also impede boating, swimming, and other recreational

Water lettuce leaves are softly hairy with parallel ridges along the leaf surface.

Left: *Mats of water lettuce block sunlight from organisms below.* Right: *Water lettuce crowds out other floating plants.*

water activities. In some areas water lettuce is itself being displaced by water hyacinth (*Eichornia crassipes*, p. 449).

HOW IT CAME TO NORTH AMERICA

Water lettuce may be native to North America, but many botanists believe it arrived here in ballast water from ships coming from South America or Africa. Pennsylvania naturalists John and William Bartram first described water lettuce in the St. Johns River during their voyage to Florida in 1765. It is often sold in the aquarium trade.

MANAGEMENT

On small ponds, plants can be harvested using a rake. Mechanical harvesters and choppers are used to manage large colonies of water lettuce. Two biological control insects help control the spread of the plant. Aquatic herbicides like endothall can also be used.

FOR MORE INFORMATION

Adebayo, A. A. et al. 2011. "Water hyacinth *(Eichornia crassipes)* and water lettuce *(Pistia stratiotes)* in the Great Lakes: playing with fire?" *Aquatic Invasions* 6:91–96. http://www.aquatic invasions.net/2011/AI_2011_6_1_Adebayo_etal.pdf.

Langeland, K. A., H. M. Cherry, C. M. McCormick, and K. A. Craddock Burks et al. 2008. *Identification and Biology of Nonnative Plants in Florida's Natural Areas*, 2nd ed. SP 257. Gainesville, FL: University of Florida IFAS.

Vandiver V. V. 1999. *Florida aquatic weed management guide.* Publication SP-55. University of Florida, Institute of Food and Agriculture Sciences, Cooperative Extension Service.

Aquatic Plants—Alternate Leaves

Water Spinach — *Ipomoea aquatica*

NAME AND FAMILY
Water spinach, water bindweed, swamp morning glory (*Ipomoea aquatica* Forssk.); morning glory family (Convolvulaceae).

IDENTIFYING CHARACTERISTICS
An aquatic morning glory, water spinach grows as a vine more than 9 ft. (2.7 m) long. Leaves are arrow shaped, 1 to 6 in. (2.5 to 15.2 cm) long and 1 to 3 in. (2.5 to 7.6 cm) wide, and grow alternately along the hollow stems. Stems have a milky sap. Flowers can be purple, pink, or white with a purple center, are 1.5 to 3 in. (3.7 to 7.6 cm) wide and funnel shaped, and grow from the leaf axils. One to six hairy seeds are held in an oblong capsule.

HABITAT AND RANGE
Grows along muddy streambanks, pond edges, and marshes in Florida. It has also been found in California. It has been cultivated in Texas for at least thirty years but has not been found outside of cultivation.

WHAT IT DOES IN THE ECOSYSTEM
Water spinach shades out submersed plants and competes with other emergent plants. Under favorable conditions it can produce 84 tons of biomass per acre

Funnel-shaped flowers and arrow-shaped leaves of water spinach.

(200 tons/ha) in nine months. Its colonies obstruct water flow in canals and can create ideal breeding habitat for mosquitoes. It can also invade rice and sugarcane fields. Spreads by fragments and by seeds.

HOW IT CAME TO NORTH AMERICA
Water spinach has been cultivated in China for over 2,000 years and is now cultivated in Africa, on Pacific islands, and in South America. In Texas, immigrants from Asia successfully won permits to cultivate the plant under special guidelines. Southeast Asian immigrants prize water spinach as an herb because it is rich in iron and easily harvested. It is usually chopped and either fried in oil or steamed, and often used as filling for spring rolls or dim sum. Despite being listed as a noxious weed by the federal government and the state of Florida, it has been introduced repeatedly since 1979.

MANAGEMENT
Small infestations can be harvested, but care must be taken to remove all plant pieces. Herbicides labeled for aquatic use show some success, but are not selective.

FOR MORE INFORMATION
Langeland, K. A., H. M. Cherry, C. M. McCormick, and K. A. Craddock Burks et al. 2008. *Identification and Biology of Nonnative Plants in Florida's Natural Areas,* 2nd ed. SP 257. Gainesville, FL: University of Florida IFAS.

Water Primrose   *Ludwigia* spp.

NAME AND FAMILY
Water primrose, creeping water primrose, Uruguay water primrose, primrose willow, seed box (*Ludwigia hexapetala* (Hook. & Arn.) Zardini, H. Y. Gu, & P. H. Raven); primrose family (Onagraceae). *Ludwigia hexapetala* is sometimes listed as *L. uruguayensis* or *L. grandiflora* ssp. *hexapetala*. Floating primrose willow, floating water primrose (*L. peploides* (Kunth) P. H. Raven) looks similar to water primrose and although it is native to the United States, an introduced subspecies, *L. peploides* ssp. *montevidensis*, is listed as invasive.

IDENTIFYING CHARACTERISTICS
Showy, yellow, five- or six-petaled flowers up to 1 in. (2.5 cm) across bloom all summer on this semiaquatic perennial plant. Flowers grow on short stalks from the leaf axils. Before flowers emerge, a cluster of spoon-shaped leaves grows on the surface of the water. Stems and leaves elongate as plants prepare to flower. Willowlike, lance-shaped leaves grow alternately along the stems, 4.5 in. (11.2

Dense mats of water primrose grow in slow-moving water.

Yellow five-petaled bloom all summer.

cm) long and 0.5 to 1 in. (1.3 to 2.5 cm) wide, becoming very crowded near the tip of the stem. Stems are sometimes hairy and by late in the growing season become reddish and woody. Woody seed capsules growing to 1.5 in. (3.7 cm) long hang on stalks from the leaf axils. Roots branch from the lower nodes of the stems in water and give the plant a feathery appearance. Plants can grow to 3 ft. (1 m) high or up to 6 ft. (1.8 m) long when floating on water. When plants are growing in mud or in dense mats, white, spongy roots often grow from the nodes. Floating primrose willow is virtually indistinguishable except by analysis of its chromosome numbers and minute reproductive characteristics.

HABITAT AND RANGE

Creeping water primrose occurs in fresh water along lake, pond, and river edges from New York south to Florida and west to Texas, as well as in California and the Pacific Northwest. Tends to prefer waterways with slow-moving, nutrient-enriched water (the nutrients often supplied by pollution). It may be

Willow-like lance-shaped leaves grow above the water before the plants flower.

native to parts of the southeastern United States, but is classified by several of these states as a noxious weed.

WHAT IT DOES IN THE ECOSYSTEM

Water primrose forms dense mats on the water surface and underwater that prevent other plants from growing. The dense stems in water also create ideal mosquito-breeding habitat, trap sediments, and impede water flow. Plants growing over water limit open water where many migratory birds forage for food. Plants are eaten by muskrats and waterfowl. Invasive populations spread mainly from stem and root fragments.

HOW IT CAME TO NORTH AMERICA

Probably introduced from South America as an ornamental plant. Herbarium specimens go back to the early 1900s.

MANAGEMENT

Plants can be hand pulled or cut repeatedly. Plants can also be killed by shading with opaque plastic. Stands can be controlled using glyphosate or triclopyr herbicides labeled for use over water.

FOR MORE INFORMATION

South Carolina Department of Natural Resources. "Water primrose." http://www.dnr.sc.gov/ invasiveweeds/primrose.html.

Washington State Noxious Weed Control Board. "Water primrose." http://www.nwcb.wa.gov/ detail.asp?weed=87.

Curly Pondweed   *Potamogeton crispus*

NAME AND FAMILY
Curly pondweed (*Potamogeton crispus* L.); pondweed family (Potamogetonaceae). One of 80 species in this aquatic family, curly pondweed most closely resembles native flat-stem pondweed (*P. zosteriformis*) and Richardson's pondweed (*P. richardsonii*).

IDENTIFYING CHARACTERISTICS
This popular aquarium perennial is easily recognized by the wavy edges of its flattened, dark-green, 1 to 4 in. (2.5 to 10 cm) long, narrow leaves that grow on flat reddish-brown stems. Leaves are bordered by very fine teeth. Curly pondweed attaches to submerged soil by shallow roots and rhizomes. In spring and summer, leaves tend to be reddish brown and more distinctly curly than in fall, when they are flatter and greener. Leaves are arranged alternately along the four-angled stem. Only a terminal flower spike appears above water, bearing inconspicuous brownish, wind-pollinated flowers in summer or early fall. In midsummer, plants begin to form vegetative buds or turions, then die off until new growth begins from the buds in fall or winter. The turions can reach 2 in. (5 cm) long and have three to seven thick leaves pointing upward from the stem. The flat-stem pondweed has flatter, smooth-edged leaves. Richardson's pondweed leaves come to a point and the leaf bases clasp the stems.

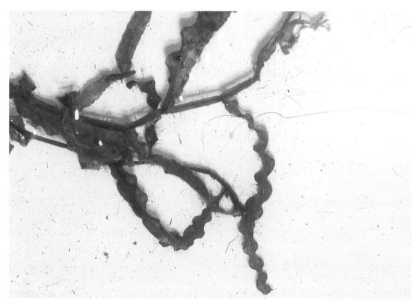

Wavy-edged leaves attach to the reddish stem.

HABITAT AND RANGE
Curly pondweed grows in fresh or slightly brackish waters. It will grow in ponds, irrigation canals, lakes, and slow moving streams. While it grows in partial shade, it does not tolerate full shade well. It has spread across the United States and southern Canada.

WHAT IT DOES IN THE ECOSYSTEM
Curly pondweed can form dense colonies of thick vegetation just below the surface of the water. That makes fishing, boating, and swimming difficult and deprives other forms of plant life of light. Summer-growing plants, however, do not have to compete with curly pondweed since it dies off in July (to dormant buds). The fact that it tends to die back in midsummer, however, when the warmer water holds less oxygen, means its rotting, oxygen-absorbing organic matter endangers other aquatic life forms by using up oxygen when oxygen levels are the lowest. New York lists it among the 20 most problematic invasives, and it is second only to Eurasian water-milfoil as a cause of concern in Midwestern waters. Since new plants form in winter, even under ice, curly pondweed has a head start on most other aquatic plants. It tolerates polluted waters and has been planted as an oxygenator. Curly pondweed is also a food for wildlife.

HOW IT CAME TO NORTH AMERICA
Curly pondweed may have come in as a hitchhiker on boats. It has been present in North America since the mid-1800s, and it was definitely popular among aquarium keepers by the late 1800s. It has often been spread by aquariums being emptied. Migrating birds spreading seeds probably play an important role in its distribution.

MANAGEMENT
Curly pondweed does not tolerate shade and it may be controlled by chemicals that darken the water, or in some cases by blanketing it with opaque fabric. Killing it before it forms buds (turions) in late spring or summer is essential for effective eradication. This also minimizes disturbance to native plants that are still dormant. Pond and lake rakes can be effective if used for several seasons, since they cut the plant at soil level, but cut vegetation should be removed from the water. Herbicides formulated for aquatic use, including diquat, endothall, or fluridone, can control pondweed. Treatments may have to be repeated a second year.

FOR MORE INFORMATION
Crowell, W. Undated. "Curly pondweed: New management ideas for an old problem." Exotic Species Program, Minnesota Department of Natural Resources. http://www.lakewashington assn.com/pdfs/curlyleaf.pdf.

Cypert, E. 1967. "The curly-leaved pondweed problem at Reelfoot Lake." *Journal of the Tennessee Academy of Science* 42:10–11.

Nichols, S. A., and B. H. Shaw. 1986. "Ecological life histories of the three aquatic nuisance plants, *Myriophyllum spicatum, Potamogeton crispus,* and *Elodea canadensis." Hydrobiologia* 131:3–21.

Stuckey, R. L. 1979. "Distributional history of *Potamogeton crispus* (curly pondweed) in North America." *Bartonia* 46:22–42.

Wisconsin Department of Natural Resources. "Curly-leaf pondweed." http://dnr.wi.gov/invasives/fact/curlyleaf_pondweed.htm.

Water Chestnut   *Trapa natans*

NAME AND FAMILY
Water chestnut (*Trapa natans* L.); water chestnut family (Trapaceae). These are not the water chestnuts used in Chinese cooking.

IDENTIFYING CHARACTERISTICS
The unmistakable fruit of this aquatic annual is a black, swollen nut more than 1 in. (2.5 cm) wide with two to four sharp spines that can penetrate shoe leather. Each nut contains a single, fleshy seed. Fruits form from four-petaled white to lavender flowers that grow singly among the upper leaves, blooming from midsummer to frost. The plant has two forms of leaves. Clusters of leaves at the water's surface are fan or diamond shaped with toothed edges, arranged alternately, and are attached to the stem by swollen stalks. The upper leaf surface is glossy, and the lower surface is hairy. Submerged leaves look feathery and may be arranged alternately or oppositely. Botanists disagree on whether

Leaves of water chestnut lie on top of the water's surface.

the submerged leaves are actually leaves or whether they are adventitious roots (roots that grow from the stems rather than in the soil). Stems are like long cords that grow to 16 ft. (4.8 m) and are rooted in the sediments.

HABITAT AND RANGE
Plants grow in ponds, lakes, slow-moving rivers, and canals in full sun from Quebec and New York south to Virginia. They grow most luxuriantly in water that is high in nutrients and has a neutral to alkaline pH.

WHAT IT DOES IN THE ECOSYSTEM
Plants form dense mats that shade out native plants and restrict water movement. When plants decay, they reduce oxygen levels in the water. Dense stands, however, can actually take up large amounts of nitrogen in rivers. Mosquitoes find favorable breeding grounds in water chestnut mats. Nuts that wash up on shorelines are hazardous to walkers and swimmers. Seeds that remain in the water can remain dormant for up to 12 years, although most germinate in 1 to 2 years. Each seed can produce 10 to 15 new stems when it germinates. It is spread to new locations most likely through fragments of plants carried by water, birds, or boats, since viable seeds tend to sink quickly. In Lake Champlain, more than $5 million was spent on water chestnut control from 1982 to 2003.

HOW IT CAME TO NORTH AMERICA
Harvard botanist Asa Gray may have been the first American to cultivate water chestnut in 1877, but the spread apparently began when a gardener cul-

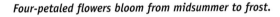

Four-petaled flowers bloom from midsummer to frost.

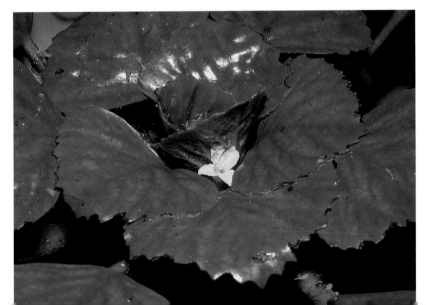

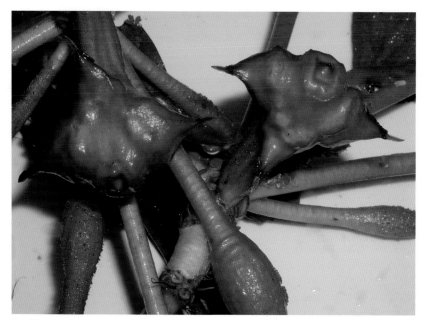

Swollen seedpods have sharp spines.

tivated water chestnut in a pond at the Cambridge Botanical Garden in Massachusetts in the late 1870s. The plant is native to Eurasia and seeds are eaten in Asia, and used medicinally and as cattle fodder. Interestingly, in Europe water chestnut became a protected species because of its recent decline in European waters.

MANAGEMENT
Plants can be pulled up or raked. Fragments will reestablish so it is important to remove all plant parts. Mechanical harvesters are often used to control large infestations; 2,4-D herbicide is licensed for use to control water chestnut in some states.

FOR MORE INFORMATION
Invasive Species in the Chesapeake Watershed Workshop. 2002. "Water chestnut." http://www
     .mdsg.umd.edu/exotics/workshop/water_chestnut.html.
Northeast Water Chestnut Web. http://www.waterchestnut.org.
Tall, L., N. Caraco and R. Maranger. 2011. "Denitrification hotspots: dominant role of invasive
     macrophyte *Trapa natans* in removing nitrogen from a tidal river." *Ecological Applications*
     21:3104–3114
Vermont Department of Environmental Conservation Water Quality Division. "Water chestnut."
     www.anr.state.vt.us/dec/waterq/lakes/htm/ans/lp_wc.htm.

Aquatic Plants—Opposite or Whorled Leaves

Alligator Weed   *Alternanthera philoxeroides*

NAME AND FAMILY
Alligator weed (*Alternanthera philoxeroides* [Mart.] Griseb.); amaranth family (Amaranthaceae). Sessile joyweed (*A. sessilis*), which looks similar to alligator weed, is a federally listed noxious weed in the United States because it is a weed of wet agricultural areas like rice and sugarcane fields.

IDENTIFYING CHARACTERISTICS
Alligator weed is a perennial that roots in wet soil or submerged soils and spreads out to form mats on top of waterways. The elliptical leaves with smooth edges grow oppositely along the stems. The leaves can be up to 4 in. (10 cm) long. Papery, ball-shaped heads of tiny white flowers bloom in summer on short stalks coming from the joints where the leaves meet the stems. Hollow stems can grow up to 3 ft. (1 m) long. Sessile joyweed has solid stems and the flowers grow in the leaf axils rather than on short stalks.

HABITAT AND RANGE
Alligator weed grows from Virginia south to Florida and west to Texas. It also occurs in California. It grows over still water and on wet banks and can withstand some salinity and immersion in floodwaters.

Papery flowers form a ball, flowering in summer.

The elliptical leaves have smooth edges and grow in pairs along the stems.

WHAT IT DOES IN THE ECOSYSTEM

Alligator weed forms mats over the surface of lakes, ponds, and canals, restricting light to plants and animals below and lowering water temperatures and oxygen content. It also creates favorable habitat for mosquitos. It discourages boating, swimming, and fishing and damages pumping and irrigation equipment. Although it was estimated to cover 97,000 acres in 1963, insects (see management, below) have reduced it to around 1,000 acres now.

HOW IT CAME TO NORTH AMERICA

Originally from South America, alligator weed arrived in the United States in the late 1800s in ballast water. It was first collected near Mobile, Alabama, in 1897.

MANAGEMENT

Starting in 1964, scientists released three biological control agents to control alligator weed. A flea beetle (*Agasicles hygrophila*), a thrip (*Amynothrips andersoni*), and a stem borer (*Vogtia malloi*) are now relied on to control alligator weed infestations, except in colder areas and swamps where the beetles cannot establish. Herbicides approved for use around water can also be used, but some populations are herbicide resistant. Mechanical removal is difficult because plants often break into pieces that can reestablish.

Alligatorweed formed huge mats over southern waterways until several biocontrol insects were introduced.

FOR MORE INFORMATION

Langeland, K. A., H. M. Cherry, C. M. McCormick, and K. A. Craddock Burks et al. 2008. *Identification and Biology of Nonnative Plants in Florida's Natural Areas,* 2nd ed. SP 257. Gainesville, FL: University of Florida IFAS.

Masterson, J. 2007. *"Alternanthera philoxeroides."* Smithsonian Marine Station at Fort Pierce. http://www.sms.si.edu/irlspec/alternanthera_philoxeroides.htm.

Spencer, N. R., and J. R. Coulson. 1976. "The biological control of alligatorweed, *Alternanthera philoxeroides,* in the Unites States of America." *Aquatic Botany* 2:177–190.

Brazilian Elodea   *Egeria densa*

NAME AND FAMILY

Brazilian elodea, giant waterweed, anacharis, egeria (*Egeria densa* Planch.); frog's bit family (Hydrocharitaceae). American elodea (*Elodea canadensis*) looks similar to Brazilian waterweed and to the invasive hydrilla (*Hydrilla verticillata*).

IDENTIFYING CHARACTERISTICS

A bushy aquatic perennial plant with leaves usually in whorls of four; the lowest leaves can be opposite or in whorls of three, while the upper leaves can be in whorls of up to eight leaves. In low light, leaf whorls are spaced further apart. Each leaf is about 1 in. (2.5 cm) long and 0.25 in. (0.6 cm) wide with finely toothed edges. American elodea generally has whorls of three leaves and leaves are less than 0.5 in. (1.3 cm) long. Hydrilla generally has a whorl of five leaves with tiny spines along the leaf edges. Brazilian elodea roots at the bottoms of waterways and the stems grow until they reach the water's surface where they spread out to form a mat. Mats can also be free floating. Showy,

Left: *The number of leaves in a whorl generally increases towards the tip of the stems.* **Right:** *Three-petaled flowers rise above the water on thin stalks.*

white, three-petaled, 0.75 in. (1.9 cm) flowers rise above the water on thread-like stalks in late spring and again in fall in the Southeast. Male and female flowers occur on separate plants, and only male plants have been found in the United States; thus no seeds are produced.

HABITAT AND RANGE
Brazilian elodea grows from New England south to Florida, in the Midwest south of Illinois and Nebraska, and in the western United States from Washington State south to California. Also found in Vancouver, Canada. Occurs in freshwater lakes and ponds and in slow-moving streams.

WHAT IT DOES IN THE ECOSYSTEM
The dense mats restrict water flow, trap sediments, and impede recreational use of waterways. Mats can become 6 to 15 ft. (1.8 to 4.6 m) deep, reducing light and outcompeting other native aquatic plants. Plants die back in fall or when water temperatures rise above 86° F (30° C). They overwinter underwater in a dormant but evergreen state.

HOW IT CAME TO NORTH AMERICA
Introduced from South America (native to Brazil, Argentina, and Uruguay) as an aquarium plant and for use in fishponds. Described as a good oxygenator, it was thought it would reduce mosquito larvae and therefore malaria. In dense mats, however, it often leads to reduced oxygen levels in water and encourages the growth of mosquito larvae. First reported in Millneck, Long Island (New York), in 1893. Known to have been sold commercially in the United States as early as 1915.

MANAGEMENT

Cutting and hand pulling tends to fragment plants and spread them. Small infestations can be shaded out using opaque cloth. Where water levels can be controlled in northern areas, draining over winter and allowing the sediments to freeze kills plants. Plants can be treated with the herbicides acrolein, diquat, fluridone, or complexed copper; note that many states have restrictions on where these herbicides can be used.

FOR MORE INFORMATION

Santos, M. J., L. W. Anderson and S. L. Ustin. 2010. "Effects of invasive species on plant communities: an example using submersed aquatic plants at the regional scale." *Biological Invasions* 13:443–457.

Schmitz, D. C., B. V. Nelson, L. E. Nall, and J. D. Schardt. 1988. "Exotic aquatic plants in Florida: A historical perspective and review of the present aquatic plant regulation program." In *Proceedings of the symposium on exotic pest plants, Nov. 2–4, 1988,* 303–326. Technical Report NPS/NREVER/NRTR=91/06. University of Miami.

Washington State Department of Ecology. 2003. "Technical Information about *Egeria densa* (Brazilian elodea)." http://www.ecy.wa.gov/programs/wq/plants/weeds/aqua002.html.

Hydrilla   *Hydrilla verticillata*

NAME AND FAMILY

Hydrilla, Florida elodea, water thyme, water-thyme, Indian star vine (*Hydrilla verticillata* [L. f.] Royle); frog bit family (Hydrocharitaceae). Hydrilla can be confused with native *Elodea* species, but hydrilla leaves have toothed leaf margins and reddish veins.

Hydrilla's whorled leaves have toothed edges.

Dense mats form at the surface of freshwater lakes and ponds.

IDENTIFYING CHARACTERISTICS

Dense mats of hydrilla often form at the surface of freshwater, growing from stems that can root as deep as 20 ft. (6 m) below (some sources say it can root as deep as 40 ft. [12 m] in clear water). Near the surface, stems branch and leaves are more frequent than in the depths and occur in whorls of three to eight leaves, about 0.75 to 1.5 in. (1.9 to 3.7 cm) in diameter, with reddish central ribs. Leaf margins are usually slightly toothed. Stems and leaves near the surface look like bottlebrushes. Plants can be either male or female. A form with both flowers on the same plant generally occurs north of South Carolina. Tiny reddish-brown male flowers grow on stalks, but when they mature they are released to float freely. Female flowers, which float on the surface of the water, have three whitish sepals and three translucent petals. Wind blows pollen from the male flowers to the female flowers.

HABITAT AND RANGE

From Maine south to the Gulf of Mexico and west to Texas, as well as on the West Coast from California to Washington, hydrilla has become a major obstacle to fishing, swimming, boating, hydroelectric generation, and irrigation in ponds, canals, and almost any slow-moving water. It has invaded about half of Florida's lakes, ponds, and waterways. It can grow in both low- and high-nutrient waters.

The tips of the stems resemble bottlebrushes.

WHAT IT DOES IN THE ECOSYSTEM

Hydrilla spreads rapidly, growing from rootstock and special buds that resemble tubers, or from a vegetative node that consists of no more than one whorl of leaves. Hydrilla out-competes native plants because it can photosynthesize in as little as 1 percent of normal sunlight. Thus it can start deep and grow vertically 1 in. a day (2.5 cm/day). When it nears the surface it branches and its stalks turn horizontally to form thick mats that rob sun from slower-growing competitors under water. The economic costs of hydrilla are in the hundreds of millions of dollars, from lost recreation opportunities to cleaning costs for irrigation works and hydroelectric plants. Hydrilla has certain benefits for largemouth bass production and some ducks eat it.

HOW IT CAME TO NORTH AMERICA

Hydrilla is probably native to India and Sri Lanka but is now found around the world. In North America its first appearance was probably as an aquarium and water garden plant in the 1950s. It was identified in a Miami, Florida, canal and in the Crystal River in Florida in 1960. Small pieces often stick to boats that then transport them to new waters. Colonies often start near boat ramps.

MANAGEMENT

Commonsense prevention of spread includes hosing down boats after leaving or before entering a lake, not dumping aquarium water into streams or ponds, and reporting any suspected colonies as soon as they are spotted. Hydrilla has become such a large problem that it has inspired a wide variety of control tactics that range from introducing grass carp to eat it (an illegal fish in some states) to dredging. For small areas, hydrilla can be eliminated by special fabric barriers pinned down over young colonies. The easiest mechanical controls are pond rakes and other cutters, but these provide only short-term reductions. Large rotovators, or tillers on hydraulic arms, till root crowns out of the bottom and regrowth after this treatment is much slower than after raking or cutting. Hydrilla can also be vacuumed off the bottom, roots and all, by a special dredge tube directed by a diver.

Herbicides, including copper sulfate, endothall, and fluridone, are a common control method. Copper sulfate can be toxic to fish, though, and copper sulfate and endothall harm most submerged aquatic plants. Fluridone is a plant growth regulator that slows growth but does not eliminate hydrilla. Biological controls include grass carp (an exotic fish) and an Asian fly that effectively diminished but did not eradicate hydrilla concentrations in some experiments.

FOR MORE INFORMATION

Batcher, M. S. 2000. "Element stewardship abstract for *Hydrilla verticillata* (L. f.) Royle." The Nature Conservancy. http://www.invasive.org/weedcd/pdfs/tncweeds/hydrver.pdf.

Langeland, K. A. 1996. "*Hydrilla vertcillata* (L.F.) Royle (Hydrocharitaceae): The perfect aquatic weed." *Castanea* 61:293–304. http://plants.ifas.ufl.edu/hydcirc.html.

Eurasian Water-Milfoil — *Myriophyllum spicatum*

NAME AND FAMILY

Eurasian water-milfoil (Myriophyllum spicatum L.); water-milfoil family (Haloragaceae). Parrot feather (M. aquaticum L.) looks similar to Eurasian water-milfoil but is not as cold tolerant. Often confused with native northern water-milfoil (M. sibiricum Komarov) and coontail (Ceratophyllum demersum L.). An invasive hybrid between Eurasian and northern water-milfoil also occurs in Minnesota. Variable leaf milfoil (M. heterophyllum Michx.) is listed as native and even endangered in some eastern states and eastern Canada but is listed by several other eastern and western states as a noxious weed.

Whorls of leaves are spaced closer together towards the tips of the stems.

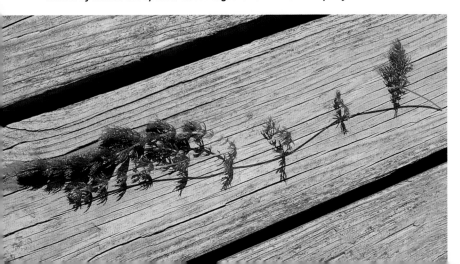

Leaves are divided into many pairs of leaflets.

IDENTIFYING CHARACTERISTICS

Bright-green, feathery leaves distinguish this aquatic plant. Whorls of three to five leaves divided into 12 to 16 pairs of thin, 0.5 in. (1.3 cm) long leaflets grow from stems 3 to 4 ft. (1 to 1.2 m) long (sometimes over 30 ft. [9 m] long). Stems are pale pink to red brown in color. At the surface, plants spread out on top of the water in dense mats. Small, four-part, yellow flowers bloom on spikes that rise 2 to 4 in. (5 to 10 cm) above the water. The hard fruits that follow the flowers have four seeds in a segmented capsule. New plants grow from nodes on the stems. Plants usually die back in late fall. The flowers of the similar-looking parrot feather can stick up out of the water by as much as a foot and are white, inconspicuous spring flowers. In contrast to the uniform thickness of stems in northern water-milfoil, Eurasian water-milfoil's stem below the flower spikes is almost twice as thick as the rest of the stem. Northern water-milfoil also tends to have fewer leaves. Coontail has toothed leaves and feels rough. Variable leaf milfoil has stems that stick above the water by 2 to 6 in. (5 to 15 cm) by late summer, with lance-shaped leaves. Leaves on its submerged stems look like those of Eurasian water-milfoil.

HABITAT AND RANGE

Versatile and adaptable, Eurasian water-milfoil has taken hold in most states and in the southern Canadian provinces of Ontario, Quebec, and British Columbia. It grows especially well in warmer, quiet waters of Mid-Atlantic reservoirs and coastal plain rivers, including tidal areas since it can tolerate brackish water. It is the most wide-ranging of nonnative aquatic plants.

WHAT IT DOES IN THE ECOSYSTEM
Like many invasives, Eurasian water-milfoil is an opportunist, taking hold most readily in disturbed areas, especially where pollutants discourage reestablishment of natives. It can root in waters as deep as 30 ft. (9 m) and when the stems grow near the surface they branch profusely to collect sunlight. These mats deny sunlight to submerged plants and animals. It spreads mainly by autofragmentation after flowering or by breakage. Fragments attached to boats and trailers carry it to new locations. Milfoil mats can increase both phosphorous and nitrogen in surrounding waters and greatly enlarge mosquito-breeding habitat. Winter kills the tops of the plants, adding to decaying, oxygen-consuming sediments that have sometimes led to fish deaths. Milfoil mats often make boating, swimming, and fishing nearly impossible. Eurasian milfoil is the target of multimillion dollar control programs by Canadian provinces, the U.S. Army Corps of Engineers, and the Tennessee Valley Authority.

HOW IT CAME TO NORTH AMERICA
This native of Europe, Asia, and North Africa probably came to North America in the late 1800s in ballast or bilgewater in the Chesapeake Bay area or for use in aquariums, but was clearly identified only in the early 1940s. Parrot feather probably arrived in the 1800s from South America to the Chesapeake Bay area.

MANAGEMENT
Pond rakes can control Eurasian water-milfoil in small areas, while larger mechanical harvesters are used on large populations. Any mechanical treatment causes breakage and can spread the population. In the early stages while milfoil is submerged, colorants that darken and filter sunlight can stunt its growth but will also impact native plants. Most herbicides that will kill milfoil also kill other plants and fish, but it is very susceptible to 2,4-D. The watermilfoil moth larvae (*Acentria ephemerella*), introduced from Europe in 1927, feeds on and in the stems, causing the leaves and stems to fall off. A native American milfoil weevil (*Euhrychiopsis lecontei*) also damages plants.

FOR MORE INFORMATION
Jacobs, J. and J. Mangold. 2009. "Ecology and management of Eurasian watermilfoil (*Myriophyllum spicatum* L.)." USDA NRCS Invasive Species Technical Note No. MT-23. http://www.plant-materials.nrcs.usda.gov/pubs/mtpmstn8523.pdf.

Langeland, K. A., H. M. Cherry, C. M. McCormick, and K. A. Craddock Burks et al. 2008. *Identification and Biology of Nonnative Plants in Florida's Natural Areas,* 2nd ed. SP 257. Gainesville, FL: University of Florida IFAS.

Moody, M. L. and D. H. Less. 2007. "Geographic distribution and genotypic composition of invasive hybrid watermilfoil *(Myriophyllum spicatum x M. sibiricum)* populations in North America." *Biological Invasions* 9:559–570.

Reed, C. F. 1977. "History and distribution of Eurasian watermilfoil in United States and Canada." *Phytologia* 36:417–436.

Brittle najad   *Najas minor*

NAME AND FAMILY
Brittle najad, also bushy, spiny or slender najad, brittle waternymph (*Najas minor* All.); water nymph family (Najadaceae). Sometimes confused with other native Najad species, coontail (*Ceratophyllum* spp.), or muskgrass (*Chara* spp.).

IDENTIFYING CHARACTERISTICS
This annual aquatic plant has brittle stems up to 4 ft. (1.2 m) long that support slender, 1 in. (2.5 cm) long, toothed leaves. The leaves usually grow opposite each other along a stem but can be whorled or clumped. The stems branch near the tips and the leaves are spaced closely together, giving the tips a bushy appearance. The leaves curve as they mature. Inconspicuous flowers bloom in the axils of the leaves from early spring through late fall. Each flower produces a single seed. Seeds germinate in spring. Compared to other najads, the teeth on brittle najad are more distinct and the base of the leaf is blocky or fan-shaped when pulled away from the stem. Coontail leaves are forked at the tips and arranged in whorls around the stem. Muskgrass is an alga with whorls of branches and a garlicky smell.

HABITAT AND RANGE
Grows from Ontario south to Florida, and west to Missouri and Oklahoma. Particularly problematic in the Southeast. Prefers calm waters but also grows in streams and rivers. Tolerates a pH range from 6.0 to 9.3 and salinity to 0.3 ppt.

Slender curved leaves have small teeth.

WHAT IT DOES IN THE ECOSYSTEM
Can outcompete other native and invasive aquatic plants, including hydrilla, by forming mats up to 12 ft. deep. Interferes with recreational activities such as boating, swimming, and fishing. Decaying mats of brittle najad can reduce oxygen levels in waterways.

HOW IT CAME TO NORTH AMERICA
Native to North Africa, central and eastern Europe, and parts of Asia. It was first found in the Hudson River in 1934. It was promoted for use as food for waterfowl, but could also have been introduced accidentally or dumped in the river from an aquarium.

MANAGEMENT
Aquatic harvesters and rotovators (like an underwater rototiller) can round up brittle najad, but fragments will often break off. Herbicides containing diquat, fluridone, and endothall are all used to treat brittle najad, but can kill other aquatic plants as well. Benthic barriers can be used to prevent seeds from germinating but can kill other organisms in the sediments.

FOR MORE INFORMATION
Global Invasive Species Database. 2010. "Najas minor." IUCN Invasive Species Specialist Group. http://www.issg.org/database/species/ecology.asp?si=1560&fr=1&sts=&lang=EN.
Cao, L. 2009. "Najas minor All." USGS Nonindigenous Aquatic Species. http://nas.er.usgs.gov/queries/factsheet.aspx?SpeciesID=1118.

Aquatic Plants—Other
(Algae, Ferns, Grasses, Rushes)

Flowering rush   *Butomus umbellatus*

NAME AND FAMILY
Flowering rush, grass rush (*Butomus umbellatus* L.); may the only member of the flowering rush family (Butomaceae). *Butomus junceus* Turcz may be a second species; it is very similar but has shorter flowers and an upright stamen. Resembles bulrushes and true rushes when not in flower.

IDENTIFYING CHARACTERISTICS
The narrow, pointed leaves grow from rhizomes, rising up to 3 feet (1 m) high. They are triangular in cross section and have smooth edges and parallel veins. The flower head, which blossoms from July into September, is a group of 20 to 50 flowers on slender stems that radiate like fireworks from the end of a supporting cylindrical stalk. The flowers are 0.75 to 1 in. (2 to 3 cm) wide with three pink sepals and three petals. In shallow water and along the shoreline, flowering rush will grow upright and stiff, but in deeper water the leaves float on the surface and move with the water. True rushes and some bulrushes have stems that are round in cross section.

This unique rush has clusters of pink and white flowers.

HABITAT AND RANGE

Flowering rush can be found on many shorelines and in slow-moving rivers, from the bank to waters up to 9 feet deep. It occurs across southern Canada and across the northern United States as far south as Illinois, Indiana, and Ohio. In the Great Lakes, flowering rush has become a serious invasive plant.

WHAT IT DOES IN THE ECOSYSTEM

Flowering rush pioneers newly exposed soils when water levels drop. It can crowd out native species like willows and cattails. Reduces open water habitat for some trout and shelters invasive fish species. Interferes with recreation. It reproduces from the pea-sized bulbils (bulb-like plant sprouts) on the rhizome when they detach and float away.

HOW IT CAME TO NORTH AMERICA

Flowering rush is native to Europe and Central Asia and was first noticed in North America along the St. Lawrence River in 1897. It was identified in 1918 on the River Rouge in Michigan and in 1929 at Lake Champlain in New York. It is sold as an ornamental plant, but is illegal in some states.

MANAGEMENT

Small infestations can be carefully dug up, taking care to leave no bulbils or pieces of fruit behind. Divers using suction dredges have also been effective, but entire rhizomes must be removed without losing the bulbils. In Michigan, ten years of cutting below water level incurred great costs and the invasion still spread. Imazapyr applied during the summer and when waters are calm can suppress populations but herbicides do not kill rhizomes.

FOR MORE INFORMATION

Jacobs, Jim and Jane Mangold, Hilary Parkinson, Virgil Dupuis, Peter Rice. 2011. *Ecology and Management of Flowering Rush* (Butomus umbellatus L.). United States Department of Agriculture.

Natural Resources Conservation Service. Invasive Species Technical Note No. MT-33. http://www.plant-materials.nrcs.usda.gov/pubs/mtpmstn10617.pdf.

Lui K, Thompson FL, Eckert CG. 2005. "Causes and consequences of extreme variation in reproductive strategy and vegetative growth among invasive populations of a clonal aquative plant *Butomus umbellatus* L. (Butomaceae)." *Biological Invasions* 7:427–444.

Killer Algae   *Caulerpa taxifolia*

NAME AND FAMILY
Killer algae, caulerpa (*Caulerpa taxifolia* [Vahl.] C. Ag.); green alga family (Ulvophyceae). The genus *Caulerpa* is native to the Caribbean and several species can be found off the coast of Florida. A Mediterannean strain of the genus known as the killer algae was found off the coast of California. A strain of sea grapes (*C. racemosa*), currently invading the Mediterranean, should also be on the watch list of invasive algaes.

IDENTIFYING CHARACTERISTICS
Featherlike fronds grow from a stolon (an aboveground stem) that runs along the bottom of the sea. Fronds of the invasive strain can be 2 in. (5 cm) long in shallow water, up to 38 in. (95 cm) in deeper water. Fronds of the native strains usually grow to only 6 in. (15.2 cm) long. Rootlike structures descend from the stolons and attach the alga to rocks or other substrates.

HABITAT AND RANGE
Native strains of the genus *Caulerpa* occur in the Caribbean and other tropical seas. In the 1970s, caulerpa began to be used as a decorative saltwater aquarium plant and in 1984 an extremely aggressive strain identical to one bred for aquariums was found in the Mediterranean near Monaco. This strain appeared in Carlsbad and Huntington Harbour, California, in 2000. It can grow in shallow coastal lagoons and in the ocean in waters up to 150 ft. (46 m) deep.

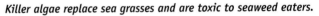

Killer algae replace sea grasses and are toxic to seaweed eaters.

Fronds attach to an aboveground stem.

WHAT IT DOES IN THE ECOSYSTEM
In the Mediterranean, one colony of killer algae spread from 1 sq. yd. to 5,400 acres in less than 10 years. It almost completely replaced all native sea grasses and greatly reduced numbers of aquatic organisms. It is toxic to fish and sea urchins that feed on other sea grasses and seaweeds, and it inhibits the growth of phytoplankton, an important part of the marine food chain.

HOW IT CAME TO NORTH AMERICA
Killer algae was probably dumped into harbors or storm drains from aquariums. Even small fragments of killer algae can form a new plant and fragments could be carried on boat anchors or fishing gear.

MANAGEMENT
Small populations may be hand pulled but small becomes big rapidly. In California, the infestations of killer algae were covered with tarps and liquid chlorine injected under the tarps killed the algae, along with all other plants and animals. More than $6 million dollars were spent to control and monitor these relatively small infestations from 2000 to 2004. Several snaillike animals are being tested as possible biological control agents in the Mediterranean.

FOR MORE INFORMATION
Jousson, I., J. Pawlowski, L. Zaninetti, F. W. Zechman, F. Dini, G. Di Guiseppe, R. Woodfield, A. Millar, and A. Meinesz. 2000. "Invasive alga reaches California." *Nature* 408:9.

Ramey, V. 2001. "*Caulerpa taxifolia* (Vahl.) C. Ag." Center for Aquatic and Invasive Plants, University of Florida and Sea Grant. http://aquat1.ifas.ufl.edu/seagrant/cautax2.html#hpconfuse.

Rock snot   *Didymosphenia geminata*

NAME AND FAMILY
Rock snot, didymo (*Didymosphenia geminata* (Lyngb.) M. Schmidt); Gomphonema family, (Gomphonemataceae). Closely resembles *Cymbella mexicana*.

IDENTIFYING CHARACTERISTICS
Rock snot are colonies of single-celled diatoms (an alga with a silica shell) that grow in freshwater streams. They begin in early summer as small 0.06 to 0.5 in. (2 to 10 mm) circles that feel like cotton balls and attach to rocks or other substrate. As the colonies expand, or "bloom," they merge into a thick tan or brown coating and put out long stalks that can eventually resemble white ropes or tissue floating in the water. When stalks of rock snot are pulled apart, there is some resistance and they feel like wet cotton balls. *Cymbella* strands feel slimy and there will be no resistance in pulling them apart. Confirmation of identification must be done using a compound microscope with at least 400x magnification to look at the shape and size of the cells.

HABITAT AND RANGE
Rock snot has been present for a long time in cold, low-nutrient freshwater streams and rivers of the far north around the globe as well as in some high-elevation streams. Since the 1980s it has become increasingly common in warmer climates and high-nutrient streams around the world. Algal blooms at

Rock snot coats a rock in a Maryland stream.

new sites where rock snot did not historically occur are considered invasive, but in areas where it has long been present the blooms are considered merely an occasional nuisance. In North America, scattered outbreaks have occurred in western streams from British Columbia, Canada, to California east through Colorado. In eastern North America, rock snot has made new appearances since 2007 in streams from Quebec to North Carolina.

WHAT IT DOES IN THE ECOSYSTEM
Rock snot forms dense mats up to 8 inches (20 cm) thick in freshwater streams, coating aquatic plants and stream bottoms. It is able to produce biomass even in low-nutrient streams by chemically attracting iron and phosphorous to its strands, where a bacteria living among the colony then converts the phosphorous into a form the alga can use to produce more cells. It changes insect larvae communities, as rock snot appears to favor small larvae like midges over large caddisfly and mayfly larvae. Rock snot does not appear to change water quality or affect human health. It causes economic losses since few people want to fish, boat, or swim in streams covered by rock snot.

HOW IT CAME TO NORTH AMERICA
Scientists are not sure why rock snot is now able to tolerate a wider range of environmental conditions. It is spread from stream to stream inadvertently on neoprene clothing, felt-soled waders, boats, and fishing gear.

MANAGEMENT
Stop the spread of rock snot by cleaning boats and equipment, disposing of algae in the trash. Thoroughly clean and soak waders and fishing gear in a 5 percent salt solution, wipe off with antiseptic hand cleaner, or scrub with dishwashing detergent. Thoroughly drying equipment or freezing for several days will also kill diatoms. Once a stream has rock snot, no effective way of killing it has been found, although high water flows may scour the algae out.

FOR MORE INFORMATION
Environmental Protection Agency. "*Didymosphenia geminata:* A nuisance freshwater alga." http://www.epa.gov/region8/water/didymosphenia.

Moen, S. 2009. "How didymo became rock snot." Minnesota Sea Grant. http://www.seagrant.umn .edu/newsletter/2009/12/how_didymo_became_rock_snot.html.

Sundareshwar, P. V., S. Upadhayay, M. Abessa, S. Honomichl, B. Berdanier, S. A. Spaulding, C. Sandvik, A. Trennepohl. 2011. "*Didymosphenia geminata:* Algal blooms in oligotrophic streams and rivers." *Geophysical Research Letters* 38 (10) DOI: 10.1029/2010GL046599

Salvinia   *Salvinia* spp.

NAME AND FAMILY
Giant salvinia, Kariba weed, African pyle, aquarium watermoss, koi kandy (*Salvinia molesta* D. S. Mitch.); and common salvinia, water spangles (*S. minima* Baker); water fern family (Salviniaceae). Giant and common salvinia are closely related and difficult to distinguish from *S. biloba* Raddi. Giant salvinia is federally listed by the United States as a noxious weed.

IDENTIFYING CHARACTERISTICS
The crumpled look of the leaves of this floating fern has led people to describe it as squashed grapes. In fact, this kind of leaves are produced when the plant grows in sunlight. Shade-grown leaves lie flat and are 1 to 1.5 in. (2.5 to 3.7 cm) long and oval, with a heart-shaped base. Tips are rounded to notched. Crowding also crumples leaves and turns them vertical. Leaves exposed to sun grow longer and begin to turn brown as they age. The other distinguishing feature of this plant is the eggbeater-shaped hairs on the leaf surface, which are clearly visible under a hand lens. They are upright, whitish, and their four tip branches curve out and then together like the cage of an eggbeater. The undersides of the leaves have brown hairs. Most plants have two leaves on the surface and one brown leaf that droops underwater. This submerged leaf often grows very long. Common salvinia (*S. minima*) has surface hairs, but their branches do not join at the tips in the eggbeater-like cage. It is also sterile and may be a hybrid, reproducing vegetatively.

HABITAT AND RANGE
Salvinias have established colonies in slow-moving to still waters of ponds, lakes, sloughs, swamps, and marshes from Virginia south to Florida and across

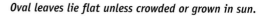

Oval leaves lie flat unless crowded or grown in sun.

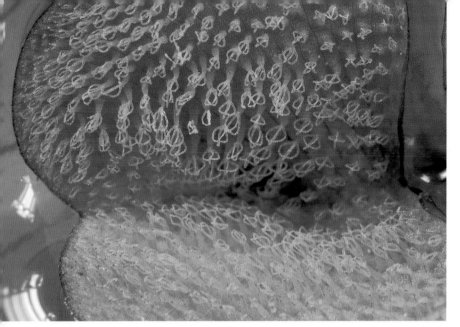

Hairs on the leaf surface are shaped like egg beaters.

the southern tier of states to California. They occupy the same general range as water hyacinth (*Eichhornia crassipes*, p. 449) and can occur in the same body of water. Salvinias tolerate both shade and sun and freezing temperatures, but they cannot survive some ice formation or brackish water.

WHAT IT DOES IN THE ECOSYSTEM

The rapidly growing, dense surface mats displace native plants and block sunlight from bottom plants and animals, and when they decay, the heavy biomass consumes large amounts of oxygen. The ferns enlarge themselves by developing new branches that emerge from buds, each side node being capable of producing five buds. Reproducing from stem fragments, common salvinia can double its population every 2 to 4 days but usually takes 4 to 10 days. One plant is reported to have covered 40 sq. mi. (104 km²) in three months. In Louisiana bayous, mats of salvinia have grown 12 mi. (19 km) long. Some mats are as thick as 9 in. (23 cm). The mats create excellent mosquito-breeding habitat and displace the native duckweeds favored by wildlife.

HOW IT CAME TO NORTH AMERICA

Native to the southern Brazilian coast but now global, giant salvinia was first detected in the United States in 1995 in South Carolina. Within a few years it had spread across the southern states to California, perhaps introduced in several places by plants sold by nurseries to aquarium and koi pond owners. Common salvinia has been cultivated since the 1880s for water gardens and continues to be sold. It is native to Central and South America.

MANAGEMENT

In coastal areas, introducing brackish water or saltwater into salvinia colonies can kill the ferns, and this technique is used to improve duck and goose habitat. Draining and drying out a colonized waterway also kills the plants. A small weevil (*Cyrtobagous salviniae*) whose adults and larvae feed on the leaf buds of salvinia has been an effective biological control in some countries, but its reproduction and appetite depend on temperature, nitrogen levels, and how much other vegetation impedes its search for salvinia. Herbicides containing glyphosate, diquat, and fluridone mixed with surfactants can be effective treatments against salvinia.

FOR MORE INFORMATION

Jacono, C. 1999. *"Salvinia molesta:* A giant among noxious weeds." *Wildland Weeds* 2(3):4–7.
McFarland, D. G., L. S. Nelson, M. J. Grodowitz, R M. Smart and C. S. Owens. 2004. Salvinia molesta *D. S. Mitchell (giant salvinia) in the United States: A review of species ecology and approaches to management.* U.S. Army Corps of Engineers, Engineer Research and Development Center, Aquatic Plant Control Research Program ERDC/ELSR-04-2. http://el.erdc.usace .army.mil/elpubs/pdf/srel04-2.pdf.
Room, P. M. 1990. "Ecology of a simple plant-herbivore system: Biological control of Salvinia." *Trends in Ecology and Evolution* 5:74–79.
U.S. Geological Survey. "Salvinia." http://salvinia.er.usgs.gov.

Smooth Cordgrass   *Spartina alterniflora*

NAME AND FAMILY

Smooth cordgrass, Atlantic cordgrass (*Spartina alterniflora* Loisel.); grass family (Poaceae). Native California cordgrass (*S. foliosa* Trin.) marshes are being invaded by several species of introduced cordgrasses and cordgrass hybrids. The most widespread invasive species is smooth cordgrass, which is native to the eastern United States. Saltmeadow cordgrass (*S. patens* [Ait.] Muhl.) from the eastern United States and denseflower cordgrass (*S. densiflora* Brongn.) from Chile are also invasive in western salt marshes.

IDENTIFYING CHARACTERISTICS

Cordgrasses are grasses of salt marshes. Smooth cordgrass grows in dense stands that gradually expand via underground rhizomes. Hollow, hairless stems tend to grow 2 to 4 ft. (0.6 to 1.2 m) tall. Leaf blades are 8 to 20 in. (20 to 50 cm) long and 1 to 8 in. (2.5 to 20 cm) wide. There is often a purplish color at the base of the stems where the leaves wrap around the stems (the leaf sheaths). A dense cluster of tan flowers made up of many closely packed spikes of flowers blooms from July to November and is 4 to 16 in. (10 to 40 cm) long. California cordgrass tends to be smaller in stature but is otherwise difficult to distinguish from smooth cordgrass. California cordgrass does not

Native to the east coast, on the west coast smooth cordgrass colonizes mudflats and hybridizes with California cordgrass.

have the purple coloration at the base of the stems or on the rhizomes as smooth cordgrass does.

HABITAT AND RANGE
Native to salt marshes along the Atlantic and Gulf Coasts; invasive along the west coast of the United States from Washington south through California, especially in San Francisco Bay, Puget Sound, and Willapa Bay.

WHAT IT DOES IN THE ECOSYSTEM
Dense stands of cordgrass often crowd into waterways and hide large expanses of mudflat in salt marshes. It spreads both by seeds and by rhizome fragments. The dense stands can choke waterways and cause mudflat elevations to increase as plants trap sediments. Smooth cordgrass and its hybrids are threatening the habitat of the endangered California clapper rail as well as the open mudflats where a million migrating shorebirds forage. Smooth cordgrass can hybridize with the native cordgrass, forming very robust clones and threatening the continued existence of California cordgrass. Smooth cordgrass and hybrids can colonize both above and below the elevations colonized by California cordgrass. Smooth cordgrass does create habitat for young salmon and is often used to stabilize estuaries and as cattle forage.

HOW IT CAME TO NORTH AMERICA

Introduced for erosion control projects in California and Washington in the early 1970s. May have been introduced to Willapa Bay with oyster shipments in the 1800s.

MANAGEMENT

Small plants can be hand pulled or dug up and left to dry above the high-tide line. Small patches can be mowed or cut, then covered with geotextile cloth or weed control fabric for two years, or with black plastic for a year, to kill plants with solar heat. Repeated mowing (eight times or more per year) can kill plants. The herbicides glyphosate and imazapyr are used to kill large infestations. A planthopper, *Prokelisia marginata*, released in 2000 in Washington, is being used as a biocontrol agent to reduce seed production.

FOR MORE INFORMATION

Ayres, D. R., and D. R. Strong. 2002. "The Spartina invasion of San Francisco Bay." *Aquatic Nuisance Species Digest.* 4(4):37–39. http://www.spartina.org/project_documents/ans_digest_Vol_4_No_4_mini.pdf.

Daehler, C. *"Spartina alterniflora."* 2000. In *Invasive Plants in California's Wildlands,* edited by C. C. Bossard, J.M. Randall, and M. C. Hoshovsky. Berkley, CA: University of California Press. http://www.cal-ipc.org/ip/management/ipcw/online.php.

San Francisco Estuary Invasive Spartina Project. *Introduced* Spartina alterniflora *hybrids (smooth cordgrass).* http://www.spartina.org/species/spartina-alterniflora-hy_v2.pdf.

PART III

More Information

BIBLIOGRAPHY

INTERNATIONAL

CABI Invasive Species Compendium. http://www.cabi.org/isc. Currently under development, this website contains information on numerous invasive species from around the world.

Global Invasive Species Database. http://www.issg.org/database/welcome. Published by the Invasive Species Specialist Group of the International Union for the Conservation of Nature, it contains a compilation of information on invasive species.

Institute for Biological Invasions. http://invasions.bio.utk.edu. A project of the University of Tennessee, founded by Dr. David Simberloff, this site includes an extensive database of literature.

Weber, E. 2003. *Invasive Plant Species of the World: A Reference Guide to Environmental Weeds.* Cambridge, MA: CABI Publishing. Basic information, including distribution, description, ecology, and control of hundreds of species occurring around the world.

NATIONAL

Agriculture Institute of Canada. *The Biology of Canadian Weeds.* http://www.aic.ca/journals/weeds.cfm. Journal series of articles on weeds of Canada contributed by different authors.

Alien Plant Working Group, Plant Conservation Alliance. *Weeds Gone Wild.* http://www.nps.gov/plants/alien. Fact sheets on many invasive plant species found in the U.S., with management information and references.

Canadian Food Inspection Agency. 2008. *Invasive Alien Plants in Canada.* Ottawa, ON: Canadian Food Inspection Agency. http://www.agrireseau.qc.ca/argeneral/documents/SIPC%20 Report%20-%20Summary%20Report%20-%20English%20Printed%20Version.pdf. Summary report of current status of invasive plants in Canada.

Center for Invasive Species and Ecosystem Health. http://www.invasive.org. Project of the University of Georgia, USDA Forest Service, and USDA APHIS. Links to many invasive species publications and information on invasive plants, insects, other animals, and diseases.

Environment Canada. *Invasive Alien Species in Canada.* http://www.ec.ga.ca/eee-ias. Environment Canada's website with basic information on invasive species.

Government of Canada. *Invasive Species in Canada.* http://www.invasivespecies.gc.ca. Canadian government website with basic information and links to national programs.

National Invasive Species Information Center. http://www.invasivespeciesinfo.gov. Information on U.S. federal efforts related to invasive species, the National Invasive Species Council, and legislation.

REGIONAL AND STATE

Alberta Invasive Plants Council. http://www.invasiveplants.ab.ca. Information on invasive plants in Alberta and link to Weeds Across Borders conference.

Bossard, C. C., J. M. Randall, and M. C. Hoshovsky, eds. 2000. *Invasive Plants of California's Wildlands.* University of California Press. http://www.cal-ipc.org/ip/management/ipcw. Information on 78 plant species that are invasive in California natural areas.

Center for Aquatic and Invasive plants, University of Florida. http://plants.ifas.ufl.edu. Photographs, information, and educational resources on native and invasive aquatic plants of the United States.

Cranston, R., D. Ralph and B. Wikeem. 2002. *Field Guide to Noxious and Other Selected Weeds of British Columbia.* Government of British Columbia. http://www.agf.gov.bc.ca/cropprot/weedguid/weedguid.htm. Provides brief species descriptions with links to additional information.

Czarapata, E. 2005. *Invasive Plants of the Upper Midwest.* Madison, WI: University of Wisconsin Press. Detailed identification information for the layperson and professional on invasive plants common to Midwest.

Gettys, L. A., W. T. Haller and M. Bellaud (eds.) 2009. *Biology and Control of Aquatic Plants: A Best Management Practices Handbook.* Marietta, GA: Aquatic Ecosystem Restoration Foundation. http://plants.ifas.ufl.edu/misc/pdfs/AERF_handbook.pdf. Information for invasive aquatic plant management.

Gray, A. N., K. Barndt and S. Reichard. 2011. *Nonnative Invasive Plants of Pacific Coast Forests: A Field Guide for Identification.* General Technical Report PNW-GTR-817. Portland, OR: U.S. Department of Agriculture, Forest Service, Pacific Northwest Research Station. http://www.fs.fed.us/pnw/pubs/pnw_gtr817.pdf. Identification, ecology, distribution and historical information for plant invaders of forests in California, Oregon, and Washington.

Hill, R. and S. Williams. 2007. *Maine Field Guide to Invasive Aquatic Plants.* Auburn, ME: Maine Center for Invasive Aquatic Plants and Maine Volunteer Lake Monitoring Program. http://www.mainevolunteerlakemonitors.org/mciap/FieldGuide.pdf. Great species descriptions that distinguish native aquatics from invasive aquatic plants.

Invasive Plant Atlas of New England. http://nbii-nin.ciesin.columbia.edu/ipane. Information on invasive plants in New England.

Langeland, K. A., J. A. Ferrell, B. Sellers, G. E. MacDonald, and R. K. Stocker. 2011. "Integrated Management of Nonnative Plants in Natural Areas of Florida." SP 242. Gainesville, FL: University of Florida IFAS. http://edis.ifas.ufl.edu/wg209.

Langeland, K. A., H. M. Cherry, C. M. McCormick, and K. A. Craddock Burks et al. 2008. *Identification and Biology of Nonnative Plants in Florida's Natural Areas,* 2nd ed. SP 257. Gainesville, FL: University of Florida IFAS.

Lassiter, B., R. Richardson, and G. Wilkerson. *Aquatic Weeds: A Pocket Identification Guide for the Carolinas.* Raleigh, NC: North Carolina State University. Aquatic weed identification; also useful for the Mid-Atlantic region.

Mabey, R. 2011. *Weeds: In Defense of Nature's Most Unloved Plants.* New York: HarperCollins.

Miller, J. H., E. B. Chambliss, N. J. Loewenstein. 2010. *A Field Guide for the Identification of Invasive Plants in Southern Forests.* General Technical Report SRS-119. Asheville, NC: U.S. Department of Agriculture, Forest Service, Southern Research Station. http://www.srs.fs.fed.us/pubs/gtr/gtr_srs119.pdf. Identification, ecology, distribution and historical information for plant invaders of forests from Virginia to Florida and west to Texas.

National Association of Exotic Pest Plant Councils. http://www.naeppc.org. Links to regional and state invasive plant councils.

Royer, F., and R. Dickinson. 1999. *Weeds of the Northern U.S. and Canada.* Renton, WA, and Alberta: Lone Pine Publishing and University of Alberta Press. An identification guide with emphasis on herbaceous plants and grasses for species growing north of Virginia and northern California.

Southwest Exotic Plant Information Clearinghouse. http://sbsc.wr.usgs.gov/research/projects/swepic/swepic.asp. Information on invasive plants in the southwestern United States.

Swearingen, J., K. Reshetiloff, B. Slattery, and S. Zwicker. 2010. *Plant Invaders of Mid-Atlantic Natural Areas,* 4th ed. Washington, DC: National Park Service and U.S. Fish & Wildlife Service. http://www.nps.gov/plants/alien/pubs/midatlantic. Basic information on selected Mid-Atlantic invasive plants and suggestions for alternatives for landscaping.

Uva, R. H., J. C. Neal, and J. M. DiTomaso. 1997. *Weeds of the Northeast.* Ithaca, NY: Cornell University Press. Mainly covers plants affecting agriculture, but has excellent photographs and identification information.

Whitson, T. D. 1999. *Weeds of the West.* Laramie, WY: Western Society of Weed Science and Cooperative Extension Services, University of Wyoming. An identification guide to more than 350 nonnative plants from New Mexico to Montana and west.

William, R.D., D. Ball, R. Parker, J. P. Yenish, T. W. Miller, D. W. Morishita, and P. J. S. Hutchinson, eds. 2006. *Pacific Northwest Weed Management Handbook.* Corvallis, OR: Oregon State University. http://pnwpest.org/pnw/weeds.

RELATED INFORMATION

Burrell, C. C. 2006. *Native Alternatives to Invasive Plants.* Brooklyn, NY: Brooklyn Botanic Garden. Noninvasive alternatives to invasive garden plants.

Community Invasive Species Network. http://protectingusnow.org. Educational resources on pests of plants.

The Nature Conservancy and the Continental Dialogue on Non-Native Forest Insects and Diseases. *Don'tMoveFirewood.org.* http://www.dontmovefirewood.org. Information on pests and pathogens affecting forests in the United States.

EatTheInvaders.org. http://eatheinvaders.org. Recipes for Japanese knotweed and kudzu among other tasty invaders.

Rotherham, I.D. and R. A. Lambert (eds.). 2011. *Invasive and Introduced Plants and Animals: Human Perceptions, Attitudes and Approaches to Management.* Washington, DC: Earthscan.

St. Louis Codes of Conduct. http://www.invasive.org/gist/horticulture/codes.html. Voluntary codes of conduct related to invasive plants.

Tallamy, D. W. 2007. *Bringing Nature Home.* Portland, OR: Timber Press. Makes the case for why native plants support more wildlife than introduced plants.

JOURNALS ON BIOLOGICAL INVASIONS

Aquatic Invasions: http://www.aquaticinvasions.net

Biological Invasions: http://www.springer.com/life+sciences/ecology/journal/10530

Diversity and Distributions: http://onlinelibrary.wiley.com/journal/10.1111/(ISSN)1472-4642

Invasive Plant Science and Management: http://allenpress.com/publications/journals/ipsm

Management of Biological Invasions: http://www.managementofbiologicalinvasions.net

NeoBiota: http://www.pensoft.net/journals/neobiota

GLOSSARY

adjuvant	a chemical added to an herbicide to increase its safety or effectiveness
adventitious root	a root that forms along the stem
allelochemical	a chemical produced by a plant that inhibits the growth or seed germination of other plants
alternate leaves	leaves that emerge on alternating sides of the stem
annual	a plant that completes its life cycle within one growing season
apomictic	able to produce seeds without fertilization
aquatic plant	a plant that grows in water
aril	a fleshy coating covering a seed; often brightly colored.
auricle	an earlike projection at the base of a grass leaf where it attaches to the stem
awn	a stiff bristle sometimes present as part of the spikelet or flowering unit of grasses or sedges
axil	the place where a leaf meets the stem
basal	from the base, as in leaves that grow from the base of a plant
berry	a fleshy fruit with many embedded seeds
biennial	a plant that completes its life cycle within two years
biological control	using natural enemies, such as insects or diseases, to control a plant species
bolt	to send up a flowering stalk
brackish	slightly salty
bract	a modified leaf found below a flower or flower cluster, not always green
bulbil	an aerial tuber
capsule	a dry fruit that opens when mature to release the seeds
catkin	a slender cluster of flowers lacking petals, usually wind pollinated; found on willows, alders, birches, and poplars
chemical control	using herbicides to control a plant species
compound leaf	a leaf divided into two or more separate sections (leaflets) arranged along a common midrib
culm	a stem of a grass or sedge
cultivar	a plant genotype selected for certain characteristics

deciduous — having leaves that fall off annually
dioecious — having male and female flowers on separate plants
disc flower — in the aster family, a flower lacking distinctive petals, usually in the center of a tight flower cluster. *See also: ray flowers.*

ecosystem — the environment and organisms comprising a certain area
ecotype — a genetically distinct geographic variety
epiphyte — a plant that grows on another plant for support
evergreen — remaining green year-round

fascicle — a bundle of pine needles
forage — food for cattle and other grazers
frond — a fern leaf that is divided into pinnae
fruit — the seed-containing structure of a plant

germinate — to begin to grow and develop
girdle — to remove a ring of bark from a tree or branch, usually done to kill the plant
gland — a structure producing nectar or another, often sticky, substance
glume — a small bract at the base of a spikelet in a grass or sedge

herbaceous — a nonwoody plant
hybrid — an offspring of two closely related species

inflorescence — a cluster of flowers
invasive plant — a nonnative introduced plant species that spreads rapidly, causing changes at the population, community, or ecosystem level

leaf blade — the wide part of the leaf
leaf scar — the mark left on a stem after a leaf falls off
leaflet — one of the sections of a compound leaf
ligule — a membrane or row of hairs on a the upper side of a grass leaf where the leaf meets the stem
lobed leaf — a leaf with deep indentations and rounded or pointed projections

mattock — a tool used to loosen the soil, like a pickaxe in appearance
mechanical control — using tools to physically remove a species, such as mowing, cutting, or burning
midrib — the central vein running the length of a leaf
midvein — the central vein running the length of a leaf
monecious — having male and female flowers on the same plant.
mycorrhizae — fungi associated with the roots of plants

native plant	a plant that occurs naturally in the ecosystem or habitat in which it evolved
nitrogen fixers	bacteria that incorporate atmospheric nitrogen into nitrogen compounds
node	the place where a leaf attaches to the stem or from which branches or roots arise
opposite leaves	leaves that emerge in pairs from the same point on opposite sides of the stem
ornamental	a plant cultivated for aesthetic purposes
palmate	a divided or lobed leaf in which all the leaflets meet at a single point and radiate outwards like the fingers on a hand.
panicle	a branching cluster of flowers
pealike flower	a flower irregular in shape, characteristic of flowers in the Fabaceae (pea) family
peltate	a leaf where the stem attaches to the center of the leaf blade
perennial	a plant that lives for three or more years
perigynium	a sacklike bract that forms around sedge flowers
petal	a part of the flower, often brightly colored
petiole	the stalk attaching a leaf to the stem
photosynthesis	the process by which plants use light and chlorophyll to turn carbon dioxide and water into carbohydrates
phyllode	a leaflike stem, typical of the eucalyptus and acacias
pinnae	the first-division leaflets of a compound frond or leaf
pistil	the female flower organ, consisting of a slender stalk (style) and the stigma, the surface to which pollen attaches
pod	a dry fruit containing the seeds
polyembryonic	having more than one embryo contained within the seed
polyploids	plants with multiple copies of each chromosone
raceme	a cluster of flowers with each flower attached to a central stalk by equal-length flower stalks
rachis	the axis of a fern frond to which the leaflets (pinnae) are attached
ray flower	in the aster family, a flower with a single petal, usually at the edge of a tight flower cluster. *See also: disc flower.*
Rhizobium	a genus of bacteria that can fix nitrogen, often associated symbiotically with plants in the Fabaceae family
rhizome	an underground stem
rosette	a cluster of basal leaves
sapling	a young tree
sepal	the outermost layer of a flower, often green and leaflike
serrated	jagged
sheath	the base of a leaf that wraps around the stem
shrub	a woody plant with multiple stems arising from the roots
spathe	a large bract surrounding a tight flower cluster

species	a group of similar organisms capable of interbreeding to produce offspring
spike	an unbranched flower cluster where flowers are attached directly to the main stem
spikelet	the flowering unit of a grass or sedge
sporangia	the spore-containing structures on a fern
spore	a reproductive cell that can develop into an adult without fusing with another cell
stamen	the male flower organ, consisting of a slender stalk (filament) and the pollen-producing anther.
stem	the axis of a plant, usually aboveground but sometimes belowground (rhizomes)
stipule	a leaflike structure, usually at the base of the leaf stalk where it attaches to the stem
stolon	a horizontal aboveground stem or runner that can root at the nodes
style	a slender stalk connecting the ovary with the stigma on the female flower organ (pistil)
subspecies	a group within a species that shares similar characteristics; abbreviated ssp.
sucker	stems that grow from underground roots, spreading from the main plant
surfactant	an additive to herbicides that helps the herbicide stick to the leaves
synconium	a type of fruit unique to figs featuring a large, fleshy surface that holds many tiny flowers
tap root	the main root, often thick
thorn	a stiff, sharply pointed modified stem
toothed	jagged, like a saw blade; teeth may be fine or coarse
tree	a woody plant, generally with one main stem arising from the roots
tuber	a thick root or rhizome used for food storage and vegetative propagation, like a potato
turion	a dormant, vegetative bud
twine	to twist
umbel	a cluster of flowers where the flower stalks originate from one point
understory	vegetation growing under the tree canopy
vascular plants	plants with specialized tissues for conducting water and nutrients
vine	a climbing or trailing plant
whorled leaves	three or more leaves that emerge from the same point on the stem
woody tissue	hard, fibrous plant tissue containing cellulose and lignin

PHOTO CREDITS

Page i
Amadej Trnkoczy

Page 44
Forest and Kim Starr
(USGS)

Page 45
Forest and Kim Starr
(USGS)

Page 47
USDA-NRCS PLANTS
Database/Herman, D. E.
et al. 1996. North Dakota
tree handbook. USDA
NRCS ND State Soil
Conservation Committee;
NDSU Extension and
Western Area Power
Administration, Bismarck,
ND (top and bottom)

Page 48
USDA-NRCS PLANTS
Database/Herman, D. E.
et al. 1996. North Dakota
tree handbook. USDA
NRCS ND State Soil
Conservation Committee;
NDSU Extension and
Western Area Power
Administration, Bismarck,
ND

Page 49
Forest and Kim Starr
(USGS)

Page 50
Forest and Kim Starr
(USGS) (left)

Mr. and Mrs. Robert G.
Young @ USDA-NRCS
PLANTS Database/USDA
NRCS. 1992. Western
wetland flora: Field office
guide to plant species.
West Region, Sacramento,
CA (right)

Page 52
John Randall/The Nature
Conservancy (top and
bottom)

Page 53
Forest and Kim Starr
(USGS)

Page 54
Forest and Kim Starr
(USGS)

Page 55
Forest and Kim Starr
(USGS)

Page 56
Forest and Kim Starr
(USGS)

Page 57
Forest and Kim Starr
(USGS)

Page 58
Forest and Kim Starr
(USGS)

Page 59
Forest and Kim Starr
(USGS) (left and right)

Page 60
Forest and Kim Starr
(USGS)

Page 61
Forest and Kim Starr
(USGS)

Page 63
Forest and Kim Starr
(USGS) (left and right)

Page 64
Forest and Kim Starr
(USGS)

Page 65
Forest and Kim Starr
(USGS) (top and bottom)

Page 67
Barry Rice/sarracenia.com
(top)
Forest and Kim Starr
(USGS) (bottom)

Page 68
Barry Rice/sarracenia.com

Page 69
John Randall/The Nature
Conservancy

Page 70
Forest and Kim Starr
(USGS) (left and right)

Page 71
Mandy Tu/The Nature
Conservancy (bottom)

Page 72
Forest and Kim Starr
(USGS) (top)
John Randall/The Nature
Conservancy (bottom)

Page 73
Forest and Kim Starr
(USGS)

Page 74
authors
Page 75
authors
Page 76
Forest and Kim Starr
(USGS)
Page 77
Forest and Kim Starr
(USGS) (top and bottom)
Page 78
Forest and Kim Starr
(USGS)
Page 79
Forest and Kim Starr
(USGS) (top and bottom)
Page 81
Jeffrey Hutchinson
Page 83
Forest and Kim Starr
(USGS) (left and right)
Page 84
Forest and Kim Starr
(USGS)
Page 86
Forest and Kim Starr
(USGS)
Page 87
Forest and Kim Starr
(USGS) (top and bottom)
Page 88
Barry Rice/sarracenia.com
Page 89
J.S. Peterson@USDA-NRCS
PLANTS Database
Page 91
Tihomir Kostadinov/
University of Richmond
(left and right)
Page 92
Bill Johnson
Page 93
authors (left and right)
Page 94
Forest and Kim Starr
(USGS) (top)
authors (bottom)

Page 95
authors
Page 96
authors (left and right)
Page 97
authors
Page 98
authors (top and bottom)
Page 99
authors (left)
Forest and Kim Starr
(USGS) (right)
Page 100
Forest and Kim Starr
(USGS)
Page 101
Chelsie Vandaveer (top)
Forest and Kim Starr
(USGS) (bottom)
Page 102
Forest and Kim Starr
(USGS)
Page 104
authors
Page 105
John Randall/The Nature
Conservancy
Page 106
authors (top and bottom)
Page 107
authors
Page 108
authors (top and bottom)
Page 109
authors
Page 111
Robert H. Mohlenbrock @
USDA-NRCS PLANTS
Database / USDA NRCS.
1995. Northeast wetland
flora: Field office guide to
plant species. Northeast
National Technical
Center, Chester, PA.
Page 112
Forest and Kim Starr
(USGS)

Page 113
Forest and Kim Starr
(USGS) (top)
John Randall/The Nature
Conservancy (bottom)
Page 115
John Randall/The Nature
Conservancy
Page 116
authors (left and right)
Page 117
authors (left and right)
Page 118
authors (left and right)
Page 119
authors
Page 120
authors
Page 121
authors (left and right)
Page 122
Tihomir Kostadinov/
University of Richmond
(top and bottom)
Page 123
authors
Page 124
authors
Page 125
authors (top and bottom)
Page 126
authors
Page 127
authors (left and right)
Page 128
authors
Page 129
authors (left and right)
Page 130
authors (top and bottom)
Page 131
USDA-NRCS PLANTS
Database/Herman, D. E.
et al. 1996. North Dakota
tree handbook. USDA
NRCS ND State Soil

Conservation Committee;
NDSU Extension and
Western Area Power
Administration, Bismarck,
ND
Page 132
USDA-NRCS PLANTS
Database/Herman, D. E.
et al. 1996. North Dakota
tree handbook. USDA
NRCS ND State Soil
Conservation Committee;
NDSU Extension and
Western Area Power
Administration, Bismarck,
ND
Page 133
USDA-NRCS PLANTS
Database/Herman, D. E.
et al. 1996. North Dakota
tree handbook. USDA
NRCS ND State Soil
Conservation Committee;
NDSU Extension and
Wesern Area Power
Administration, Bismarck,
ND
Page 134
Jeffrey Hutchinson
Page 135
Jeffrey Hutchinson (left)
Charles Bryson,
www.forestryimages.org
(right)
Page 136
James H. Miller, USDA Forest Service, Bugwood.org
Page 137
USDA-NRCS PLANTS
Database/Herman, D. E.
et al. 1996. North Dakota
tree handbook. USDA
NRCS ND State Soil
Conservation Committee;
NDSU Extension and
Western Area Power
Administration, Bismarck,
ND

Page 138
John Randall/The Nature
Conservancy (top)
USDA-NRCS PLANTS
Database/Herman, D. E.
et al. 1996. North Dakota
tree handbook. USDA
NRCS ND State Soil
Conservation Committee;
NDSU Extension and
Western Area Power
Administration, Bismarck,
ND (bottom)
Page 139
USDA-NRCS PLANTS
Database/Herman, D. E.
et al. 1996. North Dakota
tree handbook. USDA
NRCS ND State Soil
Conservation Committee;
NDSU Extension and
Western Area Power
Administration, Bismarck,
ND
Page 140
John Randall/The Nature
Conservancy
Page 141
John Randall/The Nature
Conservancy (top and
bottom)
Page 142
authors
Page 143
Bill Johnson
Page 144
authors (top and bottom)
Page 145
John Randall/The Nature
Conservancy
Page 146
John Randall/The Nature
Conservancy (top)
authors (bottom)
Page 147
authors

Page 148
USDA-NRCS PLANTS
Database/Herman, D. E.
et al. 1996. North Dakota
tree handbook. USDA
NRCS ND State Soil
Conservation Committee;
NDSU Extension and
Western Area Power
Administration, Bismarck,
ND
Page 149
USDA-NRCS PLANTS
Database/Herman, D. E.
et al. 1996. North Dakota
tree handbook. USDA
NRCS ND State Soil
Conservation Committee;
NDSU Extension and
Western Area Power
Administration, Bismarck,
ND (left and right)
Page 150
authors
Page 151
Joseph Jelich (left)
authors (right)
Page 152
authors
Page 153
Unknown@USDA-NRCS
PLANTS database
Page 155
authors (left and right)
Page 156
authors
Page 157
authors
Page 158
John Randall/The Nature
Conservancy (top)
authors (bottom)
Page 159
authors (top and bottom)

Page 161
Karan A Rawlins,
University of Georgia,
Bugwood.org (left)
Charles T. Bryson, USDA
Agricultural Research
Service, Bugwood.org
(right)
Page 162
Great Smoky Mountains
National Park Resource
Management Archives,
USDI National Park Serv-
ice/www.forestryimages.org
Page 163
Bill Johnson
Page 164
authors (left)
Amy Belding Brown (right)
Page 165
authors (left)
Bill Johnson (right)
Page 166
authors
Page 167
Bill Johnson (top)
authors (bottom)
Page 168
authors
Page 169
Pat Breen
Page 170
Pat Breen (left and right)
Page 171
Pat Breen
Page 172
authors
Page 173
authors
Page 174
authors
Page 175
Jeffrey Hutchinson (left and
right)

Page 176
Jeffrey Hutchinson (top
and bottom)
Page 177
Amy Belding Brown (left
and right)
Page 178
Amy Belding Brown (left
and right)
Page 179
authors
Page 180
authors
Page 181
authors
Page 182
Steven J. Baskauf/bioimages
.cas.vanderbilt.edu (left
and right)
Page 183
Steven J. Baskauf/bioimages
.cas.vanderbilt.edu (left
and right)
Page 184
authors (left and right)
Page 185
John Randall/The Nature
Conservancy
Page 186
authors
Page 187
Jake Barnes (top)
Louis-M. Landry 2005–2006
(bottom)
Page 177
John Randall/The Nature
Conservancy (top)
Louis-M. Landry 2005–2006
(bottom)
Page 190
Bill Johnson (top)
authors (bottom)
Page 191
Bill Johnson

Page 192
Forest and Kim Starr
(USGS)
Page 193
Forest and Kim Starr
(USGS) (left and right)
Page 194
John Randall/The Nature
Conservancy
Page 195
Barry Rice/sarracenia.com
(top and bottom)
Page 197
authors (top and bottom)
Page 198
authors
Page 199
authors
Page 200
John Randall/The Nature
Conservancy
Page 201
authors (left and right)
Page 202
authors (top and bottom)
Page 204
Jeffrey Hutchinson (left
and right)
Page 205
Jeffrey Hutchinson (top
and bottom)
Page 206
Forest and Kim Starr
(USGS)
Page 207
Forest and Kim Starr
(USGS)
Page 208
Amadej Trnkoczy
Page 209
authors (top)
Amy Belding Brown
(bottom)
Page 211
Jeffrey Hutchinson

Page 212
Jeffrey Hutchinson (top, bottom)
Page 213
Jeffrey Hutchinson
Page 214
authors (top)
authors (bottom)
Page 215
Carole Bergmann
Page 216
John Randall/The Nature Conservancy (left)
Bill Johnson (right)
Page 217
Justin Thayer, Louisiana State University
Page 218
Mary Bowen, Louisiana State University
Page 219
authors
Page 220
authors (top and bottom)
Page 221
authors
Page 222
John Randall/The Nature Conservancy
Page 223
John Randall/The Nature Conservancy
Page 224
Forest and Kim Starr (USGS)
Page 225
Forest and Kim Starr (USGS)
Page 226
authors
Page 227
authors
Page 228
authors
Page 229
authors (left and right)

Page 230
authors
Page 231
authors (left and right)
Page 232
G. Narayanaraj
Page 233
G. Narayanaraj
Page 234
authors
Page 235
authors (top and bottom)
Page 236
authors
Page 237
authors
Page 238
authors (top and bottom)
Page 239
authors
Page 240
authors (top)
Eike Wulfmeyer (bottom)
Page 241
authors
Page 242
authors (top and bottom)
Page 243
authors
Page 244
Jennifer Forman
Page 245
Charles Bryson/www.forestryimages.org (top)
authors (bottom)
Page 246
authors
Page 247
John Randall/The Nature Conservancy
Page 248
Les Mehrhoff
Page 249
authors

Page 250
authors (left and right)
Page 251
authors
Page 252
authors
Page 253
authors
Page 254
Louis-M. Landry 2005–2006
Page 255
Louis-M. Landry 2005–2006
Page 256
Louis-M. Landry 2005–2006
Page 257
authors (top and bottom)
Page 258
authors
Page 260
authors
Page 261
authors
Page 262
Forest and Kim Starr (USGS)
Page 263
Kurt Stüber/www.forestry images.org (top)
Mary Ellen Harte/ www.forestryimages.org (bottom)
Page 264
John Randall/The Nature Conservancy (left and right)
Page 265
Amy Belding Brown
Page 266
Louis-M. Landry 2005–2006
Page 267
Euphorbia esula G. Narayanaraj.JPG
Page 268
authors
Page 269
authors (top and bottom)

Page 321
John Randall/The Nature
Conservancy (left)
authors (right)

Page 322
Utah State University
Archive, Utah State
University, Bugwood.org

Page 323
USDA APHIS PPQ
Archive, USDA APHIS
PPQ, Bugwood.org

Page 324
Donna Ellis/University
of Connecticut

Page 325
Donna Ellis/University
of Connecticut

Page 326
Donna Ellis/University
of Connecticut

Page 327
Art Gover (top and bottom)

Page 328
Mary Ellen
Harte/www.forestry
images.org

Page 329
Mary Ellen
Harte/www.forestry
images.org (left and right)

Page 331
Barbara Tokarska-Guzik,
University of Silesia,
Bugwood.org

Page 332
Michael Shephard, USDA
Forest Service,
Bugwood.org (left and
right)

Page 333
Michael Shephard, USDA
Forest Service,
Bugwood.org

Page 335
John Randall/The Nature
Conservancy

Page 336
Tom Heutte, USDA Forest
Service/www.forestry
images.org (left and right)

Page 337
Les Mehrhoff

Page 338
Barry Rice/sarracenia.com

Page 339
authors (left and right)

Page 340
authors

Page 341
John Randall/The Nature
Conservancy

Page 342
Gary Monroe@USDA-
NRCS PLANTS Database
(top and bottom)

Page 343
George Rembert

Page 344
George Rembert

Page 345
Jennifer Anderson@USDA-
NRCS PLANTS Database

Page 347
Tom Heutte, USDA Forest
Service/www.forestry
images.org

Page 348
authors (left and right)

Page 349
Tom Heutte, USDA Forest
Service/www.forestry
images.org

Page 350
Matthew C. Perry

Page 351
authors (top and bottom)

Page 353
Chris Evans/University
of Georgia/www.forestry
images.org (top and
bottom)

Page 354
Chris Evans/University
of Georgia/www.forestry
images.org

Page 355
authors (left and right)

Page 356
authors (left and right)

Page 357
Art Gover

Page 358
Robert H. Mohlenbrock @
USDA-NRCS PLANTS
Database / USDA NRCS.
1995. Northeast wetland
flora: Field office guide to
plant species. Northeast
National Technical Cen-
ter, Chester, PA.

Page 359
Barry Rice/sarracenia.com
(left)
authors (right)

Page 360
John Randall

Page 362
authors (left and right)

Page 363
authors (left and right)

Page 364
authors

Page 365
authors (left)
Amy Belding Brown (right)

Page 366
authors

Page 367
Chelsie Vandaveer

Page 369
authors

Page 371
authors

Page 372
authors (left and right)

Page 373
authors

Page 430
Forest and Kim Starr
(USGS) (top)
authors (bottom)
Page 432
authors
Page 433
authors (left and right)
Page 434
Kerrie Kyde (left and right)
Page 435
Forest and Kim Starr
(USGS)
Page 436
Forest and Kim Starr
(USGS) (left)
Jeffrey Hutchinson (right)
Page 438
Joseph diTomaso
Page 439
Joseph diTomaso
Page 440
authors
Page 441
authors
Page 442
John Randall/The Nature
Conservancy
Page 444
Jeffery Hutchinson
Page 445
John Randall/The Nature
Conservancy
Page 446
Jeffery Hutchinson (left
and right)
Page 447
Forest and Kim Starr
(USGS)
Page 448
Forest and Kim Starr
(USGS)
Page 449
Louis-M. Landry 2005–2006
(left and right)

Page 450
authors (top and bottom)
Page 452
Christian Fischer
Page 454
Les Mehrhoff
Page 455
Philip Thomas (left)
authors (right)
Page 456
Charles Bryson/www.forestry
images.org
Page 458
John Randall/The Nature
Conservancy (top)
Joseph diTomaso (bottom)
Page 459
Joseph diTomaso
Page 460
Robert H. Mohlenbrock @
USDA-NRCS PLANTS
Database / USDA NRCS.
1995. Northeast wetland
flora: Field office guide to
plant species. Northeast
National Technical Cen-
ter, Chester, PA.
Page 462
John Randall/The Nature
Conservancy
Page 463
Les Mehrhoff
Page 464
Les Mehrhoff
Page 465
Robert H. Mohlenbrock @
USDA-NRCS PLANTS
Database / USDA NRCS.
1995. Northeast wetland
flora: Field office guide to
plant species. Northeast
National Technical Cen-
ter, Chester, PA.

Page 466
John Randall/The Nature
Conservancy
Page 467
John Randall/The Nature
Conservancy
Page 468
Kathy Hamel (left)
Les Mehrhoff (right)
Page 469
Jeffrey Hutchinson
Page 470
Jeffrey Hutchinson
Page 471
Jeffrey Hutchinson
Page 472
authors
Page 473
Barry Rice/sarracenia.com
Page 475
Graves Lovell, Alabama
Department of Conserva-
tion and Natural
Resources, Bugwood.org
Page 477
David Perez
Page 479
Rachel Woodfield, Merkel
& Associates
Page 480
Rachel Woodfield, Merkel
& Associates
Page 481
Ron Klauda, Maryland Dept
Natural Resources
Page 483
Barry Rice/sarracenia.com
Page 484
Barry Rice/sarracenia.com
Page 486
Unknown@USDA-NRCS
PLANTS Database

INDEX